Strategies of Biochemical Adaptation

PETER W. HOCHACHKA
Department of Zoology,
University of British Columbia

GEORGE N. SOMERO
Scripps Institute of Oceanography,
University of California at San Diego

W. B. SAUNDERS COMPANY PHILADELPHIA · LONDON · TORONTO

W. B. Saunders Company: West Washington Square
Philadelphia, Pa. 19105

12 Dyott Street
London, WC1A 1DB

833 Oxford Street
Toronto 18, Ontario

Strategies of Biochemical Adaptation ISBN 0-7216-4705-7

 Made in the United States of America. Press of W. B. Saunders Company. Library of Congress catalog card number 72-88848.

Print No.: 9 8 7 6 5 4 3 2

PREFACE

Its position notwithstanding, the preface of a book should be the last section of the volume to be written. Only after all of their words have been put to paper can authors fully appreciate what they have or have not said, and, consequently, only at that time can authors write words which will prepare the reader for what is and what is not to be.

As we trust our title indicates, ours is a volume devoted largely to an overview of the fundamental mechanisms of environmental adaptation, and the word "environment" has a broad enough meaning to encompass factors in the intracellular, the extracellular, and the extraorganismic milieux. A primary goal of our writing has been to effect a liaison between recent findings of biochemistry and decades-old questions of physiology pertaining to the means by which organisms and their constituent chemical systems adapt to various environmental parameters.

Taken as it stands, the last statement might imply that we have attempted to write what might be termed a "comparative and environmental physiology" monograph. Emphatically this is not the case. We have tried to avoid an encyclopedic approach and, instead, have tried to outline what we feel are the major "strategies" of adaptation at the biochemical level. We have also attempted to erect general guidelines as to when and by whom these various strategies are employed. Thus, the underlying theme running through the book is the nature of evolutionary change at the molecular level. To the extent that we have illuminated these various aspects of organism-environment interactions, we hope that we will have elaborated stories of interest to the ecologist, the evolutionary biologist, and the population biologist, as well as to our primary audience, the comparative biochemist/physiologist.

In writing a preface it is customary to express one's intellectual debts. Again, only after the final word of a book is written does one fully appreciate the extent of his debts to the persons whose discoveries and theories have shaped his own thoughts. In our own case we sense a particular debt to those biochemists who have succeeded in presenting their findings within a truly biological frame of reference. Scientists such as Sir Hans Krebs and Daniel Atkinson, in their wide-ranging studies of biological chemistry, have unfailingly managed to place special emphasis on the biological aspects of their subject. In so doing they have generated theories and data which the biologist can readily assimilate into his own frame of reference for use in attacking long-standing problems in physiology and comparative biochemistry.

It is to the scientists who originally generated these long-standing problems that we feel another source of indebtedness. Physiologists such as August Krogh, Per Scholander, Knut Schmidt-Nielsen, C. L. Prosser, H. Precht, and F. E. J. Fry early recognized and clearly delineated the fundamental questions we treat in this volume, questions which are only now becoming approachable as a result of modern advances in biochemistry.

Our students also deserve special mention and thanks. Certain ideas we discuss either are from their own minds or have at least been finely filtered in innumerable interactions with them.

And lastly, our efforts could not have been accomplished without the assistance and companionship of Brenda, Meredith, Laika, Heidi, and Dewdney.

Art work and cover design were by H. Hirnschall. Joyce Lewis prepared the index. Our appreciation is extended to both.

P. W. Hochachka
G. N. Somero

CONTENTS

CHAPTER 1 INTRODUCTION

WHAT ARE "BIOCHEMICAL ADAPTATIONS TO THE ENVIRONMENT"?

The structural and functional characteristics of an organism often appear to be especially designed to enhance its chances for success in its particular habitat. Such characteristics are termed "environmental adaptations," and, since the advent of evolutionary theory, have served as a focal point for an important and fascinating domain of biological research. Environmental adaptations have been charted at all levels of biological organization. Behaviorally, organisms usually appear to act in ways which increase their chances of survival in, and their abilities to exploit, their particular habitats. Anatomically, the structures of an organism often appear to gear it for its particular mode of life. At the physiological level, the manners in which the vital functions are carried out often reflect the environmental conditions encountered by the species.

Much less understood are what we term "biochemical adaptations to the environment." To comprehend what we wish to denote by this phrase, we might best begin by asking: What adaptations are *not* biochemical? Since organisms are aggregates of atoms and molecules, is not every change in an organism's physiology, anatomy, and even its behavior ultimately "biochemical"? In the strictest sense of the term "biochemical," this is true. Each change in an organism can ultimately be traced to alterations occurring at the molecular level—at least in theory. However, in this volume we are only peripherally concerned with the types of biochemical changes which accompany, for example, anatomical alterations in the shape of a limb girdle or the geometry of a bird's bill. Describing these changes in molecular terms probably tells us nothing additional about their adaptive significance.

The biochemical changes which we will describe are, for the most part, adaptive at the level of basic metabolic function, and therefore are not apparent macroscopically. To illustrate what we mean, consider the apparent macroscopic attributes of two fish species, one of which lives in the Antarctic Ocean and one of which inhabits a warm tropical sea. In their behavioral, anatomical, and physiological characteristics, both species may seem very much alike. Both display comparable hydrodynamic properties and comparable abilities to swim, to metabolize, to transport gases, to osmoregulate and so forth. Above the biochemical level we may find little reason to regard one species as "polar-adapted" and one as "tropical-adapted." It is only when we begin to "dissect" the biochemical machinery of the two species that we learn of an important relationship: the capabilities of the two species to conduct the same basic functions at quite similar rates under their very different habitat conditions are dependent on the existence of basic biochemical differences—each species displaying specific biochemical adaptations to its particular environment. It is indeed the biochemical adaptations to their respective environments which make the two species so apparently similar. Adjustments which occur at the chemical level, and can be observed only indirectly, thus can obliterate observable differences at higher levels of biological organization.

THREE BASIC STRATEGIES OF BIOCHEMICAL ADAPTATION TO THE ENVIRONMENT

Since the metabolic activities of organisms are strictly dependent upon macromolecules such as enzymes and nucleic acids, adaptational processes ensure that macromolecular functions occur at the correct *rates* and are of the proper *type* to permit the vital processes of the organism to continue satisfactorily despite perturbing effects of the environment. Three mechanisms constitute the major strategies by which this general "goal" is attained:

(1) The *types* of macromolecules present in the system may be changed.

(2) The amounts or *concentrations* of macromolecules may be adjusted.

(3) The functions of the macromolecules of the cell may be *regulated* in adaptive manners.

By one or a combination of these three strategies, the organism gains a vectorial homeostasis of metabolic function. The term "vectorial homeostasis" is used to emphasize that during environmental adaptation both the *rates* and the *directions* of metabolic reactions are adjusted so as to provide the organism with a continuing supply of essential products.

To appreciate in further detail how this homeostasis—a major outcome of biochemical adaptation—is achieved, it is necessary to examine briefly the fundamental requirements which all metabolic systems must meet if life is to be maintained.

THE ESSENTIAL FEATURES OF CELLULAR METABOLISM

To the student who is bewildered (or bored) by the apparent complexity of cellular metabolism, we hasten to point out that the basic functions of metabolism are at once few in number (Figure 1–1) and capable of being comprehended without a massive ingestion of chemical formulae and chemical reactions. In essence, cellular metabolism of all organisms must, at all times, perform the following tasks:

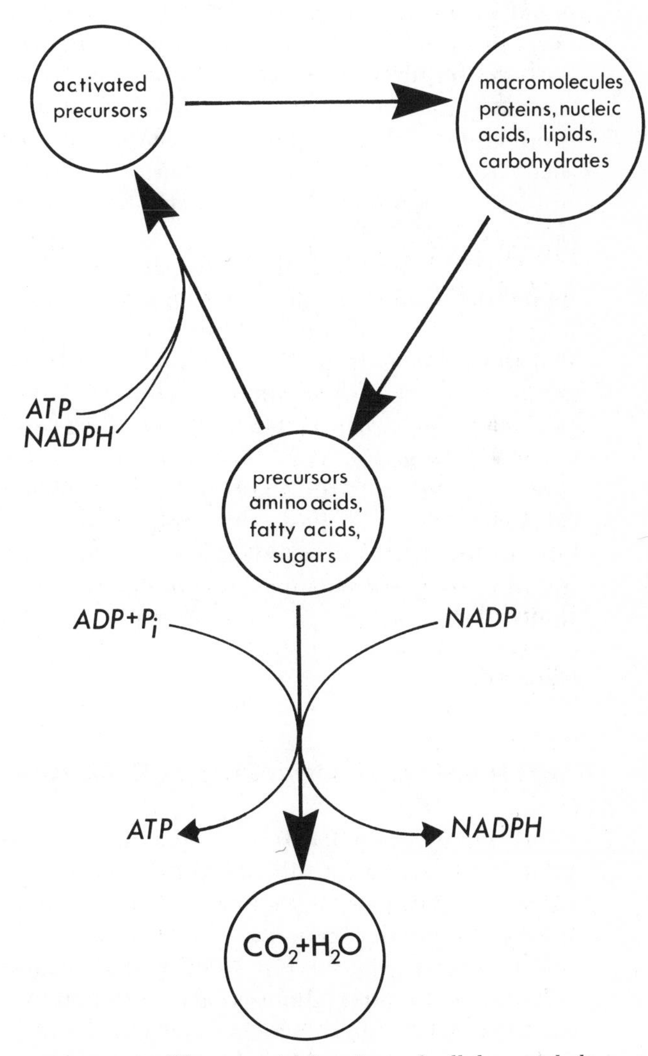

Figure 1–1 The essential functions of cellular metabolism.

(1) "High energy" compounds such as ATP (p. 22) must be generated in sufficient quantities to supply the cell with the "energy currency" it needs to keep such vital work functions as ionic regulation, contractility, and biosynthesis operating at required rates.

(2) The intermediates needed for biosynthesis must be formed, or ingested, in adequate quantities to meet the demands for the synthesis of large molecules such as nucleic acids, proteins, lipids, and carbohydrates.

(3) Biological reducing power, such as NADPH (p. 24), must be generated to support the reductions which occur in biosynthetic pathways.

(4) Lastly, through a union of the above three functions, the organism must synthesize the large molecules which serve as the basis for the distinct "biological" properties of living systems. Proteins are required for biological catalysis. Nucleic acids are needed for information transfer—from generation to generation during reproduction, and from genes to proteins during development. Polysaccharides and lipids are needed as fuel reserves. And proteins, lipids, and carbohydrates constitute important structural components of the cell.

Superimposed on all of these essential metabolic functions, or, more specifically, on the reaction sequences which conduct these functions, is an efficient and intricate control system which ensures that the participation of each metabolic process is consistent with the needs of the organism as a whole. No metabolic pathway is a free-wheeling process. The activity of each pathway and, indeed, each regulatory enzyme is subject to an array of controls which links its function to the status of the local chemistry of the cell and, ultimately, into the basic demands of the whole organism. These control functions represent important sites of interaction with the environment, and we must consider them in some detail if we are to understand how organisms biochemically adapt to their environments.

THE HIERARCHY OF METABOLIC REGULATION

Prior to about 1955, the major thrust of biochemical research aimed at elucidating the different components of the cellular metabolic "machine," with little attention being given to how the myriad reactions occurring in the cell are integrated with each other. Only during the last 10 to 15 years has biochemistry come to emphasize that metabolic reaction spans are functionally interlinked and regulated. The key fact emerging from this more recent research is that regulation is achieved through a hierarchy of control mechanisms stemming from genes to proteins. Since essentially all cellular reactions are enzyme-catalyzed, control of metabolism reduces itself to control of the rates and types of enzyme function. It is now clear that there are but two basic ways in which the rates of catalysis can be controlled: either

(1) the amounts of enzymes can be varied, and/or

(2) the activities of enzymes—the extent to which catalytic potentials are actually utilized—can be regulated.

The hierarchical nature of control arises from the simple fact that regulation of enzyme function can be achieved by placing "on-off" signals at either of these two levels—either at the level of enzyme synthesis or at the level of enzyme activity. It will be evident that the properties of these two control mechanisms differ in terms of:

(1) speed of response (the first is slow, usually requiring at least hours; the second is so fast that for practical purposes it is considered instantaneous),

(2) sensitivity of response (the first allows for only a "coarse" level of control; the second allows for a much finer "tuning" of enzyme function), and

(3) versatility of response (the first is the more versatile in that it allows the organism to modulate the relative *amounts* and the *kinds* of enzymes functioning at any given locus in metabolism, depending upon external environmental signals; in the second strategy, the qualitative composition of the enzyme battery cannot be changed, only its relative activity can be modulated).

Let us consider the basis for these differences in somewhat greater detail.

CONTROL OF ENZYME CONCENTRATION CAN OCCUR AT FOUR LEVELS: TRANSCRIPTION, TRANSLATION, ASSEMBLY, AND DEGRADATION

The amount of a particular enzyme present in a cell can be regulated at several steps in the production of the enzyme and, of course, at the stage of enzyme degradation. In the metabolic control hierarchy, the most complex mechanism for regulating enzyme concentration involves the processes of gene activation and repression. In response to specific chemical signals, the *transcription* of a given DNA sequence into messenger RNA (mRNA) may be initiated or blocked, depending on whether the signal in question is an "inducer" or a "repressor," respectively. Gene-level control can lead to (i) increased or decreased quantities of enzymes, (ii) changes in the types of enzymes which occur in the cell, and (iii) changes in the relative abundance of enzyme variants (or isozymes), each of which catalyzes the same given reaction but may display unique catalytic properties.

Control of enzyme concentrations at this highest point in the metabolic control hierarchy has obvious advantages and limitations. On the one hand, gene activation/repression is an effective way of changing enzyme concentrations in a highly specific fashion. On this count, it is a highly versatile control mechanism. However, in eucaryotic cells, with which this book is almost exclusively concerned, gene activation is a slow process. Normally, at least hours are required for an inducing or repressing signal to exhibit its effects at the level of enzyme concentration. In contrast, environmental changes may occur within seconds or minutes and, therefore, survival may depend on biochemical adaptations which can occur at the same rapid rates.

Following transcription, mRNA participates in a series of reactions which terminate in the production of "nascent" polypeptide chains. This series of reactions is termed "translation" as it is these reactions which translate the information carried by the language of the genetic code into a protein or polypeptide molecule. The control of protein synthesis at the level of translation is poorly understood. Theoretically, control could be exerted at any of the several steps in the translation sequence, including (i) the binding of mRNA to the 40S ribosomal subunit, (ii) the formation of the 80S ribosomal complex, (iii) the activation of amino acids, and (iv) the rate at which the mRNA message is read. In reality, controls at sites other than (i) probably are not of sufficient specificity to be of general importance in metabolic regulation.

Following translation, the newly formed nascent polypeptide chain usually must assume a particular higher order structure before it is functional. Thus, merely having the correct amino acid sequence (primary structure) does not assure that the protein is functional. The nascent polypeptide chain must gain the proper secondary (2°), tertiary (3°), and for most enzymes, quaternary (4°) structure* before it can be of metabolic "use" to the cell.

There appears to be precise regulation of the rate at which cer-

---

*_Primary structure_ refers to the covalent backbone of the polypeptide chain and specifically denotes the amino acid sequence. _Secondary structure_ refers to the extended or helical conformation of polypeptide chains. _Tertiary structure_ refers to the manner in which the chain is bent or folded to form the compact, folded structure of globular proteins. The general term, _conformation_, is used to refer to the combined 2° and 3° structures. _Quaternary structure_ denotes the manner in which individual chains of a protein having more than one subunit are arranged in space. Proteins possessing more than one chain are known as _oligomers_; their component chains are called _subunits_ or _protomers_.

tain levels of higher order structure are attained. Whereas for 2° structure the amino acid sequence of the protein appears to bear major, if not sole, responsibility for the geometry of the molecule, 3° and 4° structures are at least partially under the control of exogenous molecules. For example, the aggregation of subunits to form the functional holoenzyme (the 4° structure) may depend on (enzyme-catalyzed) phosphorylation of the subunits. This process, in turn, may be hormonally regulated. The equilibrium, inactive subunits $\rightleftharpoons$ active oligomer, is often also influenced by substrates and cofactors of the reaction. Thus, epigenetic control of enzyme function at the level of acquisition of 3° and 4° structure may be of significance in metabolic regulation.

Lastly, many enzymes must be integrated into definite cellular structures or joined with other large molecules before they are functional. In these processes, which represent the acquisition of quintinary (5°) structure, the enzyme combines with other enzymes, lipid molecules, or membranes and thereby becomes capable of catalysis. Thus, many enzymes are true lipoproteins: the protein is enzymatically inactive (or largely nonfunctional) in the absence of a lipid moiety. Many, perhaps most, membrane-bound enzymes fall into this category. Other enzymes function only when associated with one or more functionally related enzymes. For example, the enzyme complexes involved in pyruvate dehydrogenation (the pyruvate dehydrogenase complex) and fatty acid synthesis (the fatty acid synthetase complex) contain three and seven distinct enzymes, respectively. Any factor which enhances or interferes with the assumption of these types of 5° structures will influence the concentration of *functional* enzyme present in the cell.

ENZYME MODULATION IS THE MOST RAPID AND SENSITIVE MECHANISM OF METABOLIC REGULATION

Once an enzyme has attained its functional state, following transcription, translation, and assembly, its activity is largely determined by (i) the availability of substrate and cofactors, (ii) the influences of physical environmental parameters such as temperature and pressure, and (iii) the interaction with a class of metabolites termed "*enzyme modulators*." These latter compounds are instrumental in providing metabolism with its most *rapid, accurate, and sensitive level of regulation.*

Enzyme modulators fall into two classes. *Positive modulators*, as their name denotes, increase enzymic activity, whereas *negative modulators* have the opposite effect. The enzymes which are regulated by modulators are termed *regulatory enzymes*. Not all enzymes are regulatory in this sense, and we will see that the positioning of regulatory enzymes in metabolic sequences is a vital element in the "design" of metabolic control mechanisms.

To appreciate how enzyme modulators carry out their integrative functions, we must first "dissect" a highly generalized enzymic reaction to discern what steps in the reaction might be especially suitable for control purposes.

ENZYME CATALYSIS IS FUNDAMENTALLY A THREE-STEP PROCESS

Three major events occur during an enzymic reaction:

1. Enzyme-Substrate Complex Formation. In this critical first step of an enzymic reaction, free enzyme and free substrate bind to each other, forming the enzyme-substrate (E-S) complex. Formation of this complex is critically dependent on the geometrical and charge characteristics of the enzyme and substrate. The two molecules must have sterically complementary structures in order to "fit" together, and the molecules cannot have strongly opposing charge properties. The substrate(s) of the reaction bind at the *substrate binding site(s).*

2. Activation of the E-S Complex. As we will discuss in greater detail in the chapter dealing with temperature effects (see p. 182), the role of enzymes is to accelerate chemical transformations by reducing the "energy barriers" to chemical reactions. It is extremely important to appreciate the nature of these "barriers," for they are the primary determinants of how rapidly the chemical reactions of metabolism occur at biological temperatures. The rate at which a chemical reaction occurs is largely set by the *free energy of activation* of the reaction. This free energy value can be regarded as the amount of energy which must be "put into" the E-S complex to weaken or distort the bonds of the substrate molecule(s) and thereby facilitate the conversion of substrate(s) to product(s). Mere formation of the E-S complex does not weaken or distort bonds to the extent that an active complex results. Additional energy from the surrounding system, for example in the form of heat, must be supplied. The catalytic function of enzymes is to reduce the amount of this input energy which is required to form the active complex. Enzymes thus allow rates of physiological reactions to occur at velocities some 10^8 to 10^{12} times as fast as uncatalyzed reactions.

3. Release of Product(s) and Regeneration of Free Enzyme. Once the active complex is generated, the subsequent conversion of substrate(s) to product(s) is usually very rapid. Following product formation, the enzyme must free its substrate (and product) binding site in order to initiate a new round of catalysis. Thus, much as enzymes must efficiently bind substrate in the initial step of a reaction, they must be able to readily release the product(s) of the reaction. Biochemists often speak of "turnover" of substrate molecules when discussing these steps, and the *turnover number* of an enzyme is the number of substrate molecules converted to product per molecule of enzyme per unit of time. The higher the turnover number of an enzyme, the greater is its efficiency.

We see, therefore, that three fundamental sites exist for the control of rates of enzymic catalysis. Which of these is the most important? *A priori* we might predict that all three steps would be of importance in regulatory mechanisms, since all three steps of catalysis could affect the rate at which the enzyme "turns over" substrate. In fact, this does not occur; in almost all known cases, regulatory function is vested largely, if not entirely, at the initial step of the reaction, the formation of the E-S complex.

MEASUREMENT OF E-S AFFINITY

There are a number of approaches which can be utilized to estimate enzyme-substrate affinity. Ideally, one would like to know the three-dimensional structure of the substrate-binding site in order to determine how readily the substrate can fit onto the enzyme surface and what forces determine binding. The strength of such forces can then be obtained by direct measure of the equilibrium constant for substrate binding. A high binding constant indicates a low E-S affinity, while a low binding constant indicates a high E-S affinity. Unfortunately, this constant is not especially easy to obtain, and most workers are content to employ what is termed an apparent value for E-S affinity. This value is obtained from measures of reaction velocities at different substrate concentrations, and the criterion of affinity is *the concentration of substrate which yields half the maximal velocity*. For enzymes with hyperbolic saturation curves (Figure 1–2), the half-saturating concentration of substrate is called the apparent Michaelis constant (apparent K_m). This constant is "apparent" because it is determined graphically from velocity measurements rather than by the direct measurement of the dissociation constant of the E-S complex. For enzymes which display sigmoidal (S-shaped) saturation curves, the half-saturating substrate concentration is denoted as the $S_{0.5}$ value.

For both classes of enzyme, the affinity parameter is measured in units of substrate concentration. As in the case of the true E-S affinity, it is important to remember that a high E-S affinity is indicated by low values of K_m (or $S_{0.5}$) and vice versa.

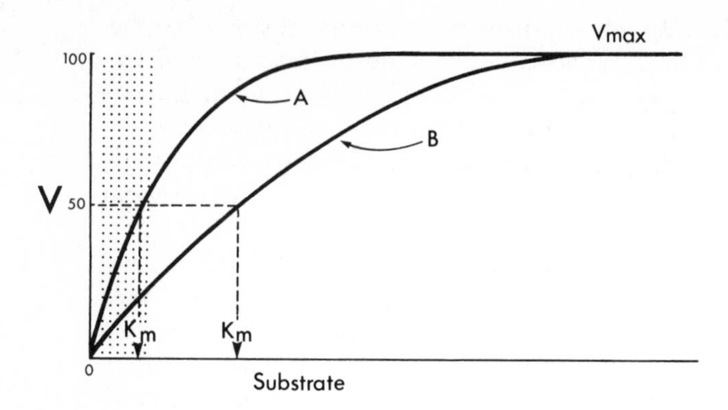

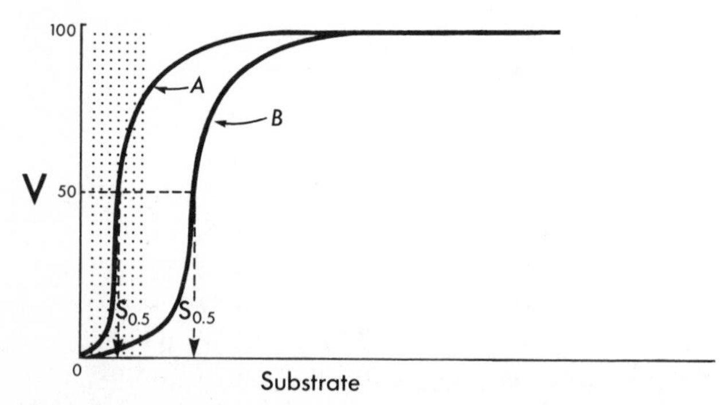

Figure 1–2 Substrate saturation curves for two kinds of enzymes —for those displaying hyperbolic saturation curves (upper panel) and those displaying sigmoidal saturation curves (lower panel). The concentration at which the reaction velocity (v) is 50% of V_{max} is defined as the K_m for hyperbolic saturation curves, or the $S_{0.5}$ for sigmoidal saturation curves. The shaded zone indicates physiological concentration ranges, which usually are somewhat lower than the apparent E-S affinity.

SIGNIFICANCE OF LOW AND HIGH E-S AFFINITIES

To gain an insight into the functional significance of E-S affinity, let us assume that two versions or "variants" of a particular enzyme are of equal efficiency in lowering the energy barrier to a given reaction, but differ markedly in their affinities for a common substrate.

In the system illustrated in Figure 1–2, each enzyme is present in equal amounts. How will their activities differ?

At sufficiently high concentrations of substrate, when there is always enough substrate to keep the substrate-binding sites of the two enzymes "saturated" with substrate, both enzymes exhibit the same maximum catalytic rate, termed the V_{max}. Under these conditions, the rate limiting step in the reaction occurs at some site other than substrate binding, most probably at the level of activation of the E-S complex.

In contrast to the V_{max} situation, at non-saturating substrate concentrations, enzyme A (the low K_m enzyme) exhibits higher activity than enzyme B. At these non-saturating conditions, the rate limiting step probably is E-S complex formation, and hence, the overall reaction is sensitive to the E-S affinity. *The lower the substrate concentration, the greater is the influence of E-S affinity upon reaction velocity,* and of course the greater the advantages of an enzyme with a high E-S affinity.

PHYSIOLOGICAL SUBSTRATE CONCENTRATIONS ARE OF THE SAME ORDER OF MAGNITUDE AS K_m AND $S_{0.5}$ VALUES

Were it not for one fact, E-S affinity might be of little biological relevance. This vital fact, that *physiological substrate concentrations are almost always too low to saturate enzymes,* means that the rate at which an enzyme functions is highly sensitive to changes in E-S affinity. This is shown in Figure 1–2, where the normal ranges of substrate concentrations *in vivo* are indicated by shading. It will be evident from this figure that *slight changes in E-S affinity can lead to large changes in catalytic rate.* That last statement is the key to much of enzyme regulation.

MODULATORS USUALLY CHANGE E-S AFFINITIES, NOT V_{max} VALUES

Figure 1–3 illustrates the manner in which most enzyme modulators exert their regulatory effects. In the absence of positive or negative modulators, a regulatory enzyme usually exhibits a substrate affinity value (K_m or $S_{0.5}$) which falls within the range of physiological substrate concentrations. The addition of a positive modulator increases E-S affinity, enabling the enzyme to operate at a velocity closer to its V_{max} potential. Negative modulators usually promote increases in the apparent K_m or $S_{0.5}$ value and, therefore, reduce the rate of catalysis. In most cases modulators have no effect on the V_{max} value of the reaction; i.e., modulators do not appear to change the enzyme's ability to lower the "energy barrier" to its reactions.

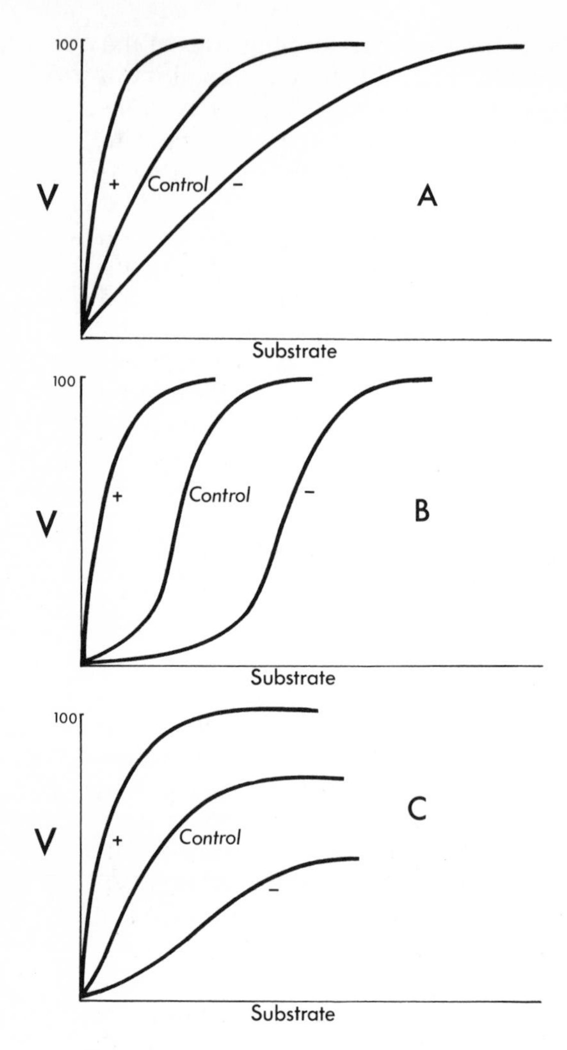

Figure 1–3 The influence of positive (+) and negative (–) modulators on three different kinds of regulatory enzymes. In (A), substrate saturation follows hyperbolic kinetics at all times and only the K_m is affected by (+) or (–) modulators. In (B), substrate saturation curves are sigmoidal and again modulators affect only the apparent E-S affinity, the $S_{0.5}$ parameter. In (C), the E-S affinity, the V_{max}, and the nature of the saturation process are affected by modulators.

ALLOSTERISM

The compounds which serve as enzyme modulators frequently bear little chemical similarity to the substrates, products, or cofactors of the enzymic reactions being regulated. This is to be expected: what an enzyme needs to "sense," if it is to gear its activity according to the demands placed by the cell for the products of the *pathway* of which it is an integral part, is the concentration of the final, key products of the pathway. For example, many of the main-line catabolic pathways have as their primary function the generation of ATP, the cell's "energy currency." The enzymes which act as the "valves" or "switches" in these ATP-generating pathways

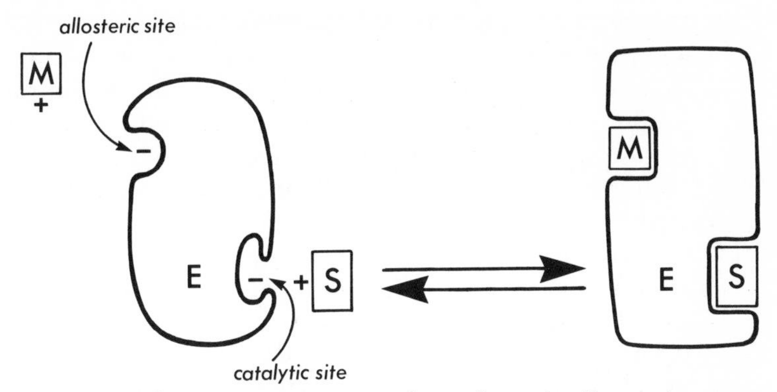

Figure 1–4 Schematic representation of a regulatory (or allosteric) enzyme, stressing the steric and spatial separation of the allosteric and catalytic sites. Binding of the positive modulator is thought to bring about a conformational change in the enzyme which increases substrate binding.

should be responsive to changes in the ATP concentration of the cell—even though ATP may be very dissimilar, structurally, to the substrates of the enzyme. In short, regulatory enzymes must exhibit affinity for modulator compounds which, for steric reasons, cannot bind at the substrate-binding site(s).

We thus find that most regulatory enzymes have distinct regions on their surfaces for binding substrates, on the one hand, and modulators, on the other. The modulator-binding sites have been termed "*allosteric sites*" (meaning sterically different) to indicate their steric and spatial separation from the site(s) where catalysis *per se* occurs. The mechanism of allosteric regulation of catalysis reveals most elegantly the level of sophistication which exists in protein "design." The most probable mechanism of allosteric modulation involves modulator-induced changes in the conformation of the protein, such that the binding of a modulator determines whether the substrate-binding site will have the proper geometry and charge configuration to permit substrate binding (Figure 1–4). We see, then, another rationale for the "macro" dimensions of biological catalysts: not only must these molecules supply complex sites for substrate attachment, but they must also have the proper degree of structure to allow the allosteric regulation of catalysis.

THE BASIC DESIGN OF METABOLIC CONTROL CIRCUITRY

It is common to hear biochemists refer to the vast interlinked network of metabolic reactions as the "metabolic map." This is an apt metaphor. In addition to there being a complex series of "routes" through which metabolites are channelled, there exist highly coordinated "traffic signalling" systems which keep the rates and directions of metabolic flow consistent with the overall interests of the cellular economy. Without these controls and their proper positioning and use, the vast catalytic potential of the "metabolic map" would be quite useless, much as a modern freeway system would be chaotic without traffic signals, off-ramps, and so forth.

The overall design of metabolic control circuitry is remarkably simple, and is based on two fundamental characteristics:

(1) The regulatory enzymes which direct metabolic flow are usually *strategically positioned* either (a) at the beginning of metabolic pathways, or (b) at vital metabolic branchpoints, where two or more pathways diverge or converge.

(2) Regulatory enzymes are, as we have already emphasized, sensitive to the cell's needs for the products of the pathway as a whole. This "sensitivity" is of two distinct classes: (i) All regulatory enzymes, whether linked to biosynthetic or to catabolic pathways, are "locked" into the *adenylate energy charge* of the cell (Figure 1–5). The importance of adenylate sensitivity can be easily appreciated, for the adenylates (AMP, ADP, and ATP) are the most important metabolic coupling agents between energy-yielding and

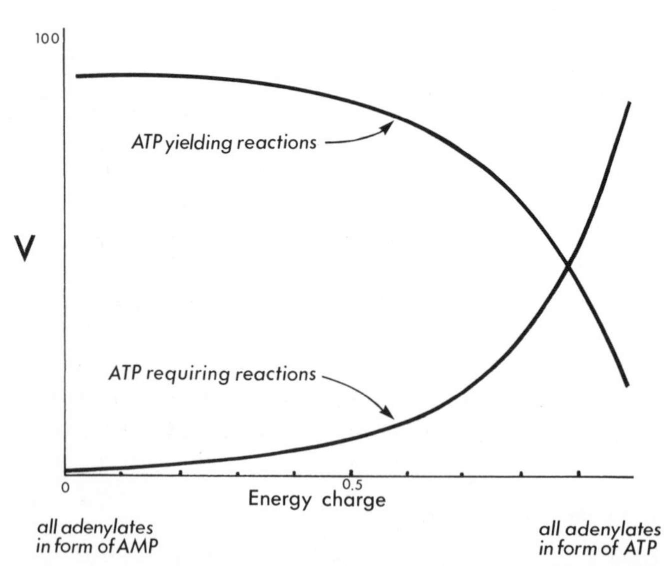

Figure 1–5 The effect of energy charge upon ATP-yielding and ATP-requiring enzyme reactions. Energy charge, defined as the concentration ratio (ATP + 1/2 ADP) / (AMP + ADP + ATP), is a measure of the degree to which AMP is "charged" with high-energy phosphate bonds. This ratio is usually maintained between about 0.7 and 0.9. In this range, small changes in energy charge cause large changes in reaction velocities.

energy-utilizing processes. (ii) In addition, regulatory enzymes are sensitive to the levels of certain metabolites which are characteristic intermediates or products of the particular pathway. An enzyme which regulates a sequence of reactions leading to the synthesis of a particular amino acid, for example, typically is feedback inhibited by this amino acid, the terminal product of the pathway.

These two types of regulatory sensitivities can be seen to offer "coarse" and "fine" control potentials for regulatory enzymes. Energy charge modulation is a relatively coarse control, whereas modulation by specific pathway products permits a fine-tuning of metabolic activity. For example, a high energy charge generally stimulates energy (ATP) utilizing pathways; specific metabolites supply the cell with the additional information required to determine which energy requiring processes are of particular need at the moment.

METABOLIC CONTROL: COMPLEXITIES IN ENVIRONMENTAL INTERACTIONS AND ADAPTATIONS

The complex, hierarchical nature of metabolic control has several important implications concerning the interactions between the parameters of the environment and the organism's biochemistry. Firstly, an environmental change may affect one or several events in the metabolic control hierarchy. Thus, the metabolic machinery of the cell is "vulnerable" at a number of sites.

Secondly, if we turn the above argument around, we can say that biochemical adaptation to the environment, to the extent that it involves changes in the rates and directions of metabolic activity, can at least in theory be effected in a number of different ways: the organism has available to it a number of different strategies of biochemical adaptation.

Thirdly, the interaction between an environmental parameter and metabolic activities may be sudden or relatively slow, depending on where in the control hierarchy the environmental parameter impinges. Altering the rate of transcription, for example, would exert a slower effect on metabolism than would a direct influence of the environment on the activity of a pre-existing regulatory enzyme.

Lastly, if we again turn around a previous point, we can state that the speed with which a biochemical adaptation to an environmental change is effected will depend on the site within the metabolic control hierarchy at which the adaptation occurs. Modulation of enzymes already active in the cells may permit a nearly instantaneous adaptation. In contrast, gene activation/repression may require hours or days. And, to take an extreme case, the accumulation of new DNA base sequences within the genome, which code for adaptive gene products, may require many generations. In short, *we would expect a strong interplay between the rate of adaptation and the strategy of adaptation.*

TIME-COURSES OF ENVIRONMENTAL ADAPTATION

Because it is apparent that the type of biochemical adaptation strategy an organism can use in responding to an environmental change is strongly influenced by the time available to the organism for its adaptive response, we must examine the different time-courses of adaptation which are observed in nature. Biologists generally recognize three fundamental time-courses of environmental adaptation:

1. Evolutionary Adaptation. The longest adaptation process is that which depends on the acquisition of new genetic information which codes for new and adaptive phenotypic traits. This is the process of evolutionary adaptation and clearly involves generations of time.

2. Acclimation or Acclimatization. There are numerous cases of adaptations which occur during the lifetime of an individual and normally require from hours to months to be completed. Good examples are the many seasonal changes which occur in both "warm-blooded" and "cold-blooded" organisms. When these phenotypic changes occur in the laboratory in response to experimental manipulation of only a single environmental variable (e.g., temperature or photoperiod), the changes are termed "acclimations." When the adaptive process occurs under natural environmental settings, the adaptations are called "acclimatizations." In the latter situation several environmental parameters may vary simultaneously, making it difficult to determine rigorously which parameter is serving as the signal to adapt or, in fact, which parameter "needs adapting to." Thus, most studies of adaptations of this time-course have focused on acclimations.

3. Immediate Adaptations. There are at least some adaptations which appear to occur essentially instantaneously. An environmental change is accommodated virtually at once by an adaptive response. At the biochemical level such "adaptations" must be inherent characteristics of metabolism even though they may appear highly perplexing and, indeed, may appear to disobey certain laws of physics and chemistry.

THE "LAST RESORT" NATURE OF BIOCHEMICAL ADAPTATIONS

While our interests are primarily in adaptations at the biochemical level, it would be unfair to give the reader the impression that adapting organisms share our particular bias. In fact, biochemical adaptations often appear to be "last resort" responses, made if and only if the organism lacks behavioral or physiological avenues of escape from an environmental stress. In general, biochemical adaptation is not "easy"; it often appears a simpler matter to migrate away from an environmental stress, for example, than to rearrange the fundamental chemistry of the cell.

"COMPENSATORY" AND "EXPLOITATIVE" ADAPTATIONS

We end this general, overview treatment of biochemical adaptations to the environment by raising one last point concerning the categories of adaptations. The biochemical adaptations we will examine in the course of this volume can be grouped under two headings: "compensatory" and "exploitative." The first of these classes of adaptations might be termed "mechanisms for righting an environmentally caused wrong." Thus, when an environmental change adversely perturbs the functionings of an organism, biochemical changes may ensue which restore the organism's functional capacities to their previous levels. Thus the term "compensatory adaptations."

Other biochemical changes may give the organism a wholly new potential for making use of its environment or for invading a new environment. We term these adaptations "exploitative." In contrast to biochemical changes which are compensatory or restorative, exploitative adaptations are, strictly speaking, not necessary. The organism could live without these new potentials; however, with them it may do significantly better in its native habitat or it may enter a new habitat previously unavailable to it.

One of the best ways of illustrating these various adaptations is to consider the manners in which organisms have adapted to one of the most critical environmental parameters: the oxygen available to the cell.

SUGGESTED READING

Books and Proceedings

Physiological and Biochemical Adaptation, in *Amer. Zoologist* (1971) *11*, 81–165.

Reviews and Articles

Atkinson, D. E., and G. M. Walton (1967). Adenosine triphosphate conservation in metabolic regulation. *J. Biol. Chem.* *242*, 3239–3241.

Atkinson, D. E. (1971). Adenine nucleotides as stoichiometric coupling agents in metabolism and as regulatory modifiers. The adenylate energy charge. In *Metabolic Pathways* (ed. H. J. Vogel), Academic Press, New York, Vol. V. pp. 1–21.

THE INFLUENCE OF OXYGEN AVAILABILITY

CHAPTER 2

THE ORIGIN OF LIFE OCCURRED IN THE ABSENCE OF O_2

Considering the absolute dependence on free oxygen of many living organisms, it may appear surprising that the origin of life, and the evolution of its most significant metabolic "inventions," could have occurred under O_2-free conditions. Nevertheless, in recent years the hypothesis of an O_2-free origin of life received convincing experimental support with the demonstration that many vital biological molecules can be produced *in vitro* under experimental conditions which simulate the essentially oxygen-free atmosphere of the prebiological earth. In these studies the gas composition of the "atmosphere" includes NH_3, N_2, hydrogen cyanide (HCN), CO_2, and water vapor. To these reagents energy is added in the forms of electrical discharges ("lightning") and ultraviolet light ("solar radiation"). The interaction between such primeval atmospheric constituents and "natural" energy sources yields amino acids, purines, pyrimidines, and other organic molecules essential for life. Interestingly, astronomers have recently detected hydrocarbon compounds (e.g., formaldehyde, formaldimine, and acetaldehyde) in the far reaches of intergalactic space. This observation, like the data from *in vitro* studies, demonstrates that the origin of many of the organic molecules requisite for life as we know it can occur under environmental conditions which, for present day organisms, would be lethal.

Not only is it possible for many "biological" molecules to originate under anoxic conditions, such an origin for these compounds may well have been *necessary*. Had free oxygen been present during the initial syntheses of these molecules, it seems almost certain that their ultimate fate would have been degradation by combustion. Only in a milieu lacking free oxygen could these precursors of living systems accumulate in sufficient concentrations to permit frequent interactions—and chemical reactions—among themselves and, thereby, allow the beginnings of primitive metabolic systems.

BIOLOGICAL EVIDENCE FOR THE PRIMACY AND PRIMORDIAL NATURE OF ANAEROBIC METABOLISM

Even if we were lacking solid geochemical evidence about the chemical composition of the atmosphere of the prebiological earth, there are strong biological arguments which, in themselves, should convince us that life arose under O_2-free conditions. Otherwise, as George Wald argues in his concise paper on the origin of life, it would be extremely difficult to rationalize the ingenuity displayed by present-day organisms in conducting so many of their metabolic transformations in oxygen-independent manners. Thus, the basic skeleton of intermediary metabolism is strictly anaerobic; metabolic reactions involving the direct participation of molecular oxygen (i) are rare and (ii) are later evolutionary embellishments added onto an already functional anaerobic framework.

THE EXPLOITATIVE NATURE OF METABOLIC EVOLUTION

The evolution of metabolism must be viewed as a slow, stepwise process through which new abilities for exploiting the chemical and physical environments have appeared sequentially over the past 3 or 4 billion years. In fact, the exploitative nature of biochemical adaptation can find no better illustrations than in the manners whereby metabolic reaction systems have succeeded in utilizing new sources of substrate or energy as these have become available in the environment. A brief summary of Wald's scheme of metabolic evolution is presented in Table 2–1. This scheme clearly illustrates the interplay between organisms and their environment: The activities of organisms lead to environmental changes, which in turn are exploited via the evolution of new metabolic potentials.

The types of metabolic transformations open to an organism are strictly dependent on two factors: (i) the chemicals (metabolites) available in the external environment, and (ii) the catalytic potentials of the enzymic machinery possessed by the organism. In the evolution of metabolism, it is clear that selection favored the retention of those enzymes, arising through random mutations of the genome, which are capable of new forms of exploitation of the chemical environment.

TABLE 2–1 THE MAJOR STAGES IN CHEMICAL AND BIOLOGICAL EVOLUTION

Millions of Years Ago	Approximate Time of Origin
	Multicellular Organisms
	Eucaryotic Cells
1,000	
	First aerobic bacteria
	$C_6H_{12}O_6 + 6\ H_2O + 6\ O_2 \rightarrow 6\ CO_2 + 12\ H_2O + 30\text{–}40$ ATP
2,000	
	Green Plant Photosynthesis:
	$6\ CO_2 + 12\ H_2O \rightarrow C_6H_{12}O_6 + 6\ H_2O + 6\ O_2$
	Bacterial Photosynthesis:
	$6\ CO_2 + 12\ H_2A \rightarrow C_6H_{12}O_6 + 6\ H_2A + 12\ A$
	Photophosphorylation: light energy into ATP
3,000	
	First Cells (Procaryotic)
	Pentose Phosphate Cycle: reducing power, pentoses
	$6\ C_6H_{12}O_6 + 6\ H_2O + 12$ ATP $\rightarrow 12\ H_2 + 5\ C_6H_{12}O_6 + 6\ CO_2$
	Fermentation: a chemical source of energy; CO_2 release
	e.g., $C_6H_{12}O_6 \rightarrow 2\ C_2H_5OH + 2\ CO_2 + 2$ ATP
4,000	
4,600	Formation of the Solar System

THE NATURE AND ORIGIN OF FERMENTATION

When enzymic catalysis first appeared, the number of metabolites present in the environment was relatively small, and most of the important metabolic reactions of present-day organisms could not occur. In addition to being relatively simple, early forms of metabolism were also highly inefficient since, in the absence of molecular oxygen, complete combustion (respiration) of foodstuff molecules could not occur. Initially, therefore, energy metabolism was based entirely on fermentative sequences. In these processes foodstuffs are oxidized (hydrogens removed) anaerobically; i.e., the ultimate hydrogen acceptors (oxidizing agents) are molecules other than molecular oxygen.

All cells, whether aerobic or anaerobic, obtain the energy required for their vital functions from oxidation-reduction reactions. In these reactions the passage of hydrogen (electrons + protons) from reducing agents to oxidizing agents is accompanied by a release of energy, since the electrons lose energy as they pass from a molecule with low electron affinity to one with a higher electron affinity. *One of the most significant achievements of biochemical evolution was the acquisition of an ability to trap a portion of this energy in biologically useful form,* for instance in the bonds of organo-phosphate compounds such as ATP or the thiolester bonds of CoA derivatives (Figure 2–1).

A key feature of biological oxidation-reduction reactions is the cyclical manner in which cofactors are reduced and oxidized. In most oxidation reactions, the foodstuff molecule reduces a pyridine nucleotide cofactor (NAD or NADP, shown in Figure 2–2). If the foodstuff oxidation process is to continue, a resupply of oxidized cofactor is obviously a necessity. The cell could maintain a steady supply of NAD or NADP by continuously synthesizing new cofactor molecules and supplying these to the oxidative enzymes. However, considering the energy required to achieve these syntheses, this solution to the cofactor supply problem is thermodynamically impossible. Instead, the cofactor molecules are cyclically reoxidized. The hydrogen (electrons) donated to the cofactor by the foodstuff molecule are ultimately donated to another molecule, a *terminal electron acceptor,* which has a higher electron affinity than the cofactors. The distinguishing feature of fermentative metabolism is that the *terminal electron acceptor is an organic molecule, and is never oxygen.*

Two major fermentative sequences exist in present-day organisms (Figure 2–3). *Glycolytic fermentation* consists of the stepwise breakdown of hexoses to a pair of three-carbon pyruvate molecules. Two molecules of NAD are reduced to NADH in the process (Figure 2–2), and a net yield of two ATP molecules per molecule of hexose is obtained. To regenerate NAD from NADH, pyruvate is reduced to lactate, the end product of glycolysis.

Alcoholic fermentation utilizes the same set of reactions as glycolysis, down to the level of pyruvate. Beyond this point the two fermentation sequences differ in that alcoholic fermentation terminates with two reactions, one of which yields acetaldehyde and CO_2, and a final step in which acetaldehyde is reduced to ethanol, with the concomitant regeneration of NAD. Alcoholic fermentation likely represents the more primitive pattern of fermentation. It also represents an example of environmental modification by metabolic activity, namely the release of CO_2 into the water and atmosphere.

It is worth our effort to consider in somewhat more detail the precise transformations involved in these primitive—and still basic—anaerobic oxidative sequences, to emphasize how centrally these reaction sequences are positioned in intermediary metabolism. The overall process of anaerobic fermentation can be divided into two stages. In the first (Figure 2–3), which can be termed a "collection" or "preparatory" process, a number of different hexoses enter the fermentative sequence, following their phosphorylation. The further metabolism of all the hexose phosphates yields a common, three-carbon intermediate, glyceraldehyde-3-phosphate.

The second stage of fermentation begins with this compound

(Text continued on p. 26.)

General structure

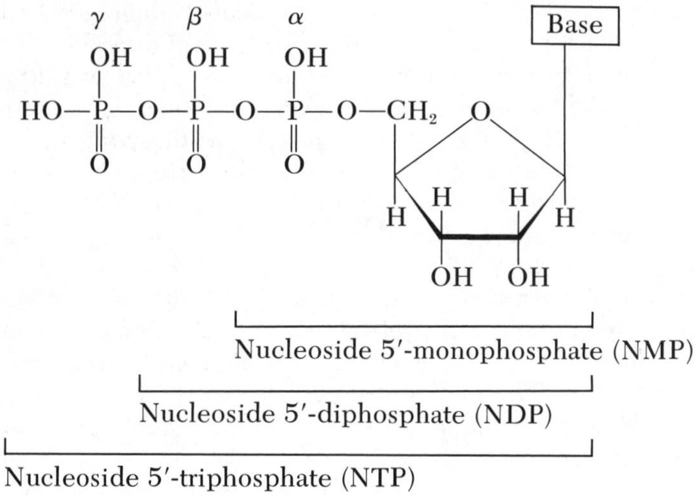

γ β α

OH OH OH

HO—P—O—P—O—P—O—CH_2 O Base

O O O

H H H H

OH OH

Nucleoside 5′-monophosphate (NMP)

Nucleoside 5′-diphosphate (NDP)

Nucleoside 5′-triphosphate (NTP)

Abbreviations

Ribonucleoside
5′mono-, di-, and triphosphates

Base	*Abbreviations*		
Adenine	AMP	ADP	ATP
Guanine	GMP	GDP	GTP
Cytosine	CMP	CDP	CTP
Uracil	UMP	UDP	UTP

Figure 2–1 (A) General structure of the nucleoside mono-, di-, and triphosphates. Of these ATP is the primary "currency" of metabolism. ADP can accept phosphate from any NTP to form ATP; hence, all of the triphosphates are metabolically equivalent. Energy released during oxidation of foodstuff in part is recaptured by the reaction, ADP + P_i → ATP. Energy-requiring processes in turn are "driven" by being coupled to the hydrolysis of ATP. This mechanism of a common intermediate between energy-yielding and energy-requiring processes in effect *quantizes metabolic energy, and the quantum is one unit of ATP or its equivalent.* ATP is often called a high-energy compound. This is a useful term, but the student must clearly understand that it refers only to the difference between the energy contents of reactants and products. The term does *not* refer to the bond energy of covalent linkage between phosphorus and oxygen. The ΔG^0 of hydrolysis of ATP is fairly large and negative (−7.3 kcal/mole); it is unique in occupying an intermediate position in the thermodynamic scale of phosphate compounds. 1, 3-DPG hydrolysis, for example, proceeds with a $\Delta G^0 = -11.8$ kcal/mole, while 3-PGA hydrolysis proceeds with a free energy drop of only −2.2 kcal/mole. Hence, the ADP-ATP system is ideally suited to act as an intermediate linking system for transferring phosphate groups between such compounds of high and low "group transfer potential." Thus, 1,3-DPG can transfer its phosphate to ADP, forming ATP and 3-PGA, a process which constitutes one example of a substrate-level phosphorylation.

(Illustration continued on opposite page.)

Acetyl Coenzyme A

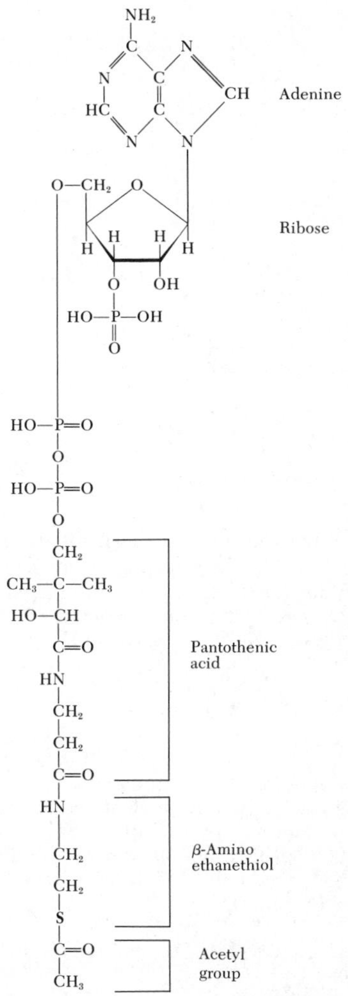

NH_2 — adenine ring (C, N, C, CH, HC, C, N, N) — Adenine

$O—CH_2$ — ribose ring (O, H, H, H, H, O, OH) — Ribose

HO—P—OH (=O)

HO—P=O

O

HO—P=O

O

CH_2

$CH_3—C—CH_3$

HO—CH

C=O

HN

CH_2

CH_2

C=O — Pantothenic acid

HN

CH_2

CH_2 — β-Amino ethanethiol

S

C=O

CH_3 — Acetyl group

Figure 2–1(B) *Continued.* The metabolism of many compounds depends upon conversion to activated Coenzyme A (CoA) derivatives. The most important of these is acetylCoA, whose structure is shown. In a sense, CoA is a carrier of acetyl groups just as ATP is a carrier of phosphate groups. CoA has a structure reminiscent of ATP because it contains adenine, ribose, and a pyrophosphate bridge. The thiol (—SH) group of CoA serves to bind acetic acid covalently through a thiolester linkage, *which is in fact a high energy bond.* The ΔG^0 of hydrolysis (acetylCoA + H_2O → acetic acid + CoASH) is sufficiently large (−7.5 kcal/mole) to be harnessed to the synthesis of ATP. Other compounds that are commonly converted to their CoA activated derivatives are succinic acid, propionic acid, and other fatty acids.

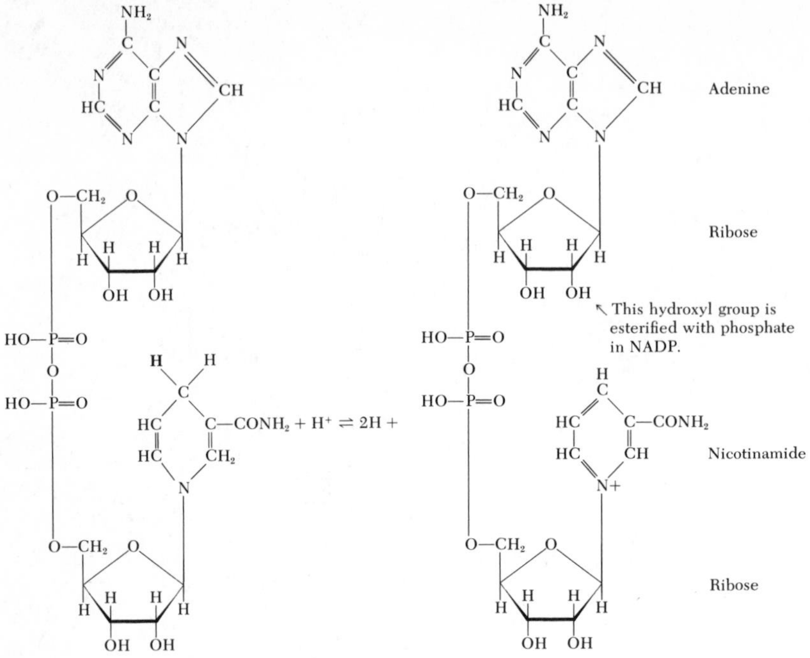

Adenine

Ribose

This hydroxyl group is esterified with phosphate in NADP.

Nicotinamide

Ribose

$$\text{C—CONH}_2 + H^+ \rightleftharpoons 2H +$$

$$NADH + H^+ \rightleftharpoons 2H + NAD$$

Figure 2–2 Nicotinamide adenine dinucleotide (NAD), which is also called diphosphopyridine nucleotide (DPN). In nicotinamide adenine dinucleotide phosphate (NADP), also called triphosphopyridine nucleotide (TPN), there is a third phosphate group as shown. In a sense, NAD is a carrier of H^+ and e^- just as ATP is a carrier of phosphate groups.

Figure 2–3 The two basic glycolytic schemes, one leading to lactate, and the other leading to ethanol, as the major end product. Note that, per mole of glucose mobilized, a net gain of 2 moles of ATP is realized. Shown also is the so-called pentose shunt or the phosphogluconate cycle. Its principal physiological functions are (1) the generation of NADPH for reductive biosyntheses, and (2) the generation of 5-carbon sugars for nucleic acid synthesis.

Metabolite abbreviations used here and throughout the text:

G1P	glucose-1-phosphate
G6P	glucose-6-phosphate
F6P	fructose-6-phosphate
FDP	fructose-1,6-diphosphate
G3P	glyceraldehyde-3-phosphate
DHAP	dihydroxyacetone phosphate
1,3 DPG	1,3 diphosphoglycerate
3 PGA	3 phosphoglycerate
2 PGA	2 phosphoglycerate
PEP	phosphoenolpyruvate
6 PG	6 phosphogluconate
Ru5P	ribulose-5-phosphate
R5P	ribose-5-phosphate
X5P	xylulose-5-phosphate
S7P	sedoheptulose-7-phosphate
E4P	erythrose-4-phosphate

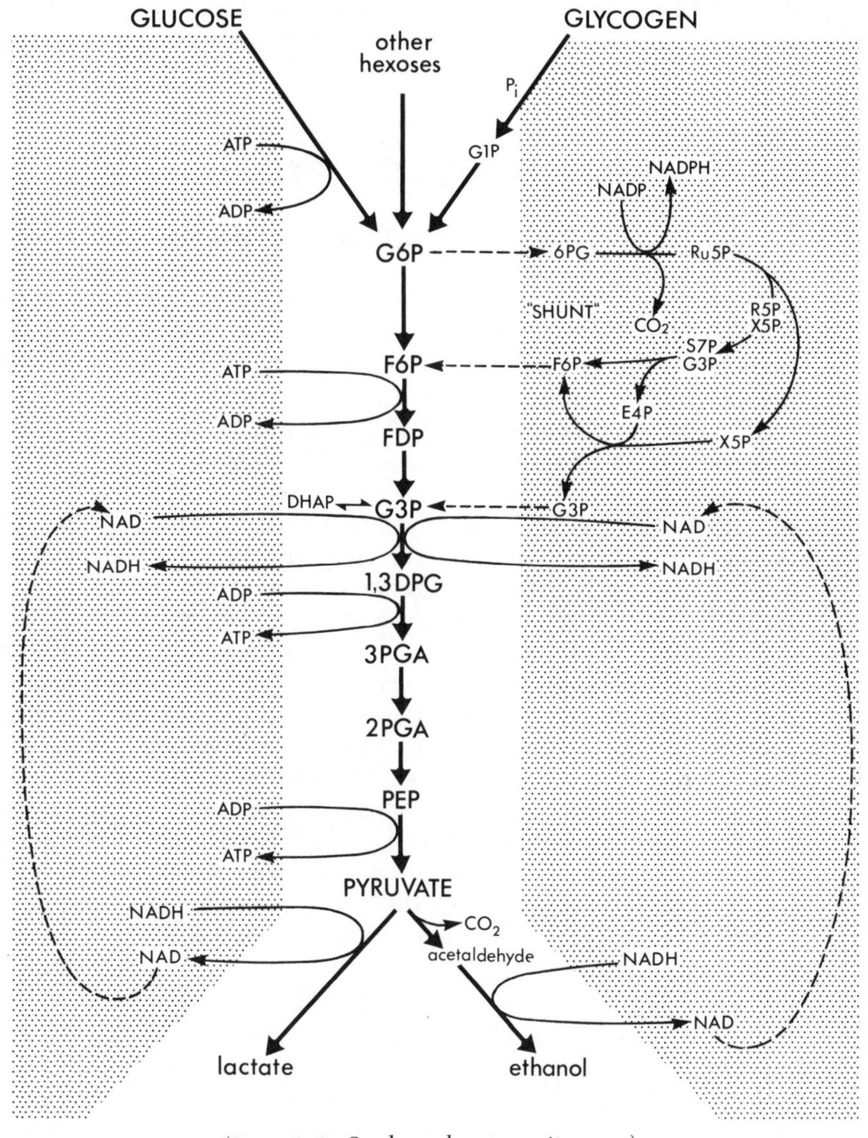

(**Figure 2–3** *See legend on opposite page.*)

and includes the important reactions in which oxidations and ATP formation occur (Figure 2–3). At the end of the second stage of fermentation, the organism has gained two molecules of ATP per hexose, plus the ability to process another hexose molecule due to the regeneration of two molecules of NAD. It should be noted that the oxidation state of the carbohydrate end product, lactate or ethanol, is the same as that of the starting hexose molecule.

THE PENTOSE PHOSPHATE PATHWAYS AND REDUCING POWER FOR BIOSYNTHESES

Fermentative pathways supplied early metabolizing systems with adequate quantities of "high energy" compounds like ATP. This trapping of foodstuff bond energy in biologically useful form can be viewed as the primeval "metabolic invention." And, as long as early living systems remained chemically simple and as long as the environment was rich in the organic molecules necessary for life, the metabolic systems were able to function largely or solely as producers of "high energy" compounds.

As the concentrations of organic molecules decreased and/or the biochemical complexities of living systems increased, a need for biosynthetic potential arose. In particular, the synthesis of macromolecules requires a high rate of supply of reducing power: molecules which function efficiently as hydrogen donors. At first glance one might predict that this need could be met efficiently by employing the reduced NADH formed in fermentative metabolism. However, this does not appear to be the strategy employed by most organisms. One reason why this strategy was avoided is that the regeneration of NAD, and thereby the continued ability to supply ATP, is not dependent on rates of biosynthesis. By means of employing a second cofactor (NADP/NADPH) primarily for biosynthetic purposes, and by generating this reduced cofactor via a reaction sequence separate from the mainline fermentative pathway, the supply of high energy compounds and biosynthetic reducing power were cleanly separated.

The pathway which supplies the larger portion of the cell's reducing power is the pentose phosphate pathway, also known as the hexose monophosphate pathway or the pentose "shunt" (Figure 2–3). This pathway has several key features which represent major "breakthroughs" in metabolic design. Firstly, the pathway can operate in a cyclical manner during which the entry of six glucose-6-phosphate molecules leads to the regeneration of five hexose molecules, with the net catabolism of only a single glucose residue. Secondly, the fact that this pathway competes for glucose-6-phosphate with the mainline fermentative sequence raises the need for metabolic control at the glucose-6-phosphate "branchpoint." Thus, the elaboration of this new pathway led to selection for important regulatory controls for the competing enzymes. Thirdly, the pentose phosphate pathway adds the significant "innovation" of the metabolic splitting of water; half the hydrogen added to NADP is ultimately derived from water. Perhaps of greatest importance is the fact that this pathway gained for the organism a substantial degree of freedom from the external environment. Pentose sugars could now be synthesized and used in nucleic acid metabolism and, as mentioned, an "in house" supply of reducing power was now available. Lastly, the pentose phosphate pathway provided a further source of environmental CO_2, a waste product which was to have far-reaching evolutionary consequences.

THE ORIGIN OF PHOTOSYNTHESIS

The appearance of photosynthesis was attendant on the occurrence of several important changes in the biochemical machinery of living systems and in the appearance of utilizable quantities of carbon dioxide in the environment. Among the former events, the capacity to trap and then transduce solar radiation into biologically useful forms could not be developed prior to the evolution of pigmented compounds like the chlorophylls and cytochromes. Chlorophylls absorb solar radiation and, as a result, electrons are raised to a high energy state. These electrons then flow through cytochrome electron transfer chains, and during this process the energy of the electron is stepwise and sequentially lowered, with a major portion of this released energy being trapped in biologically useful form (ATP and NADPH). This sequence of events, the absorption of solar radiation by one pigmented compound followed by the transduction of this absorbed energy into a form which is metabolically useable, represents the first pattern of photosynthesis to evolve: *photophosphorylation.*

The subsequent origin of carbon-fixing photosynthesis, wherein light energy is used to reduce CO_2 to the oxidation level of carbohydrate, resulted from the integration of the reactions of photophosphorylation and the pentose phosphate cycle (see Chapter 3). The latter sequence is, in essence, reversed during carbon fixation. The reducing power necessary for carbon fixation comes largely from the NADPH formed during photosynthetic electron transport.

The ultimate source of the hydrogen added to NADP varies among different photosynthetic organisms. In all cases a donor molecule is split to yield hydrogen plus an oxidized component. The hydrogen reduces NADP, and the oxidized moiety is usually released into the environment. In procaryotic organisms capable of photosynthesis ("bacterial" photosynthesis), the hydrogen donor molecule is varied; it is never water, however. Only green plant photosynthesis involves the splitting of water to obtain hydrogen. The consequent "waste product" of this reaction, molecular oxygen, is released into the environment. In terms of biological evolution, no case of environmental contamination has had such a far-reaching effect: because of its high electron affinity, oxygen has provided a necessary substrate for an entirely new pattern of metabolism, respiration. Respiratory metabolism led to high levels of metabolic efficiency which, in turn, permitted the evolution of organisms capable of vastly increased levels of physiological activity.

THE ORIGIN AND ESSENTIALS OF AEROBIC METABOLISM

Once the liberation of oxygen by green plant photosynthesis began, it probably took at least a half-billion years for oxygen-utilizing metabolic reactions to assume a major role in metabolism. The limiting factor in the evolution of aerobic energy metabolism (respiration) was, of course, the oxygen content of the atmosphere. Biochemically, organisms were essentially pre-adapted for oxygen utilization, since respiration is basically a reversal of the reactions of green plant photosynthesis. Wald estimates that aerobic metabolism could have arisen once the oxygen content of the air reached 10^{-3} to 10^{-2} atmospheres, i.e., levels of approximately 0.02 to 0.20 per cent of present day values. As indicated in Table 2–1, this may have occurred as long as 1.5 to 2.0 billion years ago.

The basic skeleton of aerobic metabolism is shown in Figure 2–4. As stated above, the net effect of these reactions is to degrade

---

Figure 2–4 The basic flow chart of cellular respiration. AcetylCoA, formed from pyruvate as well as from amino acid and fatty acid metabolism, is the activated form in which acetate is "fed" into the Krebs citrate cycle. OXA initiates the reaction cycle and OXA is regenerated by the cycle, which is thereby seen to play an overall catalytic function. H atoms (electrons and protons) are transferred to NAD or flavoprotein at the reactions indicated. For each two electrons transferred from NADH to O_2, three ATP molecules can be synthesized. For each two electrons transferred from reduced flavoprotein to O_2, only two ATP molecules can be made. NADH equivalents formed in glycolysis are transferred into the mitochondria as malate, a process which yields four ATP molecules per pair of electrons. Thus, for each two moles of pyruvate oxidized to completion, 28 moles of ATP are formed during electron transport, two moles are formed at the succinic thiokinase step, and four are formed by the oxidation of NADH generated glycolytically. When these are added to the two moles of ATP formed in glycolysis, a total of 36 moles of ATP/mole of glucose is gained by these processes.

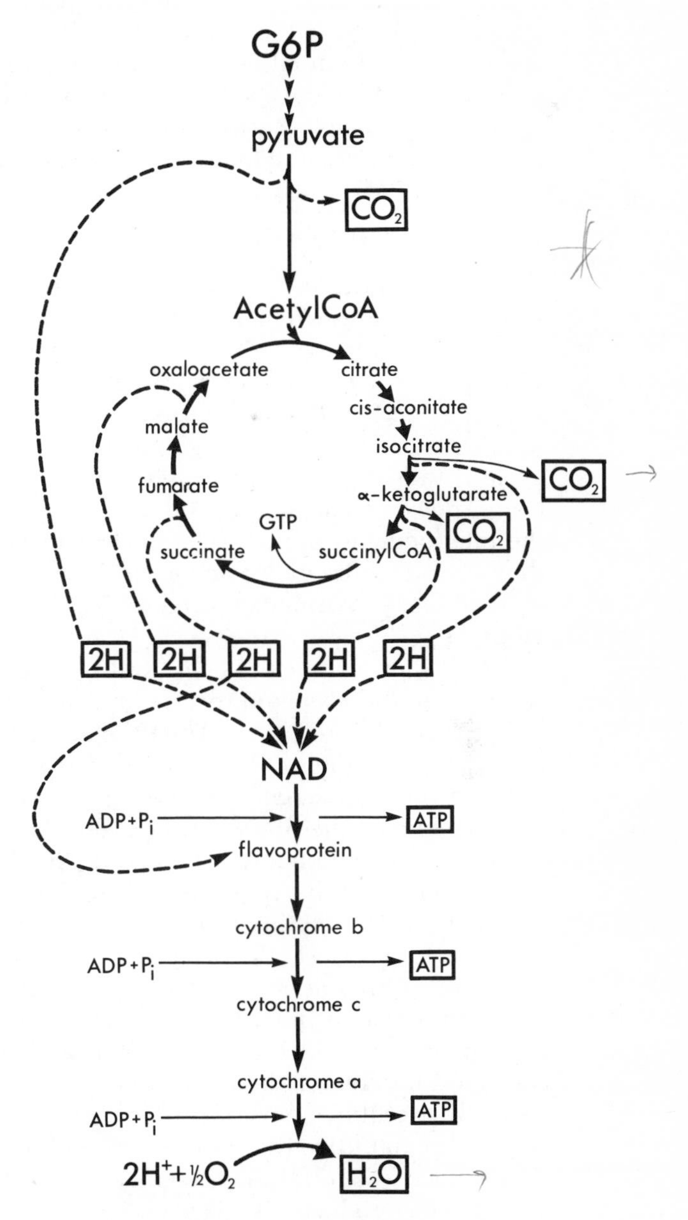

ELECTRON TRANSFER CHAIN

(Figure 2–4 *See legend on opposite page.)*

foodstuff molecules to water and carbon dioxide, using oxygen as the final electron acceptor.

The initial stage of respiration, the reactions of the Krebs citrate cycle, begins with the condensation of oxaloacetate (OXA) and acetyl CoA to form citrate. Acetyl CoA is the common breakdown product formed in the catabolism of carbohydrates, lipids, and some amino acids. Thus, the Krebs cycle is a common final processing system for all classes of foodstuffs. The overall reaction catalyzed by the enzymes of this cycle can be written:

$$CH_3COOH + 2\ H_2O \longrightarrow 2\ CO_2 + 8\ H$$

This decomposition of acetate fragments is accomplished by a series of reactions which is cyclic: the 4-carbon OXA unit utilized in the initial condensation reaction is regenerated at the end of each turn of the cycle. In other words, there is no net removal of OXA from the cycle, and, in theory, one molecule of OXA can enable an infinite number of 2-carbon fragments to enter the Krebs cycle. This cycle is thus catalytic in two senses of the term. Each separate reaction does, of course, depend on enzymic catalysis. In addition, the intermediates of the cycle function in catalytic manners in that they enable an overall reaction to occur, the oxidation of acetate, without these intermediates themselves being used up in the overall process.

Although we speak of the Krebs cycle as the initial phase of respiration, there is no direct involvement of molecular oxygen in any of the cycle's reactions. Thus, the cycle itself can be regarded as an anaerobic pathway, in the strictest sense of the term. However, unlike fermentative schemes, the regeneration of oxidized cofactors does depend on the presence of oxygen. There are no fermentative reactions for completely reoxidizing the NADH formed in the Krebs cycle, and the continued operation of this cycle is thus dependent on its coupling to the reactions of the electron transport scheme (Figure 2–4). In this latter set of reactions, the terminal phase of respiration, not only are the cofactors needed for the continued function of the Krebs cycle reoxidized, but in addition the eight hydrogens added to the cofactors are processed in such a way that the energies of their electrons are tapped and used in biologically useful manners, i.e., to form ATP. In terms of energetics, this ability to utilize the energies of electrons of reduced cofactors, rather than wasting these electrons in the formation of metabolic "dead-end" products like lactate, represents the major advance of respiration over anaerobic reactions.

The electron transport chain functions in the same manner as the pigments and enzymes involved in photophosphorylation. Through a series of steps, which terminates in the formation of water from O_2 and metabolically generated electrons and protons, the energies of these electrons are reduced and "high energy" ATP molecules are formed. Each electron carrier has a higher electron affinity than the carrier preceding it in the series, and in the transfer of an electron from a relatively low to a higher affinity carrier, a fraction of the electron's energy is given off and trapped, as "high energy" phosphate, by the enzymes associated with the transport pigments.

THE COMPARATIVE ENERGETICS OF FERMENTATION AND RESPIRATION

Careful examination of the metabolic scheme in Figure 2–4 indicates that the exploitation of oxygen for the complete degradation of foodstuff molecules led to nearly a 20-fold increase in the energetic efficiency of intermediary metabolism. The comparative efficiencies of glycolytic fermentation and respiration are documented in Table 2–2. The most important point of this comparison lies in the relative yields of ATP per mole of starting hexose in the two schemes. In glycolysis, a net yield of 2 moles of ATP per mole of glucose is obtained, whereas the complete combustion of glucose to CO_2 and water yields 36 moles of ATP per mole of glucose. In short, an organism which relies on respiration needs to ingest a significantly lower quantity of foodstuff starting material to maintain a given rate of metabolism. Furthermore, the final products of respiration, CO_2 and H_2O, are comparatively non-toxic, relative to the acidic end product lactate.

DIFFERING OXYGEN DEPENDENCIES IN PRESENT DAY ORGANISMS

In light of the advantages of respiration, it is not surprising that on a planet in which the atmosphere and hydrosphere abound in O_2

TABLE 2–2 COMPARATIVE EFFICIENCIES OF GLYCOLYSIS AND RESPIRATION

Overall Reaction	Free Energy Change (ΔG^0)	moles ATP/mole glucose
GLYCOLYSIS:		
Glucose $\longrightarrow$ 2 Lactate	−47.0 kcal/mole	2
$C_6H_{12}O_6 \longrightarrow 2\ C_3H_6O_3$		
RESPIRATION:		
Glucose $\longrightarrow 6\ CO_2 + 6\ H_2O$	−686.0 kcal/mole	36

much of the animal kingdom evolved an absolute dependence on aerobic metabolism. However, not all organisms, and not all tissues of a single individual, display the same level of dependence on oxygen. For example, vertebrate skeletal muscle may rely largely on glycolysis during rapid bursts of activity, when O_2 cannot be supplied rapidly enough to permit all of the pyruvate formed to enter the Krebs cycle. The medulla region of the kidney may also rely heavily on anaerobic metabolism. Other tissues, notably the heart and central nervous system, are strictly dependent on respiration. Most tissues fall between these two extremes and can tolerate at least short periods of anaerobiosis.

Just as O_2 dependence differs greatly among different tissues within a single organism, major differences in reliance on respiration are found among different species. One extreme class of organisms, by the criterion of dependence on respiration, are the *strict aerobes.* These organisms cannot survive without O_2. At the other extreme are organisms which are *obligate anaerobes.* Their survival is dependent on the complete absence of O_2. An intermediate class, to which many invertebrate species belong, consists of *facultative anaerobes.* These organisms utilize O_2 when it is present, but can survive indefinitely in its total absence.

STRATEGIES OF ADAPTATION TO DIFFERING O_2 AVAILABILITY

From our more functional view, the above categories can be reclassified according to the biochemical strategies which natural selection has harnessed in dealing with the specific problem of environmental O_2 supply. We can identify three basic strategies of biochemical adaptation to O_2 availability:

1. Compensatory Strategies: Some organisms compensate for the temporary depletion or unavailability of O_2 by maintaining high capacities for anaerobic generation of ATP. The strategy depends upon the ultimate return to aerobic conditions, and the metabolic adaptations occurring are typified by vertebrate white muscle.

2. Exploitative Strategies: Other organisms, such as benthic invertebrates, intertidal bivalves, and parasitic helminths, are able to invade anoxic environments for extended or indefinite periods of time. In these organisms, the substrate-level phosphorylations of glycolysis are linked to other substrate-level phosphorylations, allowing an increased yield of ATP during anoxia. This strategy does not depend upon the ultimate return of the organism to aerobic conditions, and hence allows the exploitation of anoxic environments.

3. Minimizing or Avoidance Strategies: In some instances, even the most extreme physiological demands for energy are met through highly effective capacities for O_2 delivery to aerobic cellular metabolism. In effect, these organisms "minimize" or "avoid" the physiological problem of limited O_2 availability during maximum work performance. This strategy, which reaches its zenith in certain insects, in consequence commits the organisms to an aerobic mode of metabolism.

What are the essential differences in these three biochemical strategies?

THE ANAEROBIC METABOLISM OF VERTEBRATE WHITE MUSCLE: PHYSIOLOGICAL NATURE OF SELECTIVE FORCES

Since most tissues and cells of metazoans are not in direct contact with the external environment, their O_2 requirements can only be satisfied through the use of accessory O_2 delivery systems. The design of these systems varies widely, from simple diffusion, through the open (low pressure) circulation systems of invertebrates, to the closed (high pressure) circulation of vertebrates. From our point of view in this chapter, the essential feature of the vertebrate O_2 delivery system is its limitations: under a variety of stressful conditions ("burst" activity such as diving, flying, running, and swimming), the rate of O_2 delivery cannot meet the total "demands" of all the tissues. The vertebrate solution to this "weak-link" situation is the regulation of blood circulation so as to favor some tissues, notably heart and brain, at the expense of peripheral ones in general, and *white skeletal muscle in particular.* Vertebrates in consequence exhibit an impressive capacity for supporting muscle work by anaerobic glycolysis.

GLYCOGEN AS AN ENERGY SOURCE FOR ANAEROBIC MUSCLE FUNCTION

During such limiting O_2 conditions, glycogen serves as the primary carbon and energy source. The release of glucose, as the 1-phosphate ester, does not entail the cleavage of an ATP molecule, in contrast to the situation when free glucose is shunted into the glycolytic scheme. Thus, although glycogen synthesis (occurring under energy-rich conditions) does use a UTP molecule (energetically equivalent to an ATP) per glucose added to the polymer, this energy is retained in phosphorylytic cleavage, which occurs under conditions when ATP is at a premium.

CONTROL OF GLYCOGEN PHOSPHORYLASE

Glycogen mobilization is initiated by glycogen phosphorylase, a regulatory enzyme *par excellence.* It is strategically situated between glycogen and the metabolic machinery required for its degradation, and is under the control of a variety of hormonal, ionic, and metabolite chemicals. Activation of glycogen phosphorylase (Figure 2–5) can be triggered hormonally via the action of epinephrine, norepinephrine, glucagon, and other agents, and ionically by Ca^{++}. Both activators lead to the formation of an active tetramer, phosphorylase *a*, from an inactive dimer, phosphorylase *b*.

Hormonal activation is indirect and complex. The primary action of epinephrine or norepinephrine is the activation of adenyl cyclase, which catalyzes the reaction, ATP → 3′,5′-AMP (cyclic AMP) + PP_i. Cyclic AMP, in turn, activates the enzyme protein kinase by dissociating an inactive complex (with catalytic and regulatory subunits) to an active kinase free from its regulatory subunits. Protein kinase subsequently activates phosphorylase *b* kinase and, finally, the latter enzyme activates glycogen phosphorylase by promoting a dimerization of two phosphorylase *b* molecules into one phosphorylase *a* molecule. The activations of phosphorylase kinase and glycogen phosphorylase involve the ATP-dependent phosphorylation of the enzyme proteins. Deactivation of these enzymes is accomplished by specific phosphatases.

Regulation by Ca^{++} (Figure 2–5) involves a direct activation of phosphorylase kinase, without the involvement of the initial protein kinase step. In frog muscle, Ca^{++} activation completely converts phosphorylase *b* to phosphorylase *a* in about three seconds, while hormonal activation is much slower.

Calcium ion, released into the sarcoplasm upon initiation of muscle contraction, is seen to play a dual and pivotal role in coupling mechanical and metabolic events in muscular contraction (Figure 2–5). By stimulating phosphorylase kinase, Ca^{++} increases the amount of glucose which can be degraded glycolytically. In addition, Ca^{++} activates ATP hydrolysis by the thick and thin muscle filaments during contraction. *These events set the stage for a glycolytic activation which in its precision and extent (over a 10-fold change in rate occurring within a minute) far exceeds that of all other tissues in the vertebrate body.*

ENZYMIC BASIS FOR THE GLYCOLYTIC POTENTIAL OF MUSCLE

Enzyme Concentrations. The high glycolytic capacity of muscle stems primarily from the presence of high steady-state levels of those enzymes which are unique to glycolysis. Thus, when com-

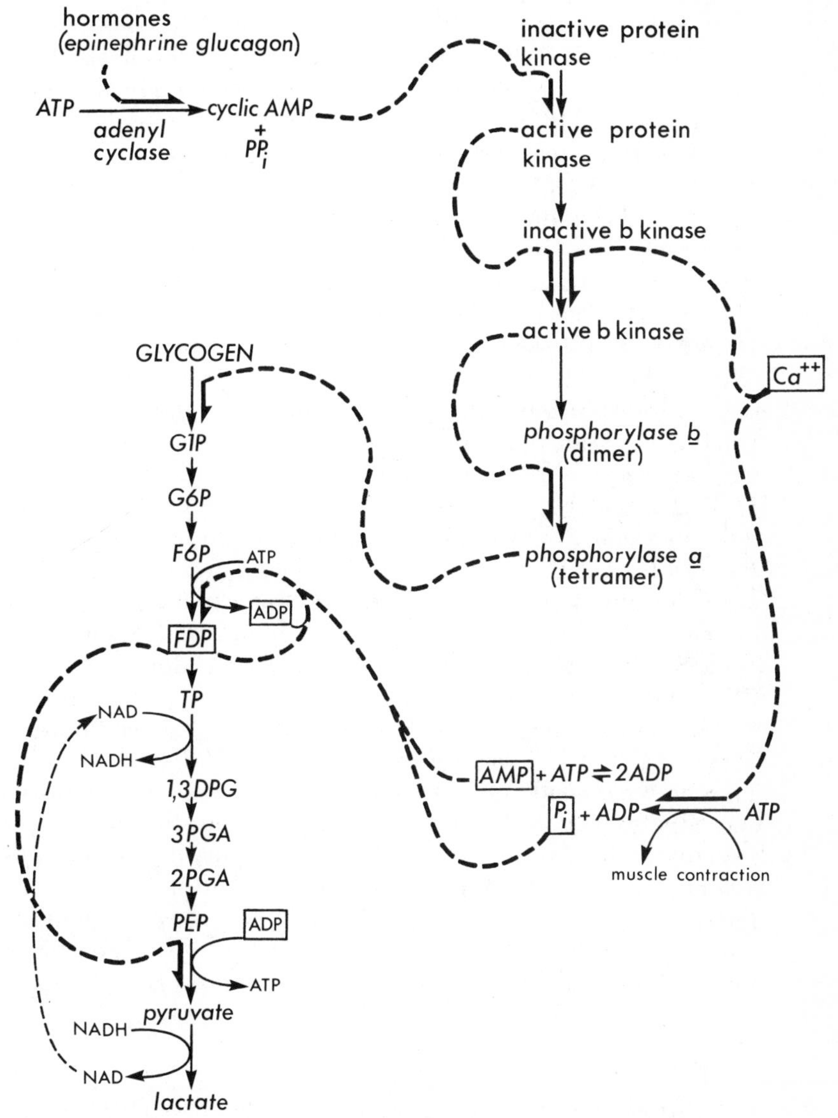

Figure 2–5 The control of glycogen mobilization in vertebrate muscle. Activation of reaction steps is indicated by a thick arrow.

pared to tissues such as liver, kidney, brain, or heart, skeletal muscle displays the highest levels of glycogen phosphorylase, hexokinase, phosphofructokinase (PFK), phosphoglycerate kinase (PGK), pyruvate kinase (PK), and lactate dehydrogenase (LDH). Recent studies by Kemp, for example, comparing liver and muscle PFKs, indicate 100-fold greater specific activities of the enzyme in muscle than in liver, and similar (though smaller) differences are documented for other glycolytic enzymes. Such high enzyme activities correlate well with the high glycolytic capacities of skeletal muscle. But of itself this property does not account for two important characteristics of muscle glycolysis: (1) its capacity for function in the glycolytic direction, and (2) its capacity for sudden changes in glycolytic rate. The basis for these two characteristics is to be found not in the amount of any given enzyme but rather in the kind of enzyme (the isozyme type) catalyzing each of the glycolytic reactions.

Tissue-Specific Isozymes. In vertebrate tissues, isozymic forms of phosphorylase, phosphoglucomutase, hexose isomerase, hexokinase, phosphofructokinase, aldolase, triose phosphate dehydrogenase, enolase, pyruvate kinase, and lactate dehydrogenase are now known (see for example, Annals N.Y. Acad. Sci., Vol. 151, Art. 1). It is clear, therefore, that many, and probably all, of the glycolytic enzymes in muscle occur in isozymic forms which are fairly specific to that tissue. Detailed documentation of their "adaptedness" for function in the microenvironment of muscle cells cannot be made for all of these, but entirely convincing arguments can be made for some of them. PFK is one of these. It, along with PGK and PK, catalyzes reactions which proceed with large free energy drops: the PFK step proceeds with a $\Delta G^0 = -3.4$ kcal/mole; PGK, with a $\Delta G^0 = -4.5$ kcal/mole; PK, with a $\Delta G^0 = -7.5$ kcal/mole. Thermodynamically, these reactions are essentially irreversible, and they therefore "poise" the flux of carbon through the multienzyme path in the catabolic direction. Thus, these reactions are essentially "pre-adapted" for the assumption of control function. Indeed, in most organisms and tissues these three enzymes have been identified as the major control sites in glycolysis; of these, PFK usually plays the predominant role in glycolytic activation.

THE NATURE OF GLYCOLYTIC ACTIVATION DURING MUSCLE CONTRACTION

As shown in Figure 2–5, the ATP cleavage which occurs during muscle contraction triggers a "cascade" pattern of activation of glycolysis. Central to this event is the overall energy charge of the cell. The fall in ATP concentrations, and the concomitant rise in

ADP, AMP, and P_i levels, potently activate muscle PFK which is the first important control enzyme situated "below" glycogen phosphorylase.

The activation involves several additional components (Figure 2–6):

(1) F6P reverses any residual ATP inhibition of the enzyme by reducing the ATP affinity of the allosteric site on the enzyme. Thus, F6P increases the $K_{i(ATP)}$ but does not affect the $K_{m(ATP)}$.

(2) F6P substrate saturation follows sigmoidal kinetics; thus, F6P is both a substrate and a positive modulator, as the binding of the first F6P molecule facilitates the binding of subsequent ones.

(3) Functionally, by far the most important aspect of the "flare-up" glycolytic activation in muscle is product activation of PFK. As shown in Figures 2–5 and 2–6, both FDP and ADP product activate muscle PFK. Since exponential "flare-ups" of glycolysis do not occur in other tissues, it is not surprising that this regulatory property is either absent or less pronounced in PFKs from tissues such as liver, red blood cells, or intestine.

(4) The final aspect of PFK catalysis which contributes to the

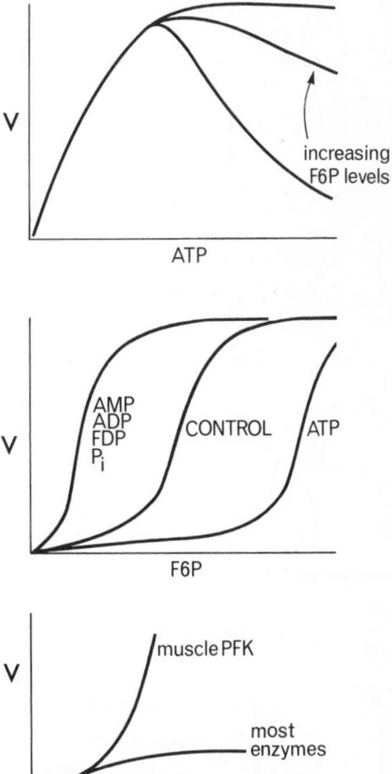

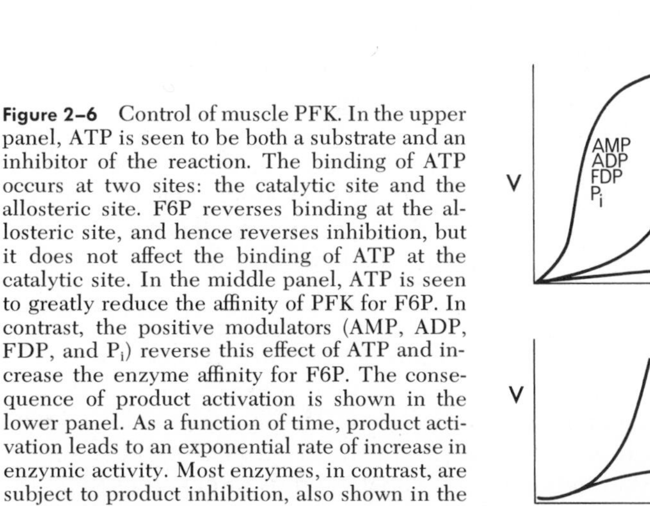

Figure 2–6 Control of muscle PFK. In the upper panel, ATP is seen to be both a substrate and an inhibitor of the reaction. The binding of ATP occurs at two sites: the catalytic site and the allosteric site. F6P reverses binding at the allosteric site, and hence reverses inhibition, but it does not affect the binding of ATP at the catalytic site. In the middle panel, ATP is seen to greatly reduce the affinity of PFK for F6P. In contrast, the positive modulators (AMP, ADP, FDP, and P_i) reverse this effect of ATP and increase the enzyme affinity for F6P. The consequence of product activation is shown in the lower panel. As a function of time, product activation leads to an exponential rate of increase in enzymic activity. Most enzymes, in contrast, are subject to product inhibition, also shown in the lower panel.

precision of PFK control in muscle concerns enzyme-ligand affinities. Thus, muscle PFK displays a 2- to 3-fold greater affinity for its substrate, F6P, than do PFKs from other tissues; under physiological conditions, it will therefore compete for limiting F6P with a 2- to 3-fold greater ability. Similar differences in enzyme-substrate affinities indeed are known for other glycolytic isozymes specific to muscle; all of these presumably contribute to overall efficiency of carbon flow in the glycolytic direction. In addition, muscle PFK shows higher affinities for key regulatory metabolites, such as the adenylates, than do PFK forms in other tissues, a situation which accords well with the tighter control of PFK which is needed in muscle.

To complete the "cascade" activation of muscle glycolysis, the activities of PFK must be integrated with the next major control site in the pathway: pyruvate kinase. This integration is achieved in two ways. Firstly, the ADP formed in the PFK reaction is, of course, a substrate for the PK reaction. And secondly, the other product of the PFK reaction, FDP, serves as a feed-forward activator of muscle PK, at least in the case of the lower vertebrates. The mode of FDP control at this site is complex and involves (1) a direct activation by decreasing the $K_{m(PEP)}$ and increasing the V_{max}, and (2) an indirect or apparent activation by reversing any residual ATP inhibition of the enzyme. By these means, the activity of PK is closely geared to that of PFK. As in the latter, muscle PK displays a much higher affinity for its glycolytic substrate, PEP, than do PKs from other tissues; hence, under limiting PEP conditions, it is able to convert PEP to pyruvate with 5- to 6-fold greater efficiency than is observed in other tissues.

Finally, the pyruvate formed by the PK reaction is converted to lactate in the LDH reaction with the concomitant regeneration of NAD, which can then act as the oxidizing agent for an additional molecule of glyceraldehyde 3-phosphate (Figure 2–5).

These events are so closely integrated that in anoxia *lactate production is directly proportional to muscle contractile work,* and the muscle sustains larger accumulations of lactate than ever occur in other tissues. Indeed, the final functional requirement of muscle glycolysis is some provision for the accumulation of large quantities of lactate and for its subsequent metabolism.

MECHANISMS OF TOLERANCE TO HIGH LACTATE LEVELS

Three specific mechanisms are available which can be viewed as adaptations to this problem, and again at least two of these appear to rely upon the elaboration of specific kinds of enzymes to do the job at hand:

(1) The LDH isozyme typically predominant in white muscle of vertebrates (designated the M_4 tetramer) is highly insensitive to pyruvate inhibition and hence allows large accumulations of lactate from pyruvate. Heart type LDH (the H_4 tetramer) functions in more aerobic tissues, where sensitivity to pyruvate inhibition prevents lactate accumulation and thus favors channelling of pyruvate into the Krebs cycle (Figure 2–7). In addition, the M_4 LDH affinity for pyruvate is rather low compared to the H_4 LDH. In consequence, the muscle enzyme becomes saturated at higher pyruvate concentrations than are required to saturate the H_4 type LDH, a situation which presumably functions to prevent pyruvate concentrations from rising without limit in muscle.

(2) Under normal circumstances, most of the lactate formed in muscle is flushed out into the blood circulation. Change in blood pH is prevented by a bicarbonate buffering system; the buffering capacity is increased in athletic individuals, who are able to tolerate higher levels of lactate. Parenthetically, it might be added that the buffering capacity of blood is remarkably high in vertebrates such as aquatic turtles, which can sustain extreme anoxia (p. 45).

(3) Perhaps the most important mechanism of tolerance of high lactate levels is the development of a means for metabolizing the lactate produced in the muscle. Upon termination of anoxia, a por-

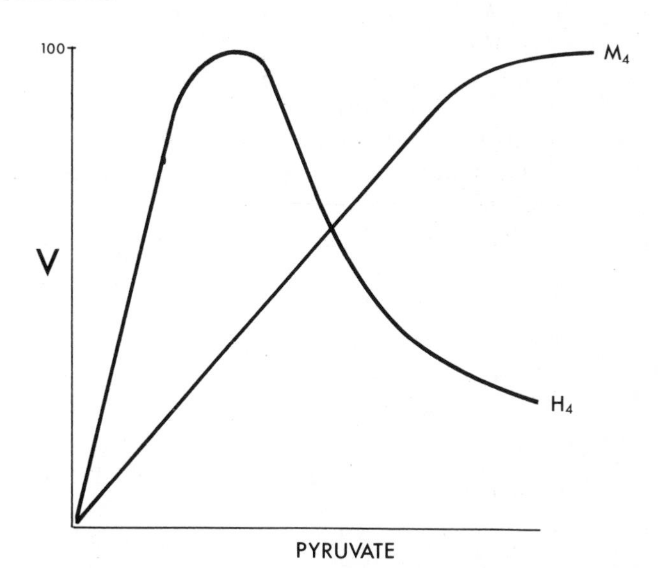

Figure 2–7 Comparison of H_4 and M_4 type LDHs in tissues of most vertebrates. The H_4 enzyme is subject to potent pyruvate inhibition and displays a very much lower K_m for the substrate.

tion of this lactate is reconverted to pyruvate in the muscle and is then "burned" in aerobic metabolism. However, by far the largest portion of the lactic acid is delivered by the blood to glucogenic tissues (liver and kidney), where it is nearly quantitatively converted to glucose or glycogen via the gluconeogenic pathway. In these terms, gluconeogenesis in liver (and to a lesser extent in kidney) can be viewed as a mechanism for dealing with the unusual lactate accumulations in muscle during anaerobic work. (It should be stressed that gluconeogenesis plays other important inter-tissue functions, chief among these being the synthesis of glucose for the brain.)

GLUCONEOGENESIS AS A MEANS FOR METABOLIZING MUSCLE LACTATE

Lactate arriving at the liver is converted to pyruvate by reversal of the LDH reaction. Although this reaction is normally far in the direction of lactate, it is reversed in the liver because of high concentrations of lactate and because pyruvate is being withdrawn for glucose synthesis. Most of the reactions from pyruvate to glucose are catalyzed by enzymes of the glycolytic scheme, and thus proceed by reversal of steps employed in glycolysis. However, there are four irreversible steps in the normal "downhill" glycolytic pathway which cannot be utilized in the "uphill" conversion of pyruvate to glucose. These are the reactions catalyzed by (1) pyruvate kinase, (2) phosphoglycerate kinase, (3) phosphofructokinase, and (4) hexokinase. During glucose synthesis, these steps are bypassed by alternative reactions which are thermodynamically favorable in the glucogenic direction.

Three such bypass reaction routes are now known (Figure 2–8):

(1) Pyruvate kinase is bypassed by the concerted action of at least two enzymes:

Pyruvate carboxylase catalyzes the reaction

$$CO_2 + ATP + \text{pyruvate} \xrightarrow[\text{acetyl CoA}]{g^{++}} OXA + ADP + P_i$$

PEP carboxykinase (PEPCK) catalyzes the decarboxylation of OXA

$$OXA + ITP \xrightarrow{Mg^{++}} IDP + PEP + CO_2$$

The net effect of these two reactions is the synthesis of PEP from pyruvate.

(2) PFK is bypassed by FDPase, which catalyzes the hydrolysis of FDP to F6P + P_i.

(3) Hexokinase is bypassed by glucose-6-phosphatase.

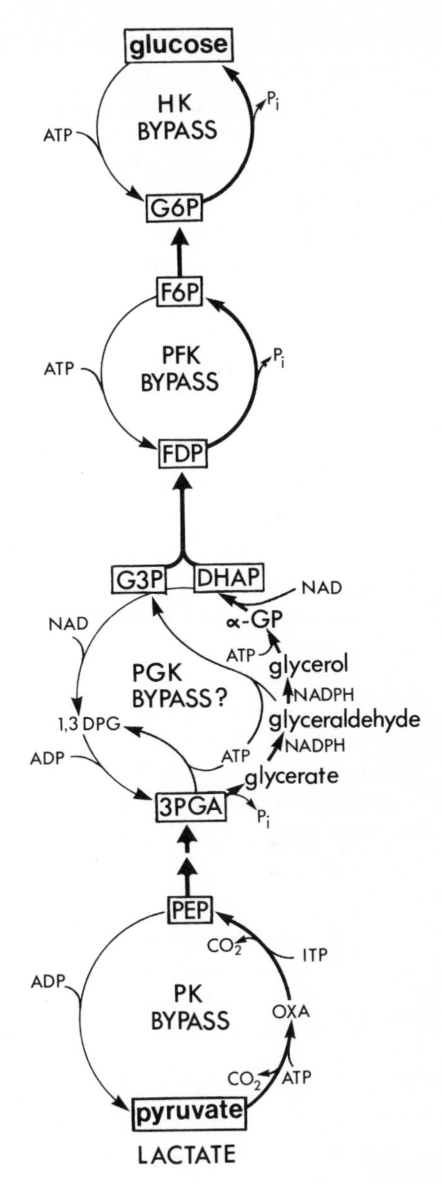

Figure 7–8 Bypass reactions in glucose synthesis in the vertebrate liver. A PGK bypass is not yet known with certainty. Three possibilities are indicated, one of which is a "kinetic bypass" and simply assumes a reversal of the PGK step. Because this would represent a strongly "uphill" reaction on both kinetic and thermodynamic grounds, it is considered the least likely of the three alternatives.

The mechanism for "physiologically reversing" the phosphoglycerate kinase step is unknown. Two properties of this reaction appear to necessitate a bypass process of some sort: in the first place, the reaction proceeds in the forward, glycolytic direction with a large (−4.5 kcal/mole) free energy change, and secondly, the affinity of the enzyme for 1,3 DPG is over 100 times greater than for 3 PGA. Hence, even though it is possible to "reverse" the reaction under test-tube conditions, a "kinetic bypass" would not appear to be physiologically efficient if indeed feasible. Two alternative bypasses are suggested in Figure 2–8.

It is important to stress that in physiological terms the known bypass reactions achieve the reversal of the "forward" glycolytic reactions, but in chemical terms these are, of course, entirely different. The occurrence of both FDPase and PFK, for example, in a single compartment raises a potential short-circuit in both carbon and energy metabolism at this point, as the simultaneous function of both enzymes would lead to a futile carbon cycling with a net hydrolysis of ATP. It is therefore evident that, in glycogenic tissues, regulation of the two enzymes must be tightly integrated. An entirely analogous problem of cycling arises whenever and wherever two oppositely-directed reaction paths occur within a single cell. Thus, the glucose ↔ G6P interconversion and the PEP ↔ pyruvate interconversions constitute further examples of this problem of carbon and energy cycling within tissues such as liver and kidney. In principle, all of these problems have been solved in the same manner: *intracellular conditions which favor catalysis in the catabolic direction are highly unfavorable for catalysis in the anabolic direction,* and vice versa.

In the case of the F6P ↔ FDP interconversion, it is now widely accepted that AMP serves as the major control signal for integrating the activity of FDPase with that of PFK. Thus, as we have already noted, under glycolytic conditions, increased AMP levels lead to an activation of PFK; at the same time, AMP, which is a specific allosteric inhibitor of FDPase, reduces FDP hydrolysis to F6P and P_i. When the system swings over to net gluconeogenesis, which would be favored by a high ATP/AMP ratio, the reduced AMP levels lead to a *deinhibition* of FDPase while the raised ATP levels inhibit PFK.

These conditions in effect *temporally* separate the two enzymes so that both do not compete simultaneously for the same metabolite pool. In the case of the glucose ↔ G6P interconversion, the two enzymes are *spatially* separated as well. G6Pase, a lipoprotein, is bound to the microsomes, while hexokinase probably is either loosely attached to the mitochondria or is free in solution. This spatial separation undoubtedly contributes to the prevention of cycling at this locus in metabolism.

INTEGRATION OF ANAEROBIC MUSCLE GLYCOLYSIS WITH LIVER GLUCONEOGENESIS

As we have indicated, an important aspect of the adaptation of muscle energy metabolism to low O_2 supplies is the development of a tolerance for unusually high lactate accumulations. One component of that tolerance is a means for metabolizing the lactate upon termination of anaerobic muscle work. Most of the lactate is flushed out of the muscle and delivered to the liver, where it is converted to glucose. What signal integrates these specific metabolic functions in these different tissues? It appears that the signal may be a mild acidosis induced by the high amounts of lactic acid appearing in the blood and delivered to glucogenic tissues.

It is now well established that mild acidosis, such as may be expected by flooding the circulation and the glucogenic tissues with muscle lactic acid, leads to an increase in gluconeogenesis (Figure 2–9). At least two enzymic mechanisms could contribute to this activation.

(1) PEPCK, which functions as the major control site in gluconeogenesis, appears to be remarkably sensitive to pH changes. At pH 7.0, the enzyme affinity for OXA is about 10-fold greater than at pH 8.0. Thus, a mild acidosis would be tantamount to a potent increase in the competitive ability of PEPCK for its substrate, OXA. At the same time, mammalian pyruvate kinases display a reduced affinity for PEP as the pH is lowered. Hence, at the same time as it

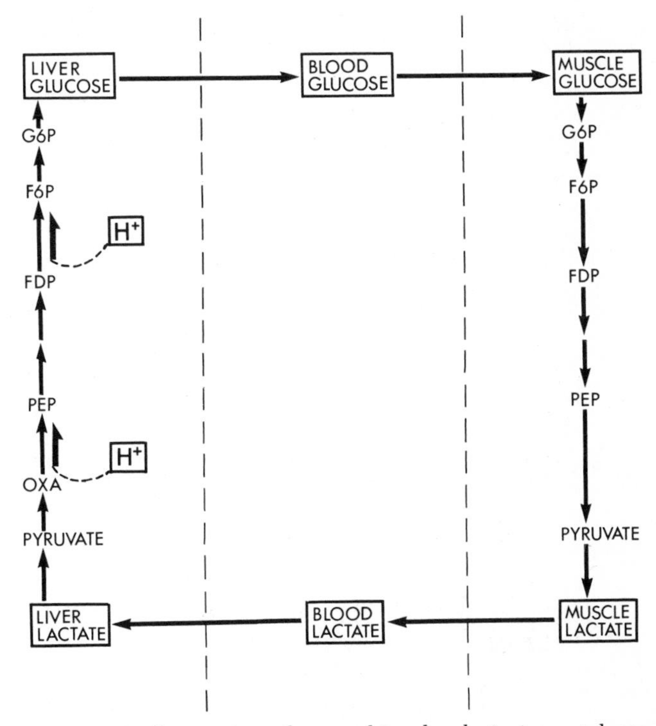

Figure 2–9 Integration of anaerobic glycolysis in vertebrate muscle with liver glucogenesis.

would activate the PEPCK catalyzed conversion of OXA to PEP, a mild acidosis would inhibit PEP conversion to pyruvate; that is, the system is well integrated to avoid cycling of carbon at this point. The PEP formed is then made available for gluconeogenesis.

(2) Another regulatory effect of H^+ may occur at the FDP ↔ F6P interconversion. Both PFK and FDPase are known to be highly sensitive to pH changes. Recently, it has become evident that the optimum pH for FDPase is about pH 7.0, while PFK displays a much more alkaline pH optimum (about pH 8.2). Moreover, the allosteric ATP inhibition of PFK is much more potent at pH values below pH 7.5. Thus, at this interconversion site as well, mild acidosis would favor decreased glycolytic flow simultaneously with increased flow of carbon towards glucose. The glucose formed in the liver is then available for redelivery to muscle for further energy metabolism.

COMPENSATORY NATURE OF ANOXIA ADAPTATION IN VERTEBRATE MUSCLE

From these considerations, we can conclude that anaerobic muscle work in the vertebrates is supported by three fundamentally adaptive characteristics:

(1) a high glycolytic potential stemming from high concentrations of glycolytic enzymes;

(2) the occurrence of specific muscle isozymes kinetically attuned (a) for function in the glycolytic direction and (b) for exponential rate of change from low-activity to high-activity states; and

(3) a tolerance of high lactate accumulation with provision for its subsequent metabolism.

These mechanisms in the vertebrates compensate for the *temporary absence of molecular oxygen in muscle* by allowing for an impressive increase in the anaerobic ATP-generating capacity. *In the most skilled of vertebrate divers (aquatic turtles, marine mammals, and diving birds), it is precisely these loci which we might expect to find even further adjusted.*

THE DEPENDENCE ON ANAEROBIC GLYCOLYSIS DURING PROLONGED DIVING IN TURTLES

On the basis of direct calorimetric measurements of metabolism, Jackson has divided the diving period of turtles into three metabolic phases:

(1) During phase I, lasting about a half-hour at 24°C, the metabolic rate persists at the same rate as in the prediving condition, but O_2 tensions in the blood and the lungs rapidly fall.

(2) During phase II, also lasting about 30 minutes, the metabolic rate falls to 40% of the initial rate and remaining O_2 reserves are exhausted.

(3) The remainder of the dive (phase III), which can last for many hours (and in some species, at low temperature, for many days), is *totally anoxic.* All maintenance and work functions in phase III are sustained by anaerobic metabolism. On the basis of heat measurements, Jackson has estimated that this metabolism can yield 20% of the total energy which is available in the prediving state (Figure 2–10). Apparently, anaerobic glycolysis accounts for all of the energy generated during this time. The unusual activity of this pathway in these species leads to unusually large lactate accumulations in the tissues and in the blood, and accounts for the extreme insensitivity of turtles to metabolic poisons such as cyanide, which specifically block oxidative metabolism but do not affect glycolysis. Because even "aerobic-type" tissues such as the heart have a high glycolytic capacity, the usual kinetic differences between H_4 and M_4 LDHs are absent. In these organisms, as well as

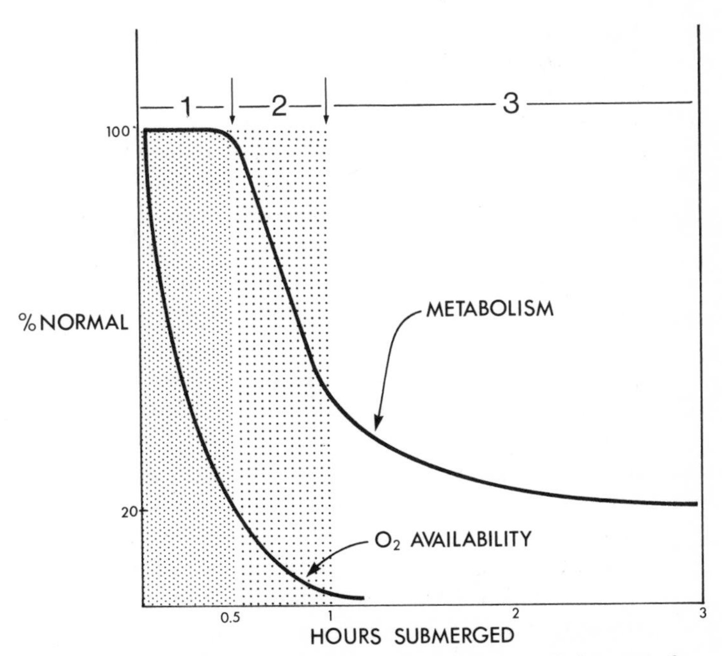

Figure 2–10 Decrease in O_2 availability and in metabolic rate of an aquatic turtle during a 3 hour dive at 24° C. (Modified after Jackson, 1968.)

in diving ducks and seals, H_4 LDH does not show the usual sensitivity to substrate inhibition. These LDHs maintain pyruvate levels at 0.1 mM, while lactate levels can rise to over 60 mM!

Belkin first provided an unequivocal demonstration of the critical importance of anaerobic glycolysis to the anoxic turtle. He selected a metabolic poison, iodoacetate, whose chief locus of action is known to be the triose phosphate dehydrogenase step in glycolysis. If anoxic survival depended upon glycolysis, inhibition of this enzyme by injection of iodoacetate into the turtle should lead to a greatly reduced tolerance to anoxia. This prediction was verified. Jackson then observed, by direct calorimetry, that the injection of iodoacetate leads to a predicted drop in the remaining energy metabolism of the anoxic organism. These experiments emphatically underline the pivotal role of glycolysis in the adaptation to anoxia; by this means, enough energy is generated to sustain aquatic turtles for periods of many hours—possibly, under certain conditions, for periods of many days—in a state of total anoxia. Yet sooner or later even the aquatic turtle surfaces and "repays" his O_2 debt, for here, as in other vertebrates, the biochemical strategy of anoxia adaptation depends upon an ultimate return to aerobiosis. Hence, this strategy is a poor one for the exploitation of O_2-free environments on a sustained basis.

EXPLOITATIVE NATURE OF ANOXIA ADAPTATION IN INVERTEBRATE FACULTATIVE ANAEROBES

In contrast to the vertebrates, massive exploitation of O_2-free environments has been achieved by many invertebrate organisms. Three well-studied groups of organisms that belong to this category are the helminths (particularly parasitic helminths), intertidal bivalve molluscs, and benthic invertebrates which burrow into the bottom sediments. Within these groups, a spectrum is encountered ranging from obligate anaerobes to facultative anaerobes. Mere mention of their habitats indicates that quite different (environmental vs. physiological) selective forces are at work here, and that "exploitative" strategies allowing indefinite anoxia are more likely to succeed than are "compensating" strategies which only allow temporary excursions into O_2-free environments.

Three processes seem to have been favored in the evolution of energy metabolism in facultative anaerobes:

(1) The deletion of certain enzymes, in particular, the "deletion" of LDH to avoid metabolic *cul de sacs* such as lactate production.

(2) The modification of the kinetic properties of certain key

branchpoint enzymes to allow an efficient transition from aerobic to anaerobic metabolism.

(3) The coupling of other substrate-level phosphorylations to the glycolytic reactions, thus increasing the potential yield of high energy phosphate compounds. The situation can be well illustrated by considering the anaerobic metabolism of adductor muscle in intertidal bivalve molluscs such as the oyster, but the basic principles are probably valid for all invertebrate facultative anaerobes.

BASIC PATTERNS OF ENERGY METABOLISM IN FACULTATIVE ANAEROBES SUCH AS THE OYSTER

The basic pathways of carbohydrate catabolism in oyster muscle are similar to those described in Figure 2–5. Like vertebrate skeletal muscle, oyster muscle is a highly glycolytic tissue which derives much of its energy from glycogen. Under aerobic conditions the metabolism of glycogen, via glucose-6-phosphate, occurs in a manner comparable, if not identical, to that of vertebrate muscle. The PEP formed is directed towards pyruvate, which is then fed into the Krebs cycle. The evidence for this scheme comes from three kinds of studies: (1) studies demonstrating the existence of appropriate intermediates, (2) studies demonstrating the existence of appropriate enzymes, and (3) studies demonstrating the flow of C^{14} of glucose or other precursors through the intermediates of both glycolysis and the Krebs cycle.

Under anoxic conditions, the breakdown of glycogen (or glucose) to the level of phosphoenolpyruvate (PEP) is also similar to the process in vertebrates, but in contrast to the latter (which converts PEP to pyruvate and accumulates lactate), the main end products of anaerobic glucose catabolism in intertidal bivalves are succinate and alanine (Figure 2–11). Succinate also accumulates in parasitic helminths such as *Ascaris lumbricoides*. In these forms, *muscle succinate production during anoxia is directly proportional to muscle work;* that is to say, in these organisms the metabolic production of succinate is the major energy-yielding mechanism during anoxia.

Although the metabolic pathways leading from PEP to succinate have been in dispute, current evidence strongly indicates that during the aerobic-anaerobic transition, pyruvate kinase activity is reduced, while PEP carboxylation to oxaloacetate (OXA) is favored. The latter reaction is catalyzed by PEP carboxykinase, and the newly-formed OXA is subsequently converted to succinate.

In helminths such as the adult *Ascaris*, which is in effect an *obligate* anaerobe, this pathway is strongly favored by the "deletion" of pyruvate kinase and lactate dehydrogenase. However, in

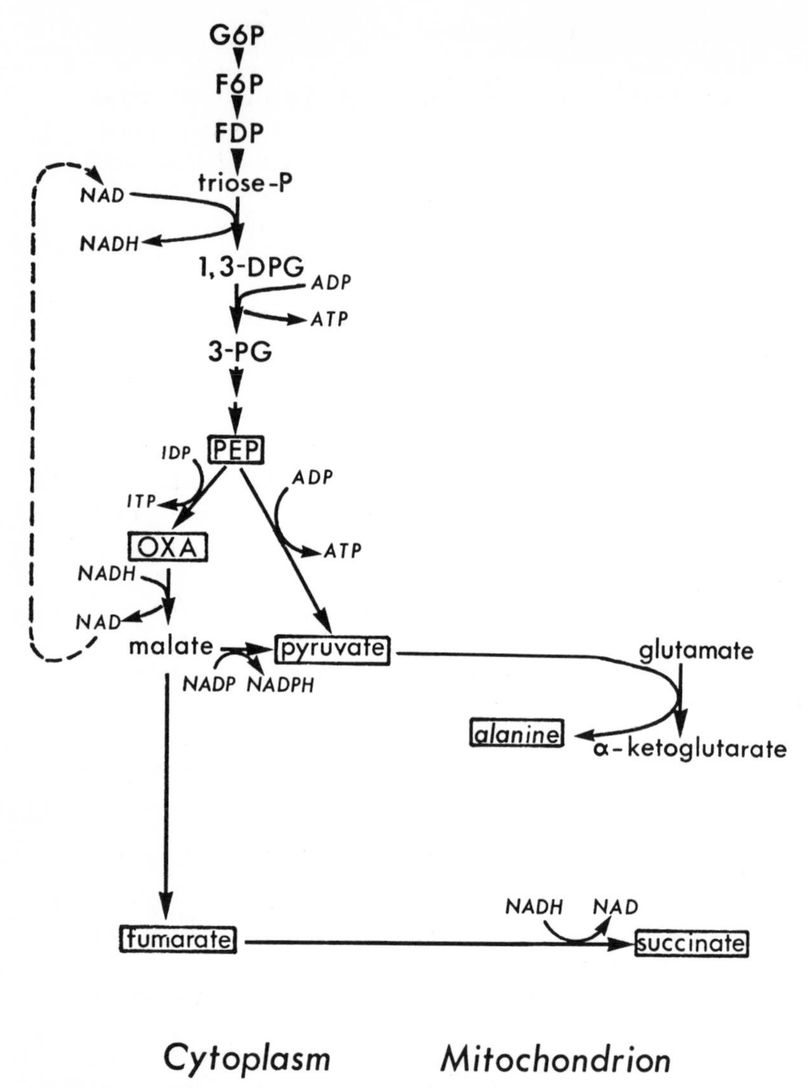

Figure 2–11 Pathways of anaerobic glucose metabolism in the oyster adductor muscle. Alanine and succinate are the two end products of the reaction scheme.

intertidal bivalves, the aerobic-anaerobic transition may occur over time periods that are as short as a single tide-cycle. These organisms maintain both PK and PEPCK. Both enzymes occur in the "soluble" fraction, and consequently both are competing for the same PEP pool during aerobic-anaerobic transition (Figure 2–11). The question which arises is what controls the flow of PEP towards pyruvate during aerobiosis but towards OXA during anoxia.

CONTROL OF THE PEP BRANCHPOINT

From demonstrable properties of PK and PEPCK, it appears that the two enzymes are not able to function at significant rates simultaneously; rather, their catalytic requirements and control properties appear to be arranged for function on a reciprocal, *either/or* basis (Figure 2–12). The nub of the argument favoring this conclusion can be summarized as follows.

The activities of many enzymes involved in cellular energy metabolism are governed at least in part by the energy status of the cell, with the adenylates being the primary metabolic "signals." In oyster muscle, both the PK and the PEPCK reactions generate a high energy phosphate compound (ATP in the case of PK; ITP or GTP in the case of PEPCK) and both are subject to product inhibi-

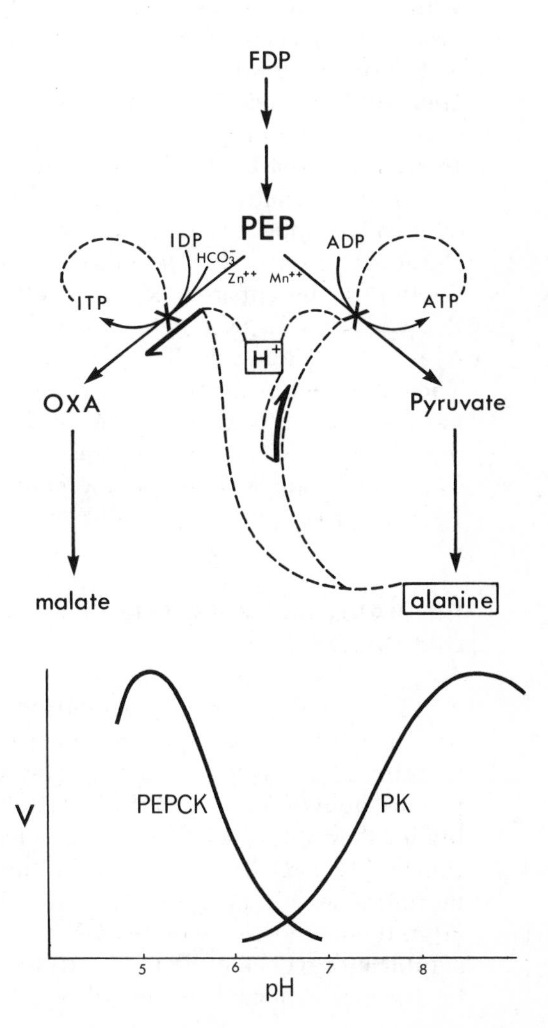

Figure 2–12 Known control interactions at the PEP branchpoint in oyster adductor muscle. Activation or deinhibition of a process is shown as a dark arrow; inhibition is indicated by a dark cross. In addition to the regulatory interactions shown, FDP is an established feed-forward activator of PK, and this mechanism could maintain some PK activity during anoxia. In the lower panel, the non-overlapping pH profiles for the two branchpoint enzymes, PEPCK and PK, are illustrated. (After Hochachka and Mustafa, 1972.)

tion by these compounds. In this sense, both enzymes behave in accordance with Atkinson's energy charge concept. Whereas these mechanisms undoubtedly contribute to the physiological "poise" of the PEP branchpoint, they do not, of themselves, display adequate specificity to account for transition from aerobic to anaerobic metabolism. The specificity required can be supplied by H^+ and L-alanine.

It is widely held that under anaerobic conditions, bivalve molluscs sustain substantial acidification of their tissues and fluids. This drop in pH appears to us to play a pivotal role in the channelling of PEP away from the PK reaction and towards PEPCK, because the *pH profiles for PEPCK and PK are essentially nonoverlapping* (Figure 2–12). In consequence, in the absence of any other factor, decreasing pH leads to an automatic inhibition of PK with a concomitant activation of PEPCK. At the same time L-alanine, which accumulates along with succinate under anaerobic conditions, potently inhibits PK (by increasing the $K_{m(PEP)}$ and decreasing the maximum catalytic rate). It is particularly instructive that the *L-alanine inhibition is potentiated by decreasing pH.* Indeed, low pH likewise potentiates ATP inhibition of PK.

In sharp contrast (Figure 2–12), the primary effects of L-alanine on PEPCK appear to be (i) a reversal of ITP inhibition and (ii) a slight activation at low *p*-enolpyruvate concentrations due to a reduction in the apparent $K_{m(PEP)}$. Both these effects of L-alanine on PEPCK occur at pH ranges in which PK activity is very low and in which L-alanine and ATP inhibition of PK is unusually extreme. The net effect of decreasing pH and increasing L-alanine concentration is an autocatalytic increase in the PEPCK activity concurrent with an exponential decrease in the PK activity (Figure 2–12). This would appear to be an adequate arrangement for channelling PEP towards oxaloacetate.

THE MAJOR METABOLIC FATE OF CYTOPLASMIC OXALOACETATE

The major fate of OXA formed from PEP carboxylation is reduction to malate. This appears to be true both in molluscs and in helminths. In tissues of these organisms, cytoplasmic NAD-linked malate dehydrogenase (MDH) activities are high, usually much higher than either PEPCK or PK activities. The equilibrium position for the MDH reaction is far in the direction of malate, and it is therefore widely accepted that the high MDH activities function (a) in the maintenance of low OXA concentrations, thus preventing significant reversal of PEPCK activity, and (b) in regenerating NAD for the triose phosphate dehydrogenase (TDH) reaction. If all the carbon of glucose is diverted into this pathway during anaerobiosis (i.e., if PK is fully blocked), for each two reducing equivalents formed at the TDH reaction, two are oxidized by the MDH reaction, and to this point the system is in perfect redox balance (Figure 2–11). (Probably this is another reason why PK and PEPCK do not function simultaneously.) The student will recognize that lactate dehydrogenase serves an identical function in the anaerobic glycolysis of vertebrate tissues.

THE METABOLIC FATE OF MALATE

Malate can be metabolized by two alternate routes, and from available data it appears that each route competes approximately equally for cytoplasmic malate. In one reaction scheme, malate is converted to fumarate by a reversal of the well known fumarase reaction. In bivalves the fumarase appears to be localized in the cytosol, but in helminths it is positioned in the mitochondria. The fumarate in turn is reduced to succinate, a reaction catalyzed by fumarate reductase. In helminths and bivalves, this enzyme appears to be kinetically adapted for function in the direction of succinate production. All previous workers in this field have assumed that this pathway is the major source of the succinate which accumulates during anaerobiosis (Figure 2–11). As we shall see, it is only one of at least two available routes.

A second route of malate metabolism also begins in the cytosol, for in bivalves "malic enzyme" competes directly for cytoplasmic malate. The reaction catalyzed by "malic enzyme"

$$\text{malate} + \text{NADP} \rightleftharpoons \text{NADPH} + CO_2 + \text{pyruvate}$$

is freely reversible, but in the oyster adductor muscle, the relative affinity for malate is so high that *in vivo* the enzyme probably functions only in the direction of pyruvate production, and this is also the established direction of function in *Ascaris*. This pyruvate is the starting point for a second path to succinate.

METABOLIC FATE OF PYRUVATE AND A SECOND ROUTE TO SUCCINATE

Whatever the predominant route of pyruvate formation during anaerobiosis, the primary metabolic fate of pyruvate is conversion to alanine according to the transamination reaction

$$\text{pyruvate} + \text{glutamate} \rightarrow \alpha\text{-ketoglutarate} + \text{alanine}$$

On this, the evidence in helminths and molluscs is unequivocal.

Since bivalves and many helminths possess the enzymes capable of converting α-ketoglutarate to succinate, this reaction span *represents another major pathway for the accumulation of succinate during anaerobiosis* (Figure 2–13). This pathway may be particularly important since a variety of amino acids could "feed" into it via transamination reactions:

$$\text{amino acid} + \alpha\text{-KGA} \rightarrow \text{glutamate} + \text{keto acid}$$

In most organisms, α-KGA produced during the glutamate-alanine transamination is reconverted to glutamate by glutamate dehydrogenase (GDH), a reaction which utilizes NADH. In facultative anaerobes, the specific activity of GDH is very low while α-ketoglutarate dehydrogenase activities are often high. For these reasons, the α-ketoglutarate dehydrogenase reaction would probably outcompete the GDH reaction for the common substrate, α-KGA. Under these conditions, *the glutamate-alanine transaminase serves to channel α-KGA directly towards α-ketoglutarate dehydrogenase.*

FUNCTIONAL SIGNIFICANCE OF TWO ROUTES TO SUCCINATE

The facultative anaerobe gains a critical energetic advantage by utilizing this route: the overall α-ketoglutarate dehydrogenase reaction

$$\alpha\text{-KGA} + \text{CoASH} + \text{NAD} \rightarrow \text{succinylSCoA} + CO_2 + \text{NADH}$$

sets the stage *for conversion of thiolester bond energy into nucleoside triphosphate.* The reaction, catalyzed by succinic thiokinase, is exergonic and can utilize either GDP or IDP as cosubstrate, generating GTP or ITP:

$$\text{succinylSCoA} \xrightarrow[\text{GDP} + P_i \;\curvearrowright\; \text{GTP}]{Mg^{++}} \text{succinate} + \text{CoASH}$$

This energy yielding reaction span is utilized as an anaerobic mechanism for supplanting aerobic metabolism in certain mammalian tissues, and presumably has been selected for an analogous function in facultative anaerobic invertebrates. In both systems, for two ATPs made in glycolysis, a third is made by the succinic thiokinase reaction. In both systems, however, some provision must be made for the regeneration of the NAD required for α-KGA dehydrogenase activity. In facultative anaerobes the most likely candidate for this job is fumarate reductase, which couples the oxidation of NADH with the reduction of fumarate to succinate (Figure 2–13). It is for this function that the fumarate → succinate reduction was

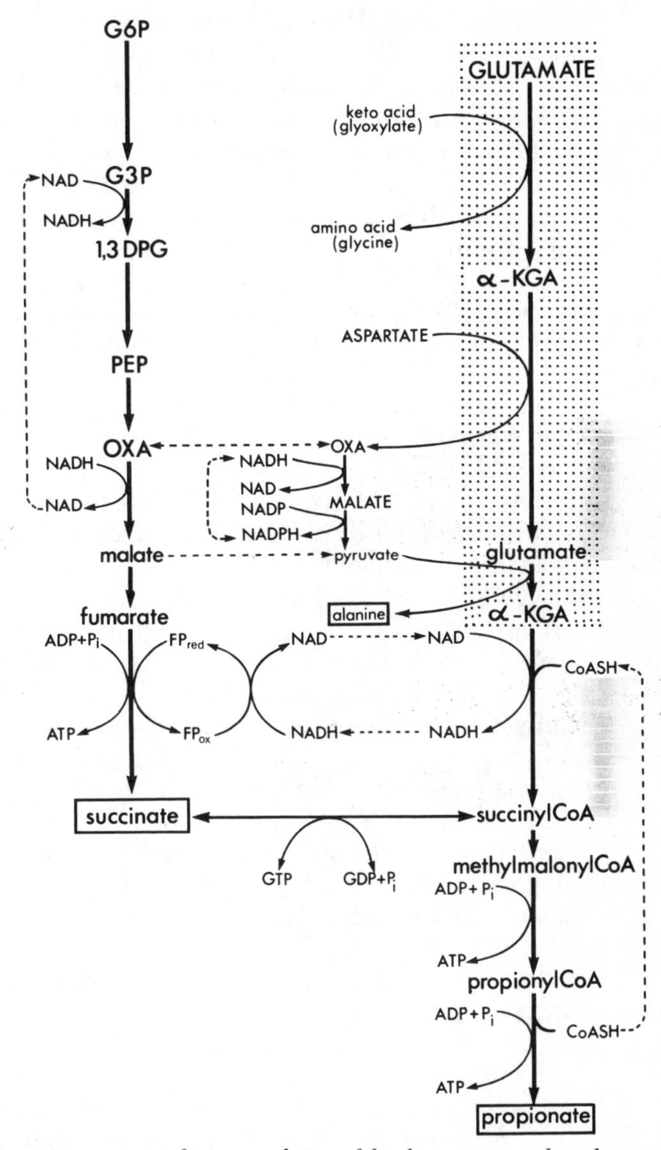

Figure 2–13 A metabolic map accounting for accumulation of the three major end products during anoxia in facultatively anaerobic animals. Most of the data are from parasitic helminths, annelids, and intertidal bivalves. Coupling between carbohydrate and amino acid catabolism is achieved primarily by the redox couple forming between fumarate reductase and α-KGA dehydrogenase. *Because 2 moles of PEP are formed per mole of G6P, 2 moles of aspartate and 2 moles of glutamate must be mobilized simultaneously to maintain redox balance during the sustained production of alanine, succinate, and propionate, the three major end products of anaerobiosis.* Energy yield: 8 moles of ATP/mole of G6P plus 2 moles of aspartate plus 2 moles of glutamate.

selected in facultative anaerobes. It is this function which explains the unique kinetic features of the enzyme (a relatively high affinity for fumarate and a relatively low affinity for succinate), as well as its high activity in these organisms. In addition, fumarate reductase is properly positioned in the mitochondrion for the delivery of NAD to the α-ketoglutarate dehydrogenase reaction (Figure 2–13).

From these considerations, one can view the pathways of anaerobic glucose metabolism in facultative anaerobes as a means for "priming" the flow of glutamate $\rightarrow$ α-KGA $\rightarrow$ succinylSCoA $\rightarrow$ succinate, by (1) supplying pyruvate for the transaminase reaction and (2) regenerating NAD through fumarate reduction for the α-ketoglutarate dehydrogenase reaction.

SOME FACULTATIVE ANAEROBES ACCUMULATE SUCCINATE AND PROPIONATE

A number of parasitic helminths and free living annelids accumulate propionate as well as succinate as major end products of anaerobic metabolism. Three well studied cases are *Ascaris* and *Fasciola*, parasitic helminths, and *Alma*, a swamp-dwelling annelid All current information suggests that in these organisms glucose is again converted to succinate and alanine. Succinate may accumulate (as it does in *Fasciola* under anoxic conditions) or it may be converted to propionate, since all the reactions between succinate and propionate are reversible (Figure 2–13). SuccinylCoA, also undoubtedly formed in large quantities from α-KGA, is the immediate precursor of propionate. Again, it will be evident that, if for each mole of glucose metabolized, 2 moles of aspartate and 2 moles of glutamate are mobilized, the system will remain in redox balance indefinitely (Figure 2–13).

NATURE OF ENERGY YIELDING REACTIONS IN FACULTATIVE ANAEROBES

In the scheme depicted in Figure 2–13, four kinds of processes are potential sources of ATP (or its metabolic equivalent):

(1) One group of reactions depends upon a substrate of high phosphate-transfer potential and includes the reactions catalyzed by phosphoglycerate kinase (1,3 DPG + ADP $\rightarrow$ 3PGA + ATP), by pyruvate kinase (PEP + ADP $\rightarrow$ pyruvate + ATP), and by PEP carboxykinase (PEP + CO_2 + IDP $\rightarrow$ OXA + ITP). Pyruvate kinase is usually inactive in facultative anaerobes, particularly during anoxia; hence, only two of these reactions are utilized for production of high energy phosphate compounds.

(2) In addition, one biotin-dependent enzyme, propionylCoA carboxylase, is utilized in ATP synthesis. During catalysis of the

reaction, methylmalonylCoA + P_i + ADP → CO_2 + propionylCoA + ATP, a carboxybiotin-enzyme complex is formed. Cleavage of this complex proceeds with a large enough free energy drop to be used in ATP synthesis.

(3) Another kind of ATP-generating process harnessed by facultative anaerobes appears to involve *a step in the electron transfer system that utilizes fumarate as an electron acceptor and NADH as an electron donor.* The basic concept here emphasizes that in normal oxidative phosphorylation, the first site of ATP synthesis occurs during the transfer of electrons from NADH to a flavoprotein (see Figure 2–4). Since fumarate has a sufficient electron affinity to serve as the ultimate electron acceptor for this step, the facultative anaerobe not only reoxidizes NADH, but also gains a mole of ATP per mole of fumarate reduced to succinate by the process.

(4) Thiolesters are a final and most important source of energy in these organisms. As we indicated in Figure 2–1 (B), thiolester compounds such as succinylCoA and propionylCoA are high energy compounds and hence can be utilized to "drive" the synthesis of ATP. This is an important insight, since a number of minor end products of helminth anaerobiosis (isovalerate, isobutyrate, methylbutyrate, and even acetate) can be produced from their CoA derivatives with the concomitant formation of ATP:

$$\text{acylCoA} + P_i + \text{ADP} \rightarrow \text{ATP} + \text{acid end product} + \text{CoASH}$$

Since CoASH is released in the process, each thiokinase must be functionally linked to a ketocarboxylate dehydrogenase (such as α-ketoglutarate dehydrogenase), so that there is neither a depletion nor an accumulation of CoASH. Note that in Figure 2–13, the CoASH utilized by α-KGA dehydrogenase is regenerated by propionate thiokinase; hence, the role of CoASH in this process is a catalytic one.

YIELD OF ATP IN FACULTATIVE ANAEROBES COMPARED TO CLASSICAL GLYCOLYSIS

An insight into the energetic efficiencies of succinate/propionate production compared to lactate production can be obtained by comparing the number of moles of ATP which can be gained by the two processes per mole of glucose utilized. In classical glycolysis, 2 moles of ATP are produced per mole of glucose (Figure 2–3) and the system remains in redox balance so long as a 1:1 TDH/LDH activity ratio obtains. In the facultative anaerobe, the situation is more complex. Since 2 moles of OXA are formed from a mole of

G6P, a net of 2 moles of ATP are synthesized and to this point the system is comparable to glycolysis. However, two additional moles of ATP are gained at the fumarate → succinate reduction. In order to remain in redox balance, 2 moles of α-KGA are converted to succinylCoA, setting the stage for the synthesis of four more moles of ATP during propionate production. Thus, in a species accumulating succinate and propionate in a 1:1 ratio (as, for example, in *Fasciola* under anoxia), the overall yield is 8 moles of ATP per mole of glucose + 2 moles of α-KGA (Figure 2–13). *Compared to glycolysis, the scheme quadruples the high energy phosphate yield.* On a percentage basis, these are astonishing advantages over classical glycolysis.

Lastly, it should be mentioned that in helminths, under certain conditions, pyruvate has a different metabolic fate, accumulating ultimately as acetate. If pyruvate dehydrogenase is the first step in the reaction pathway, the scheme is formally analogous to that for the metabolism of α-ketoglutarate:

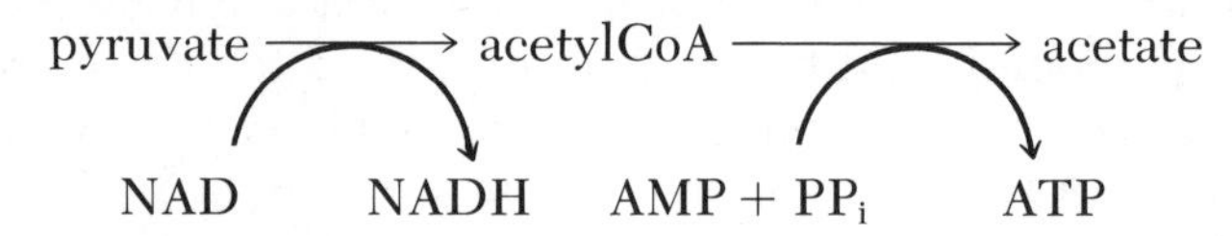

This route, therefore, supplies the organism with another potential mechanism for the anaerobic generation of ATP. As in the case of α-KGA dehydrogenase, the NAD for pyruvate dehydrogenase may be regenerated by fumarate reductase. This reaction pathway probably would be favored in those facultative anaerobes that rely solely upon glucose as a carbon and energy source. In contrast, in the bivalve molluscs, concentrations of free amino acids are up to 100 times greater than those occurring in mammalian tissues, and these are widely recognized as important potential energy sources. In these groups, the α-KGA pathway probably is favored because it couples glucose and amino acid catabolism. Indeed, it is the *ready availability of amino acids* which may have selected an alternative role for pyruvate dehydrogenase: the generation of acetylCoA for condensation with OXA to form citrate.

ORIGINS OF THE KREBS CYCLE

The latter statement is tantamount to suggesting a significant anaerobic function for the first span of the Krebs cycle—the span catalyzed by citrate synthase, aconitase, and isocitrate dehydrogenase. The last reaction in this span of course generates α-KGA, and if our considerations are correct, any anaerobic scheme which produced α-KGA in these organisms would be strongly selected since it sets the stage for an efficient substrate-level phosphorylation at a time when high energy phosphate compounds are at a premium.

The detailed exploration of this question is beyond the scope of this chapter. It suffices to mention that with the elaboration of enzymes catalyzing the first span of the Krebs cycle, all the enzymes of the cycle were available for serving *anaerobic* functions. Of course, these are precisely the preconditions needed to explain the origin of an *aerobic* metabolic scheme as complex as that of the Krebs cycle. With these preconditions satisfied, the subsequent utilization of O_2 as a terminal electron acceptor "freed" these reactions from their anaerobic functions; they could now be "hooked up" in a different sequence to subserve a different physiology.

THE BASIC STRATEGY IN FACULTATIVE ANAEROBIOSIS

From the above perspective, it is evident that the basic strategy of invertebrate facultative anaerobes is to couple other substrate-level phosphorylations to the glycolytic reactions, thus increasing the potential yield of high energy phosphate compounds. Currently, two important coupling sites can be identified:

(1) Fumarate reductase catalyzes fumarate reduction to succinate. The reaction ultimately regenerates NAD, and thus supplies coenzyme for α-ketoglutarate dehydrogenase.

(2) Glutamate-alanine transaminase catalyzes the formation of alanine and α-ketoglutarate, and thus supplies substrate for α-KGA dehydrogenase. SuccinylSCoA formed by the α-ketoglutarate dehydrogenase reaction can be utilized to "drive" the substrate-level phosphorylation of GDP to GTP.

Substrates for both of the above coupling reactions, fumarate for (1) and pyruvate for (2), arise from glucose. Hence, the two coupling reactions can be viewed as a means for achieving the *simultaneous mobilization of carbohydrates and amino acids during anoxic excursions.*

These mechanisms are commonly exploited by many animals; some estimates include representatives of 12 different phyla! By and large, they do not, however, appear to be utilized by vertebrates, with the possible exception of certain groups of fishes.

FACULTATIVE ANAEROBIOSIS IN FISHES

Because of the use of classical laboratory animals for most biochemical studies, the degree to which some of the lower vertebrates utilize anaerobic mechanisms to sustain periods of anoxia is not widely appreciated. Nevertheless, it is now clear that some fishes are true facultative anaerobes. During winter conditions the European carp, for example, often becomes "ice-locked" in small ponds which gradually grow anoxic and remain O_2 free for two to three months until the spring thaw. The carp shows no apparent ill effects after this extreme exposure to anoxic conditions. Blazka was probably the first to recognize the fundamental consequences of this habit in carp. Unlike fishes such as the salmonids, which depend upon an aerobic metabolism, Blazka found that the resting carp at low temperature does not accumulate an O_2 debt. Although it has been known for many years that fishes produce lactate as an end product of glycolysis during "burst" activity, lactate apparently does not accumulate in anoxic carp under resting conditions, particularly at low temperatures.

An important enzymic basis for this observation may be found in the kinetic properties of fish LDHs, for pyruvate inhibition of LDH activity is highly temperature dependent. For example, at high temperatures, muscle LDH from *Gillichthys* shows no substrate inhibition even at high (2 mM) pyruvate concentrations. At lower temperatures, marked pyruvate inhibition occurs (Figure 2–14). Similar results have been obtained for several poikilothermic

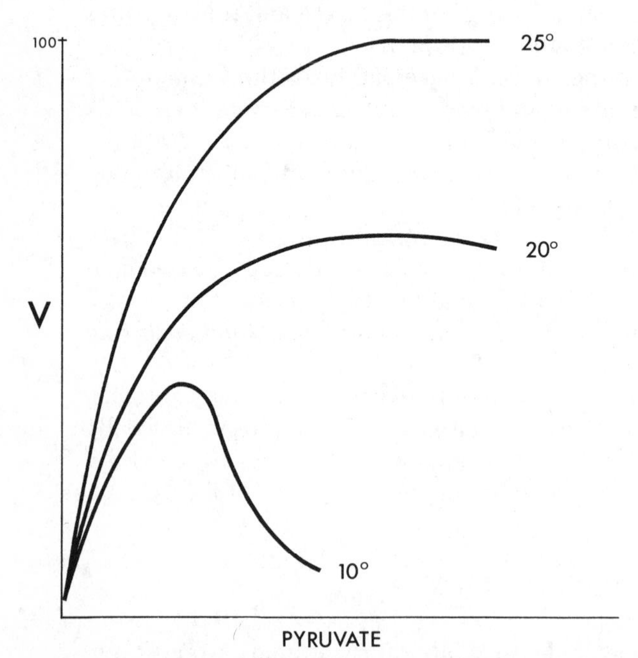

Figure 2–14 Pyruvate saturation of muscle LDH from the mudsucker fish, *Gillichthys*, showing the extreme temperature dependence of pyruvate inhibition. (After Somero, 1973.)

LDHs. This property could account for the absence of lactate accumulation in resting carp at low temperatures.

Parenthetically, it should be stressed that the physiological significance of these LDH properties in most fishes would be the favoring of anaerobic glycolysis at high temperatures. Usually, at high temperatures the organism is at once more active and exposed to reduced oxygen tensions, since the solubility of oxygen drops sharply as temperature rises. Under these conditions the organism may have to depend more highly on an anaerobic form of muscle metabolism. Conversely, at low temperatures activity is less and oxygen content is higher. Under these conditions the fish can likely "burn" all or at least a large share of its pyruvate aerobically. The LDH enzymes of these fishes are thus adapted to direct the flow of pyruvate in the proper direction, under conditions of varying oxygen content and temperature.

From these kinds of enzyme studies and from direct measurements of blood and tissue lactate levels, there is no doubt that fishes such as the carp are able to produce lactate under various physiological and environmental conditions. The controversy initiated by Blazka's original 1958 study, then, has been to some extent spurious. From a more holistic view of metabolism, the fundamental question is not whether lactate can be produced during anaerobic metabolism, *but whether any additional end products are formed.* One such additional end product—CO_2 of metabolic origin—is known with certainty.

The data here derive from studies of the goldfish and carp, two closely related species which are highly tolerant to extended anoxia. The goldfish can, for example, tolerate CO poisoning indefinitely and is remarkably resistant to CN^-. When goldfish injected with glucose-1-C^{14} or acetate-1-C^{14} are placed in O_2-free water at low temperature, they are able to produce $C^{14}O_2$ at rates which are as high as 1/3 the rates in the presence of high O_2 tensions. Kutty has shown that at oxygen concentrations near 15% of air saturation the goldfish sustains a respiratory quotient (CO_2/O_2) of about 2 for week-long periods. Using manometric techniques, Ekberg directly measured metabolic CO_2 produced by gill slices from goldfish. From these studies it is clear that goldfish and carp have mechanisms for the production of metabolic CO_2 even in the total absence of O_2. At low temperatures these mechanisms apparently supply all the resting energy demands of the organism.

At this time it is difficult to ascertain the frequency of these anaerobic mechanisms among fishes. Coulter lists some 10 species of benthic fishes in Lake Tanganyika which appear to live in, or at least tolerate extended excursions into, deep anoxic water. In warm swamp waters of tropical regions, O_2 tensions often become critically reduced; one common adaptation to this condition has been the development of the air-breathing habit in many of the fishes of the area. It would appear that other species which have not taken to air breathing in this environment must rely heavily upon an anaerobic metabolism.

Nothing is known of the mechanisms of CO_2 production during anoxia in these fishes. Two obvious possibilities are the pyruvate dehydrogenase and α-ketoglutarate dehydrogenase reactions, which appear to be so effectively coupled to the acetic and succinic thiokinase reactions, respectively, in helminths and molluscs. The α-KGA pathway is also established as an important anaerobic decarboxylating scheme in mammalian kidney cortex under conditions when aerobic metabolism must be supplanted with anaerobic ATP production. Cohen has shown that on a weight basis, the ATP yield from this pathway is about equal to that of glycolysis in the kidney tubules; hence, this scheme significantly contributes to the energy required for various transport functions of the kidney tubule cells. Similar anaerobic decarboxylations may account for the high CO_2/O_2 ratio observed in goldfish during anoxic stresses and in other fishes during excursions into anoxic waters. If so, succinate or acetate or both should be produced in stoichiometric quantities as additional end products of anaerobiosis (Figure 2–15). As far as we know, these have not been looked for, and it is evident that understanding of the nature of biochemical adaptation to anoxia in lower vertebrates must await further studies.

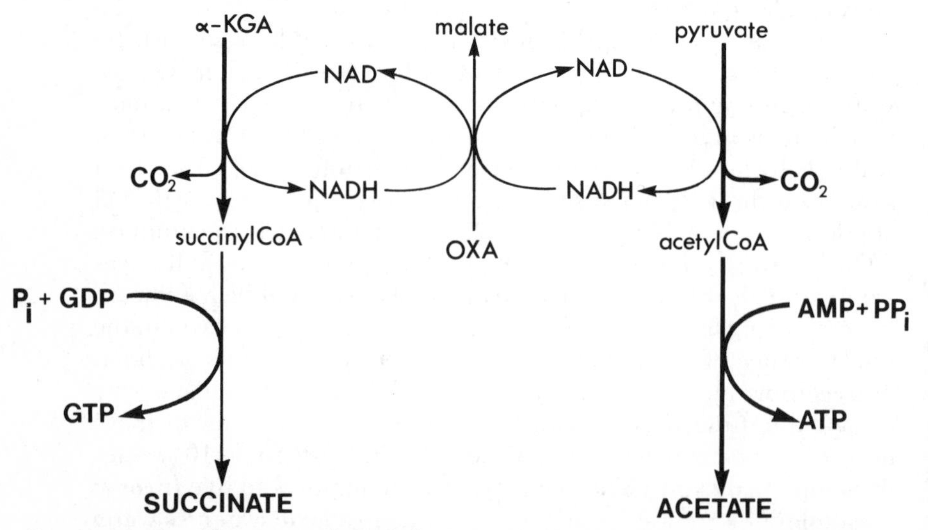

Figure 2–15 Possible sources of metabolic CO_2 and of high energy phosphate compounds during anoxia in goldfish and carp.

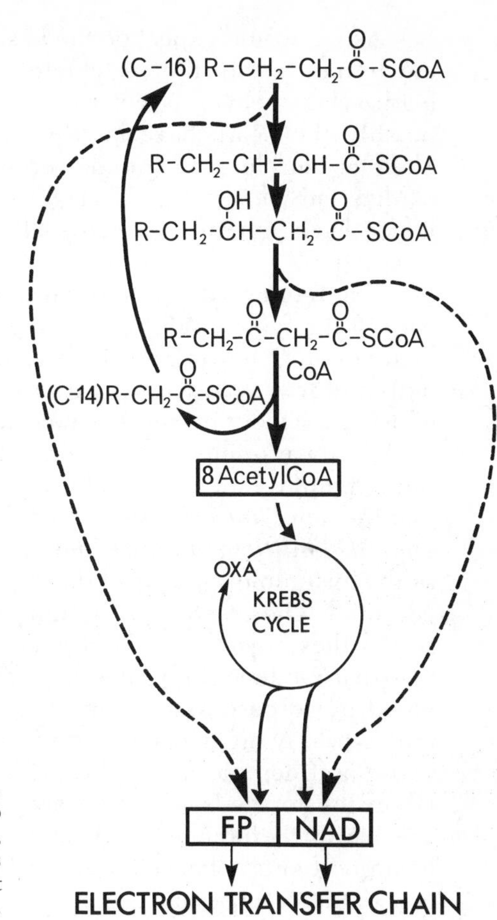

Figure 2–16 The β-oxidation spiral. Oxidation of C-16 palmitate leads to the production of 8 moles of acetylCoA, which is fed directly into the Krebs cycle. Per mole of palmitate, a net yield of 130 moles of ATP is achieved.

INTERACTIONS BETWEEN GLUCOSE AND LIPID METABOLISM: LIPIDS AS AN AEROBIC ENERGY SOURCE

To this point in our discussion we have focused our attention largely on carbohydrate and amino acids as carbon and energy sources. Whereas sudden, large scale bursts of muscular activity most commonly rely on glycolytic sources of ATP, long term muscular activity of vertebrates is generally sustained by the metabolism of fatty acids. Since the breakdown of fatty acids via the β-oxidation scheme (Figure 2–16) yields acetylCoA as the terminal product, which is then fed into the Krebs cycle for complete combustion to CO_2 and water, long term muscular activity which relies on fatty acid fuel is linked in an absolute manner to O_2 availability.

As we would expect on the basis of their different roles, different tissues within the vertebrate body display varied abilities to oxidize fatty acids. Tissues which are adapted for aerobic function, notably the heart, have large capacities for fatty acid oxidation. Skeletal muscle, if well endowed with red muscle fibers, can also oxidize fatty acids at high rates. White muscle fibers, on the other hand, have minimal capacities for this aerobic pattern of ATP production.

The rate and duration of muscular activity is a major determinant of the type of ATP generating scheme which will function to maintain muscle work. As the organism begins to exercise, both the output of free fatty acids (FFA) from adipose tissue and the uptake of FFA by working muscle increase. This increased turnover of FFA is always much greater in trained (fit) individuals than in untrained (unfit) individuals. Thus, as a first approximation, it appears that the *aerobic work capacity of the vertebrate organism depends upon its ability to mobilize and utilize FFA*, in much the same way as the anaerobic capacity of the organism depends on its glycolytic potential. There are, indeed, important regulatory interactions between these two channels of muscle ATP supply: lactic acid, the terminal product of glycolysis, inhibits the release of FFA into the blood by adipose tissue (Figure 2–17). Thus, when the muscle is forced to rely on glycolysis due to an insufficient supply of oxygen to support aerobic metabolism, the fuel reserves for aerobic metabolism, storage lipids, are spared.

During aerobic metabolism, when sufficient oxygen is present to support entry of acetylCoA residues into the Krebs cycle, citrate concentrations rise and feedback inhibit phosphofructokinase, the major control valve in the glycolytic pathway. Thus, under aerobic conditions, the carbohydrate fuel reserves of the muscle which are needed under anaerobic conditions are spared. Interestingly, the mechanisms for mobilizing both lipid and glycogen share two key properties: both are cyclic AMP-activated, and both display a "cascade" activation pattern (compare Figures 2–5 and 2–17).

WHY MANY ANIMALS USE FAT AS A FUEL FOR SUSTAINED WORK

We can gain an insight into the utility of fatty acids as a fuel for continuous work by comparing the energy yield available from the complete biological oxidation of a mole of glucose and a mole of palmitic acid. In the case of glucose, we indicated that the process

$$\text{glucose} + O_2 \rightarrow CO_2 + H_2O$$

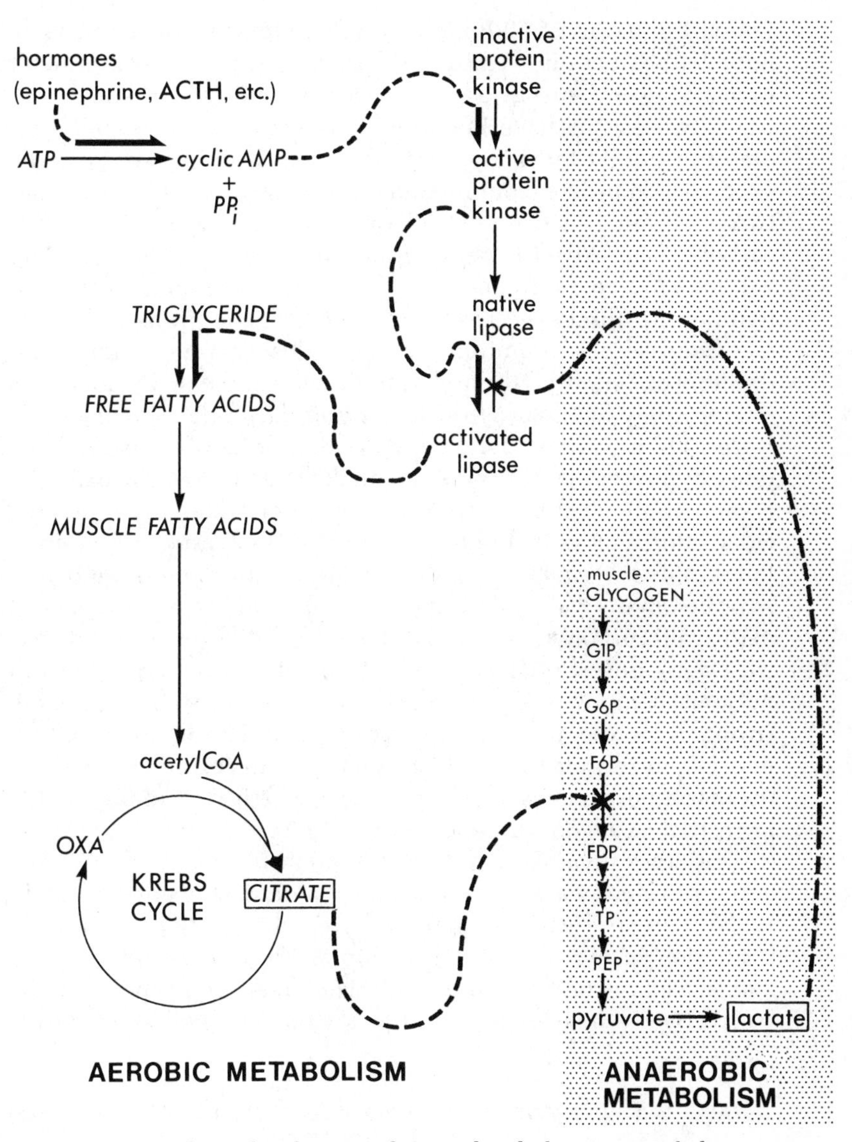

Figure 2–17 Relationship between fatty acid and glycogen metabolism in vertebrates.

yields 36 moles of ATP/mole of glucose. In comparison, an examination of Figure 2–16 will indicate that the process of palmitate oxidation

$$\text{palmitate} + O_2 \rightarrow CO_2 + H_2O$$

yields 130 moles of ATP/mole of palmitate. On a molar basis, then, fatty acid oxidation "captures" more than three times as much biologically useful energy as does carbohydrate catabolism.

Physiologists have been aware of the larger potential yield of energy from fat for many years. Because fat is in a more reduced state, its complete combustion yields, on a weight basis, about twice as much heat energy (Table 2–3) as can be obtained from carbohydrate. All else being equal, then, an organism burning fat can gain a 2- to 3-fold energetic advantage over one burning carbohydrate. All things, however, are seldom equal in nature.

Because of a premium on water, desert animals, for example, often must "choose" their metabolic fuels judiciously not for maximizing energy yield but *for maximizing the production of metabolic water.* On a weight basis, the oxidation of fat (because it contains more hydrogen) yields twice as much water as does the oxidation of carbohydrate (Table 2–4). Little wonder, then, that the desert camel stores fat in its hump! Although this fat, upon oxidation, does not supply the camel with all its water requirements, fat oxidation supplies more water than would be obtained by storing (and then oxidizing) carbohydrate; in the desert this difference could be critical. Many desert rodents appear to be able to survive indefinitely without drinking water. In these groups, there would appear to be strong selective pressure for fat (vs. carbohydrate) as a carbon, energy, and water source. The same conclusions are equally valid for marine organisms such as the seal which live under conditions of restricted availability of osmotically free water.

In other organisms, a great premium is placed on "lightness," and for these, on a weight basis, fat is a more convenient fuel depot to carry than is carbohydrate. Whereas this may be a minor consideration in the case of many animals, in the case of flying birds and insects weight is often minimized and a light fuel may be selected over a heavier one, and this would be particularly true for

TABLE 2–3 AVERAGE METABOLIC ENERGY DERIVED FROM THE THREE MAJOR FOODSTUFFS

Energy Source	kcal/gm
Carbohydrate	4.1
Lipid	9.3
Protein	4.1

TABLE 2–4 FORMATION OF OXIDATION WATER, OR METABOLIC WATER, WHEN DIFFERENT FOODSTUFFS ARE METABOLIZED

Energy Source	gm H_2O formed/ gm food	liters O_2 consumed/ gm food
Carbohydrate	0.556	0.828
Lipid	1.071	2.019

long range fliers. Weiss-Fogh has shown that if birds should store glycogen, for example, the mass of glycogen and its associated water would be *eight times that of a fat store containing the same potential amount of energy.*

For these various reasons, it is not surprising that on a gram-for-gram basis the storage of fat in the body of an organism is usually favored over the storage of carbohydrate. Thus, we find that animals from many different phylogenetic levels store fat prior to, and oxidize it during, prolonged migrations. This is as true for migrating caribou in the Arctic tundra as it is for migrating whales in the Antarctic seas, but at the biochemical level, the most impressive adaptations for fat catabolism are to be found in the flight muscle of flying birds and insects.

UTILIZATION OF LIPIDS FOR BIRD FLIGHT

The greatest capacities for lipid-fueled muscle metabolism are found in those birds which are adapted for particularly strenuous flight. Ducks, for example, can fly at about 45 miles per hour for many hours at a time. The golden plover migrates from the Aleutian Islands to Hawaii without an opportunity to feed, an over-water flight of about 2500 miles! Small birds are also known to fly some 1000 miles across the Sahara Desert without feeding stop-overs. These long range flights are fueled by fat; in preparation for the migration, huge amounts of fat (up to 50% of the total body weight) are laid down and most of this fat is gone at the termination of the flight.

The exceptional capacities of avian flight muscle probably reach their zenith in the hummingbird. This tiny creature, with a wing frequency of 80 to 100 wing beats per second, performs flight migrations across the Gulf of Mexico, a non-stop journey of over 500 miles. It possesses the incredible ability to hover (common among the insects, but unique among the birds), and is able to fly both forwards and backwards! The ratio of maximum/basal metabolic rate in this tiny bird is probably the highest to be found in all the vertebrates.

An important foundation for such extraordinary performance is clearly evident in the ultrastructure and the biochemistry of flight muscle. As the photomicrographs of Figure 2–18 illustrate, the fuel (fat droplets) and machinery for fat metabolism (mitochondria) are so pronounced in this tissue that the myofibrils themselves appear to be a minor component of the muscle! Furthermore, the architectural arrangement of (1) the fuel reserves, (2) the ATP-generating machinery, and (3) the ATP-utilizing muscle fibers favors the rapid "transduction" of energy to the working myofibrils.

In general, flight muscle of birds contains both red fibers and white fibers. The greater the aerobic potential of the muscle, the higher the proportion of red fibers. In the supracoracoideus muscle of the hummingbird (Figure 2–18), only red fibers are present. In organisms which have both red and white fibers within the muscle mass, clear structural differences between the two fiber types are noted. White muscle fibers are of relatively large diameter, contain fewer and smaller mitochondria, have much lower levels of oxidative enzymes (and the corresponding fuel: lipid), and have higher concentrations of glycogen than red muscle. (The largest muscle fibers known, some 3 mm in diameter, are to be found in the Arctic

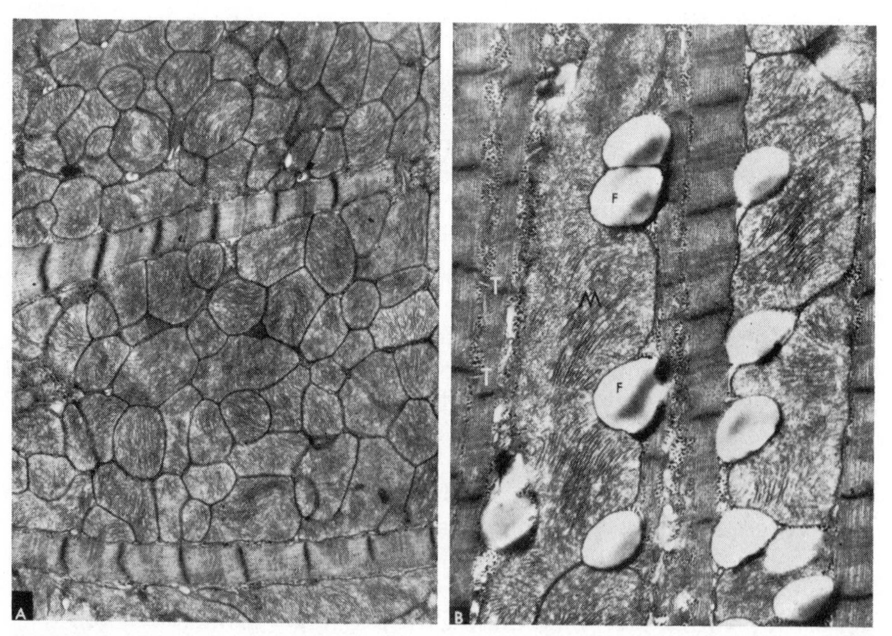

Figure 2–18 Electron photomicrographs of the supracoracoideus muscle of the ruby throated hummingbird. In the left panel, note the abundance of mitochondria. In the right panel, note the relationship between the mitochondria (M), the fat droplets (F), and the myofibrils. From Drummond (1971), courtesy of J. George.

king crab and the giant barnacle; these, too, have exceedingly high glycolytic potentials but very much reduced oxidative capacities.)

Finally, as the terms "red" and "white" imply, there is an additional important biochemical difference between these two kinds of fibers. The reddish color of "red" muscle derives from unusually high concentrations of myoglobin which occur within these cells. Myoglobin serves in part as a temporary O_2 "storehouse;" in part, it serves as a "transport" vehicle for facilitating O_2 diffusion from blood to the mitochondrial sites of utilization. Myoglobin concentrations are also quite high in heart muscle, another highly aerobic tissue. In both tissues, the regulation of myoglobin concentration contributes to the fine balance that exists between maximum O_2 demands of the muscle and the maximum O_2 delivery capacity of its copious blood circulation.

DESIGNING O_2 DELIVERY SYSTEMS TO MEET ALL POTENTIAL O_2 DEMANDS

In large measure, we have concentrated in this discussion on different biochemical strategies with which organisms have faced problems of limited O_2 availability. For heuristic purposes, we have categorized these as (1) compensatory strategies which depend upon a return to aerobiosis, and (2) exploitative strategies which do not depend upon a return to aerobiosis and which therefore allow massive exploitation of O_2-free environments. A third potential solution to the problem is to simply avoid it. This can be done behaviorally by a host of ways which we need not consider here. Biochemically and physiologically, *avoidance of O_2 depletion can only be achieved by elaborating O_2 delivery systems effective enough to balance even the most extreme O_2 demands with O_2 supply capacities.* As we have already indicated, vertebrates approach this condition for certain tissues (such as red muscle, heart, and brain) by regulating blood circulation so as to favor the delivery of O_2 to them. These tissues, however, possess a glycolytic capacity that can sustain anoxia for short periods of time. The mammalian brain, for example, can tolerate a few minutes of anoxia.

Terrestrial insects, on the other hand, balance O_2 demands with O_2 supply capacities by a tracheal system of O_2 delivery. In essence, each cell has a *direct tubular supply line to an infinite O_2 reservoir: the outside atmosphere.* Electronmicrographs show tracheoles to be in close apposition with mitochondria, forming a "mitochondrial-tracheole continuum." In small insects, diffusion through this system adequately meets the O_2 demands of the most extreme metabolism, but in large insects such as dragonflies, locusts, and wasps, the primary tracheole supply must be strongly ventilated by abdominal movements or by movements of the thoracic walls.

Insects such as the Diptera and the Hymenoptera utilize carbohydrate (glycogen, glucose, trehalose) as nearly the exclusive fuel for flight metabolism. Whereas in other organisms these are excellent sources of energy during anoxia, in these insects contractile work for even short periods of anoxia cannot be sustained. That is to say, in these insects *flight muscle contraction has become totally dependent upon an aerobic metabolism!* This suggests a very unusual interaction between glycolysis and mitochondrial metabolism, and we shall now turn to an examination of that interaction and its functional advantages.

GLYCOGEN AS AN AEROBIC SOURCE OF CARBON AND ENERGY IN INSECT FLIGHT MUSCLE

As in the case of the facultative anaerobes, the first phase of glycogen mobilization in Dipteran insect muscle is similar to that occurring in the vertebrates. With the arrival of an electrical impulse and the depolarization of the sarcolemma, Ca^{++}, which in resting muscle is sequestered in the sarcoplasmic reticulum, is liberated. The free Ca^{++} activates myofibrillar ATPase and phosphorylase *b* kinase as in Figure 2–5. Phosphorylase *b* kinase converts the inhibited phosphorylase *b* to an active phosphorylase *a*, resulting in glycogen mobilization. The decrease in ATP levels caused by myofibrillar ATPase and the consequent formation of ADP, AMP, and P_i lead to potent activation of muscle PFK (again, by mechanisms already described). As in the vertebrate case, the flare-up of glycolysis in insect muscle is due to the autocatalytic (product-mediated) PFK activation. The FDP formed by the PFK reaction is cleaved to two triose phosphates. At this point, the similarities between vertebrate and insect glycolysis end. In vertebrate muscle, all of the triose phosphate is ultimately converted to pyruvate, and, in anoxia, to lactate. In insect flight muscle, the precise situation depends upon the availability of O_2.

If the flight muscle is made anoxic, the G3P formed in the aldolase reaction is converted to pyruvate in the usual manner. However, in the insects the other product of the aldolase reaction, DHAP, is acted upon by a highly active α-glycerophosphate dehydrogenase (α-GPDH) and is converted to α-glycerophosphate (α-GP):

$$\text{DHAP} + \text{NADH} + \text{H}^+ \rightarrow \text{NAD} + \alpha\text{-GP}$$

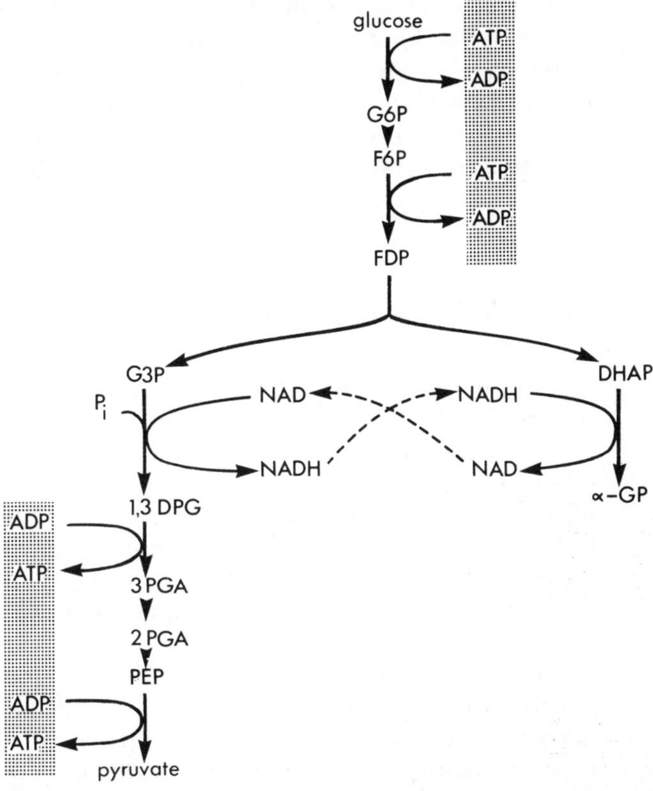

Figure 2–19 Anaerobic glycolysis in Dipteran flight muscle. End products: pyruvate and α-GP in a 1:1 ratio. Net ATP yield is 0 moles ATP/mole glucose.

During anoxia in Dipteran flight muscle, then, G6P is stoichiometrically converted to pyruvate and α-GP (Figure 2–19). Lactate does not accumulate under these conditions because of the "deletion" of lactate dehydrogenase in this tissue. It is particularly instructive that if glucose is the starting metabolite for this scheme the *net yield of ATP by anaerobic glycolysis is zero*: two ATP equivalents are utilized by hexokinase and PFK respectively, and two are regenerated by PGK and PK. If glycogen is the starting metabolite for this scheme, G6P formation is achieved without involvement of ATP. Hence, per mole of glucose residue glycolyzed, 1 mole of ATP is utilized by PFK, but 2 moles are regained at the PGK and PK steps, yielding a net gain of 1 mole of ATP/mole of glucose. Since *in vivo* trehalose (which yields glucose on cleavage by trehalase) and glycogen are both mobilized, it is evident that the actual ATP yield during anoxia would fall somewhere between 0 and 1.0 mole of ATP/mole of G6P glycolyzed. These are instructive observations, for they indicate that anaerobic glycolysis in insect flight muscle is extremely inefficient with respect to conservation of energy in the form of ATP. The need for a completely aerobic metabolism in this tissue is thereby fully appreciated, as is the observation that flight muscle work cannot be sustained during anoxia. What then is the aerobic fate and function of α-GP?

THE FUNCTIONAL SIGNIFICANCE OF α-GP FORMATION

In Figure 2–19, it will be evident that an immediate function of the α-GPDH reaction is the regeneration of NAD for the triose phosphate dehydrogenase. This function is performed by LDH in other, less active insect muscles and in vertebrate tissues; as we have seen, cytoplasmic MDH serves this role during anaerobiosis in molluscs. Of itself, then, this particular re-oxidation would not appear to supply the insect with any particular advantage. A critical energetic advantage becomes evident when we consider the subsequent fate of α-GP.

Because it moves freely across the mitochondrial membrane barrier, α-GP is quickly oxidized by a mitochondrial α-GPDH (a flavoprotein), thereby regenerating DHAP (Figure 2–20). In the process, *electrons and protons are carried from the cytosol into the mitochondria, where their subsequent transfer to O_2 by the electron transfer chain is coupled to oxidative phosphorylation.* The DHAP formed in the mitochondria is then available for further oxidation of extramitochondrial NADH. Examination of Figure 2–20 will indicate that the overall operation of this system is cyclic and catalytic, that the α-GP cycle is in fact a shuttle system in which NAD-linked substrates in reduced and oxidized states, respectively, enter and leave the mitochondria. Since it is self-generating, only a catalytic quantity of DHAP is needed to oxidize the NADH being continuously formed. This observation led Sacktor to suggest that most of the DHAP is isomerized to G3P, and that essentially all of the G6P catabolized during prolonged flight is convertible to pyruvate and is then made available for complete oxidation by the Krebs cycle. That this is in fact the case was shown in *in vivo* studies which demonstrate that α-GP does not accumulate during prolonged flight, whereas it is produced in a 1:1 ratio with pyruvate during anoxia. Glycolysis during prolonged flight therefore is completely aerobic, *with essentially 2 moles of pyruvate being formed from each mole of G6P.* Under these conditions, the situation is analogous to that in the vertebrates, but the immediate aerobic fate of pyruvate is unique to insect flight muscle.

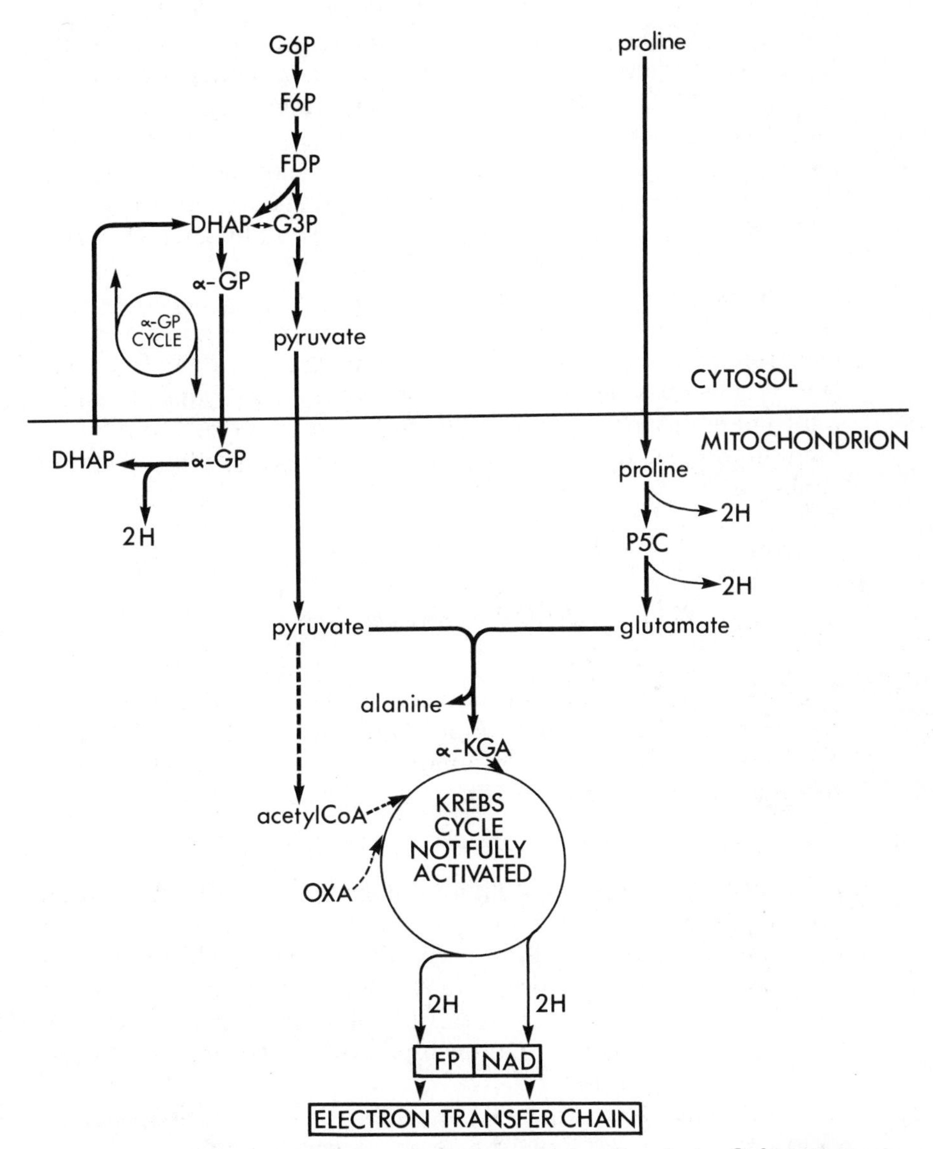

Figure 2–20 Simultaneous mobilization of glucose and proline during flight initiation in the blowfly. End products are alanine, CO_2, and water. Net ATP yield, assuming proline catabolism only to the level of OXA, is 21 moles of ATP per mole of G6P and of proline.

AEROBIC METABOLISM OF PYRUVATE DURING FLIGHT MUSCLE CONTRACTION

During the first few minutes of flight in the blowfly, pyruvate is produced by glycolysis at faster rates than it can be oxidized by the Krebs cycle. Pyruvate therefore momentarily accumulates and then is converted to other metabolites, alanine being predominant among these. The basis for this momentary limitation of the Krebs cycle is *inadequate supplies of OXA* for citrate synthase. Interestingly, these can only be generated from proline; that is, the Krebs cycle in insect flight muscle can be "sparked" *only* by proline, which occurs in resting flight muscle at remarkably high (10 mM) levels. None of the Krebs cycle intermediates, nor glutamate or aspartate, can penetrate the mitochondrial membrane barrier. Because of the strategic role of proline, it is not surprising that proline dehydrogenase (PDH), the enzyme responsible for mobilizing proline, is under tight energy-charge control. With the drop in ATP/ADP ratios at flight initiation, there is a strong activation of proline dehydrogenase:

$$\text{proline} \xrightarrow[\text{NAD} \curvearrowright \text{NADH}]{\text{PDH}} \text{P-5-C} \xrightarrow[\text{NAD} \curvearrowright \text{NADH}]{} \text{glutamate}$$

The mechanism of activation involves a large ADP-mediated increase in PDH affinity for proline. The glutamate formed has only one important route to follow, and this involves pyruvate as a cosubstrate (Figure 2–20):

$$\text{glutamate} + \text{pyruvate} \rightarrow \alpha\text{-KGA} + \text{alanine}$$

During the first minutes of flight, this pathway is strongly favored over GDH catalyzed conversion of glutamate to α-KGA because pyruvate concentrations are very high. As one would predict, as pyruvate concentrations drop, alanine levels rise; furthermore, the amino group in alanine derives ultimately from proline. Hence, there is a stoichiometric relationship between the proline utilized and the alanine formed.

The α-KGA is in turn metabolized by the Krebs cycle reactions, yielding a high-energy phosphate at the succinic thiokinase step, and ultimately generating the required OXA. The student will recognize that this scheme is identical to that in the facultative anaerobe, and it indeed gives us a clue as to why proline is the major metabolite that is "chosen" for sparking the Krebs cycle: if only proline can penetrate the mitochondrion, its further metabolism can be effectively and stoichiometrically coupled to glycolysis at the glutamate-pyruvate transaminase reaction. This assures the cell of a high-energy phosphate yield even as the stage is being set for a 100-fold activation of the Krebs cycle, and it is a strategy which we have already discovered in facultative anaerobiosis of molluscs and helminths.

Within the first few minutes of flight, the Krebs cycle becomes fully activated and essentially all of the pyruvate formed by glycolysis is oxidized by the Krebs cycle. Under these conditions, the concentrations of pyruvate are held at low, steady-state levels.

FUNCTIONAL ADVANTAGES OF THE METABOLIC ORGANIZATION IN INSECT FLIGHT MUSCLE

An insight into the overall functional advantages of the peculiar metabolic organization in insect flight muscle can be obtained if we compute the expected yield of high-energy phosphate by this system *prior to complete Krebs cycle activation*; this calculation can then be compared to that of typical vertebrate muscle. In insect flight muscle mobilizing glycogen under aerobic conditions, one ATP is utilized by the PFK step, two ATPs are regained at the PGK reaction, and two are regained at the PK step, yielding a net of 3 moles of ATP/mole of glucose residue glycolyzed. In addition (Figure 2–20),

(1) the α-GP cycle carries 4 H to the respiratory chain, yielding 4 ATPs

(2) H atoms produced at the PDH, α-ketoglutarate dehydrogenase, SDH, and MDH steps likewise feed into the electron transfer system, yielding 13 ATPs, and

(3) a final high-energy phosphate can be obtained at the succinic thiokinase reaction.

Thus, a potential yield of 21 moles of ATP per mole of G6P and proline is obtained prior to complete Krebs cycle activation. Under comparable conditions, vertebrate muscle glycolysis yields 3 moles of ATP/mole of G6P glycolyzed. By coupling glucose and proline catabolism, the insect therefore achieves a 6- to 7-fold advantage over the process that is classically operative in vertebrate muscle.

This large energetic advantage is a temporary one and occurs only during initial phases of flight. At steady state (i.e., after about 4 to 5 minutes of flight), all of the pyruvate formed by glycolysis is converted to acetylCoA, which is metabolized by the Krebs cycle in the usual manner. Under these conditions, of course, the yield of ATP/mole of glucose is similar to that in vertebrate muscle. Proline catabolism now follows a slightly different route, entering the Krebs cycle by the GDH catalyzed formation of α-KGA:

$$\text{proline} \rightarrow \rightarrow \text{glutamate} \xrightarrow{\text{GDH}} \alpha\text{-KGA} + NH_4^+$$

The α-KGA is metabolized in the usual manner and generates OXA. This scheme, however, does not generate acetylCoA. The complete oxidation of proline by this pathway therefore *requires an exogenous source of acetylCoA, which in insect flight muscle is supplied by pyruvate* (Figure 2–21). It is this mechanism, as a matter of fact, which couples proline and glucose catabolism under steady state conditions. Although the data cannot specify exact stoichiometry, it is evident that, in effect, carbohydrate supplies acetylCoA while proline supplies OXA for the 100-fold activation of the Krebs cycle that occurs during flight metabolism. At steady-state, this coupling of proline and glucose catabolism can potentially gain for the insect an additional 18 moles of ATP for each 36 moles obtained during the complete oxidation of a mole of glucose.

The increased efficiency of the metabolic scheme in insect flight muscle is evident at other levels of metabolism as well. Thus, in insects, the following advantages are gained:

(1) Aerobic glycolysis can be sustained without even a temporary accumulation of deleterious end products, such as lactate, because the system for regenerating NAD is catalytic and self-regenerating.

(2) Glucose metabolism does not wastefully accumulate end products such as lactate even during intense glycolysis at initiation of flight, when the overall flux of carbon is increased by up to 100-fold.

(3) H^+ and electrons of the NADH produced by the TDH reaction are quickly and efficiently transferred to O_2, with the consequent and immediate generation of ATP. In effect, the end products of glucose catabolism are therefore CO_2 and H_2O.

The biological cost of these apparent advantages is the total commitment of insect flight muscle to aerobiosis, a situation which may indeed give us an important insight into the kinds of evolutionary forces that shape cellular metabolism. Evidently, our initial classification of an "avoidance" strategy here may be somewhat of an oversimplification. Whereas it may be true that insects "avoid" or "minimize" the problem of O_2 depletion, it is also clear that their evolution of cellular energy metabolism was guided by a more active strategy of exploiting the nearly infinite O_2 content of the atmosphere. By elaborating tubular supply lines connecting the cell with the atmosphere, insects have gained the advantage of its high O_2 content. It was this essentially unlimited O_2 availability that led to the development of an aerobic scheme of energy metabolism unmatched in its efficiency in nature.

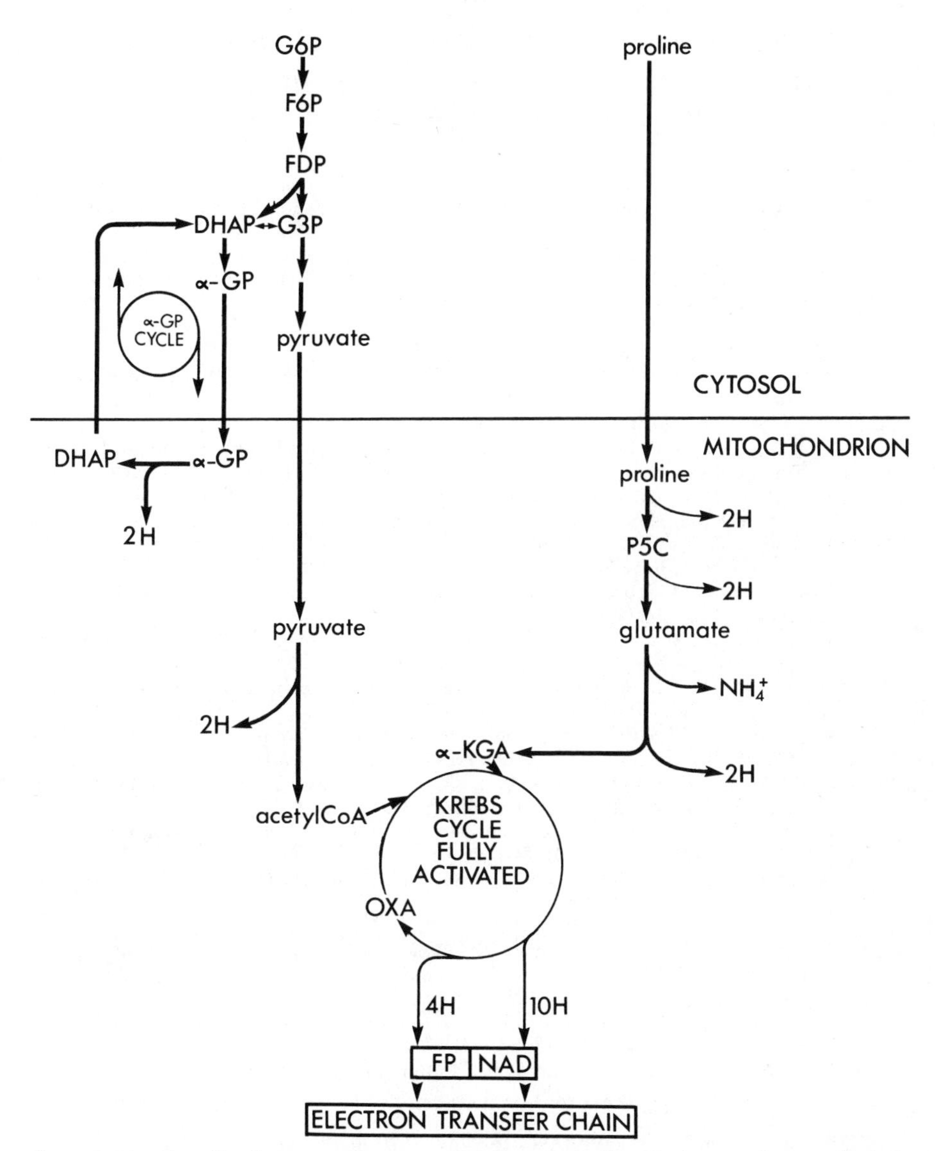

Figure 2–21 Coupling between glucose and proline catabolism during steady-state flight in the blowfly. End products are CO_2 and water. Net ATP yield, assuming proline catabolism only to the level of OXA, is an additional 18 moles ATP for each 36 moles of ATP gained per mole of glucose oxidized. In some insects, possessing active OXA decarboxylases, OXA formed from proline is decarboxylated to pyruvate, which is then "fed" into the Krebs cycle. This mechanism can supply a pathway for the complete oxidation of proline in these organisms.

SUGGESTED READING

Books and Proceedings

Biochemistry of Parasites by T. von Brand (1966). Academic Press, New York, 429 pp.

Multiple Molecular Forms of Enzymes (1968) in *Annals of the New York Academy of Sciences 151*, 1–689. (E. S. Vesell, Consulting Editor).

Reviews and Articles

Altman, M., and E. D. Robin (1969). Survival during prolonged anaerobiosis as a function of an unusual adaptation involving lactate dehydrogenase subunits. *Comp. Biochem. Physiol. 30*, 1179–1187.

Blažka, P. (1958). The anaerobic metabolism of fish. *Physiol. Zool. 31*, 117–128.

Cohen, J. J. (1968). Renal gaseous and substrate metabolism *in vivo. Proc. Intl. Union Physiol. Sci. 24*, 6, 233–234.

Coulter, G. W. (1967). Low apparent oxygen requirements of deep water fishes in Lake Tanganyika. *Nature 215*, 317–318.

de Zoeten, L. W., D. Posthuma, and J. Tipker (1969). Intermediary metabolism of the liver fluke, *Fasciola hepatica.* I. Biosynthesis of propionic acid. *Hoppe-Seyler's Z. Physiol. Chem. 350*, 683–690.

Drummond, G. I. (1971). Microenvironment and enzyme function: control of energy metabolism during muscle work. *Amer. Zoologist 11*, 83–97.

Ekberg, D. R. (1962). Anaerobic and aerobic metabolism in gills of the crucian carp adapted to high and low temperatures. *Comp. Biochem. Physiol. 5*, 123–128.

Hammen, C. S. (1969). Metabolism of the oyster, *Crassostrea virginica. Amer. Zoologist 9*, 309–318.

Hochachka, P. W. (1961). Glucose and acetate metabolism in fish. *Can. J. Biochem. Physiol. 39*, 1937–1941.

Hochachka, P. W., and T. Mustafa (1972). Enzyme mechanisms and pathways in invertebrate facultative anaerobiosis. *Science 176*, 1056–1060.

Jackson, D. C. (1968). Metabolic depression and oxygen depletion in the diving turtle. *J. Applied Physiol. 24*, 503–509.

Kemp, R. G. (1971). Rabbit liver phosphofructokinase. Comparisons of some properties with those of muscle phosphofructokinase. *J. Biol. Chem. 246*, 245–252.

Krebs, E. G., and D. A. Walsh (1972). The mechanism of action of cyclic AMP in mammalian systems. *Fed. Proceedings 31*, 14.

Kutty, M. N. (1968). Respiratory quotients in goldfish and rainbow trout. *J. Fisheries Res. Board Canada 25*, 1689–1728.

Mangum, C. P. (1970). Respiratory physiology in annelids. *Amer. Scientist 58*, 641–647.

Sacktor, B. (1970). Regulation of intermediary metabolism, with special reference to the control mechanisms in insect flight muscle. In *Advances in Insect Physiology* (ed. J. W. L. Beament, J. E. Treherne, and V. B. Wigglesworth), Vol. 7, 267–347. Academic Press, New York.

Saz, H. J. (1971). Facultative anaerobiosis in the invertebrates: pathways and control systems. *Amer. Zoologist 11*, 125–135.

Somero, G. N. (1973). Thermal modulation of pyruvate metabolism in the fish *Gillichthys mirabilis*: The role of lactate dehydrogenases. *Comp. Biochem. Physiol. 44B*, 205–209.

Wald, G. (1964). The origins of life. *Proc. Natl. Acad. Sci. (U.S.) 52*, 595–611.

Weiss-Fogh, T. (1968). Metabolism and weight economy in migrating animals, particularly birds and insects. In *Insects and Physiology* (ed. J. W. L. Beament and J. E. Treherne). American Elsevier Publ. Co., New York, pp. 143–159.

CHAPTER 3 CARBON DIOXIDE

THE TRANSPORT OF CO_2: GENERAL CONSIDERATIONS

Carbon dioxide, the terminal product of oxidative metabolism and the initial metabolite in photosynthetic carbon fixation, raises two distinct kinds of transport problems. For actively respiring tissues, in which CO_2 levels may rise to over 30 mM, a necessity exists for transporting CO_2 away from the locus of its generation. A build-up of CO_2 results in a decrease in pH, owing to the fact that CO_2 can combine with water to yield carbonic acid (H_2CO_3) which, in turn, dissociates to H^+ and HCO_3^-. Particularly in animals, the major adaptations related to CO_2 metabolism involve mechanisms for increasing the efficiency with which this metabolite can be removed from respiring tissues and released at the respiratory surfaces.

In plants we find quite a different situation, for here the major requirement is one of transporting large quantities of CO_2 into the organism. Because of the limited availability of atmospheric CO_2 (in plant leaves equilibrating with the atmosphere, CO_2 concentrations are about 7 μM), this is a formidable task. Thus, both plants and animal are faced with the problem of CO_2 transport across steep concentration gradients; in both plants and animals, solutions to this problem involve fundamentally similar adaptations in the enzyme systems responsible for CO_2 transport and metabolism. And as we shall see, many of these enzymic adaptations also reflect the types of biochemical strategies outlined in the first chapter of this volume.

CARBONIC ANHYDRASE: ITS FUNCTION IN CONTROL OF CO_2 MOBILITY

Central to control of CO_2 mobility is control of its physical state. Essentially all plant and animal tissues contain the enzyme carbonic anhydrase (CA) which catalyzes the formation of carbonic acid from CO_2 and H_2O,

$$CO_2 + H_2O \rightleftharpoons H_2CO_3 \rightleftharpoons H^+ + HCO_3^-$$

Carbonic acid dissociates rapidly, and the overall reaction is often written as

$$CO_2 + H_2O \rightleftharpoons H^+ + HCO_3^-$$

In mammalian red blood cells, these carbonic anhydrase catalyzed reactions serve as an important means for trapping metabolically generated CO_2 *inside* the cell, thus facilitating its rapid transport from the actively respiring tissues to the lungs. The mechanism of this trapping is based on the differential mobilities of CO_2 and HCO_3^-. CO_2, an uncharged gas molecule, diffuses relatively rapidly. It can enter the erythrocyte without difficulty. Once CO_2 is converted to the charged HCO_3^- ion, however, it can no longer as readily permeate the cell membrane. It is thus trapped inside the erythrocyte, and the bulk of it remains there until conditions arise which favor the regeneration of CO_2 from HCO_3^-. These conditions are met in the lungs, where CO_2 is being exchanged with the external environment; CO_2 concentrations are in consequence low, and, therefore, supply a *kinetic* "pull" on the carbonic anhydrase reaction in the direction of CO_2 formation. These kinetic conditions, of course, are just opposite to those occurring at the respiring tissues. Since the thermodynamic characteristics of the CA reaction also favor the production of CO_2, the removal of CO_2 at the lungs is rapid and efficient.

CARBONIC ANHYDRASE IN PHOTOSYNTHETIC SYSTEMS

The "engineering principles" found in animal systems which must transport relatively large amounts of CO_2 towards the environment might be expected to apply, albeit in a reverse direction, in plants which face the task of accumulating CO_2 from a CO_2-poor environment. Indeed, there is compelling evidence linking CA function to the operation of photosynthesis.

One line of evidence concerns changes in the quantities of CA which occur as ambient CO_2 levels are varied. In general, when CO_2 is limiting, CA is abundant; when CO_2 is abundant, CA levels are not maintained. For example, *Chlorella* cells grown first in a CO_2 rich medium, then transferred to a low CO_2 environment, are unable to photosynthesize for about a half-hour. A return to former rates of photosynthesis does not occur for at least 90 minutes. During this recovery period the quantities of CA in the cells increase approximately 1000-fold (Figure 3–1). Although the signals for these events are not known, CA induction at this time constitutes a clear example of an environmental adaptation involving the large scale increase of a needed enzyme species.

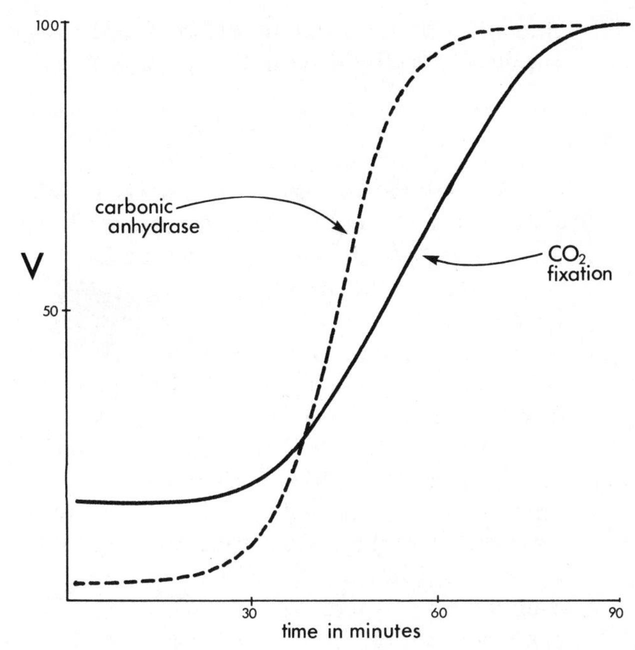

Figure 3–1 *Chlorella* cells, when grown first in a CO_2 rich medium and then transferred to a low CO_2 environment, are unable to photosynthesize for about 30 minutes. The return of full photosynthetic capacity is preceded by a 1000-fold increase in carbonic anhydrase activity. For further discussion of this area, see Hatch *et al.* (1971).

A second line of evidence implicating CA in photosynthetic processes comes from studies in which plants are treated with specific inhibitors of CA. In most cases these poisons greatly reduce the rates of photosynthesis.

While these data appear to offer unequivocal evidence that CA participates in photosynthesis, the precise mode of function of the enzyme remains unclear. Two hypotheses have been forwarded to explain its role in photosynthesis:

(1) Carbonic anhydrase facilitates the transport of CO_2 across the chloroplast membrane.

(2) Carbonic anhydrase increases the availability of CO_2 in alkaline environments, in which bicarbonate accumulation is favored.

Each hypothesis has received some experimental support and, of course, the two hypotheses are not mutually exclusive. Support for the latter hypothesis has come from studies of the pH dependence of photosynthesis and growth. For example, the marine alga *Dunaliella* requires CA activity only when growth is occurring at alkaline pH values. Carbonic anhydrase displays a distinctly alkaline pH optimum (about pH 8.5) and presumably increases the supply of CO_2 by driving the reaction towards the right,

$$H^+ + HCO_3^- \rightarrow CO_2 + H_2O$$

Support for the former hypothesis, that CA facilitates CO_2 transport across the chloroplast membrane, has come from studies of artificial membrane systems to which CA can be bound, on either or both sides of the membrane (Figure 3–2). These artificial membranes, like true cellular membranes, are hydrophobic. Consequently, charged molecules such as HCO_3^- cannot diffuse through them, whereas uncharged molecules, including CO_2, can freely cross the hydrophobic layer.

When purified CA (which is normally membrane bound *in situ*) is artificially linked only to the CO_2 donor side of the artificial membrane (Figure 3–2, middle panel), the rate of CO_2 transport across the membrane is increased about 1.5 times that of the control value (top panel). At the donor side of the membrane, the CA reaction favors the production of CO_2 from HCO_3^- because CO_2 is continuously being "bled off" across the membrane. Thus, the local environment of the membrane surface is continuously supplied with a readily diffusible molecule. When carbonic anhydrase is attached to both sides of the membrane, CO_2 transport is facilitated even further, and CO_2 transport rates are twice those observed for a CA-free artificial membrane. Presumably, the CO_2 which has crossed the membrane is bound by the "internal" carbonic anhydrase under conditions favoring conversion to HCO_3^-. This process accentuates the CO_2 gradient across the membrane, thereby increasing CO_2 diffusion rates across it (Figure 3–2, lower panel).

These model experiments suggest that carbonic anhydrase may play a comparable role in biological membranes such as those of the chloroplast. Be that as it may, the unique and fascinating aspect of this mode of CA function is that *it enables a charged species, HCO_3^-, to attain a mobility normally associated only with an uncharged gas, without the organism losing control of its concentration or its transport.*

CO_2 FIXATION: GENERAL CONSIDERATIONS

The roles of carbonic anhydrase relate entirely to the transport of CO_2 between areas of the body or between the organism and its environment. The enzyme plays no part in the fixation of CO_2 into more complicated organic compounds. CO_2 fixation or reduction is effected by another class of enzymes, which can be grouped into two distinct categories:

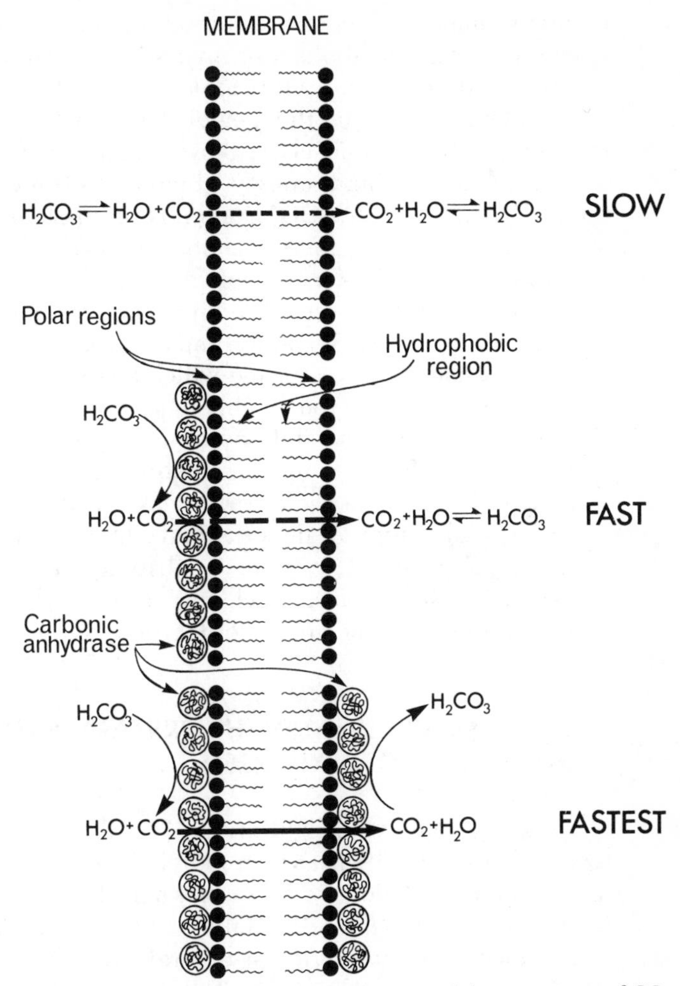

Figure 3–2 Diagrammatic representation of the movement of CO_2 through a naked hydrophobic sheet (upper panel), through a hydrophobic sheet chemically combined with carbonic anhydrase on the donor side only (middle panel), and through such a sheet with CA combined to both sides of the membrane. Neither liquid water nor ions (such as HCO_3^-) can cross such hydrophobic films, but molecular CO_2 readily penetrates them. The presence of CA on the donor side increases CO_2 transport rate 1.5-fold; CA bound to both sides of the membrane increases CO_2 transport rate over 2-fold. Data from Brown *et al.* (1970).

(1) Enzymes which fix or reduce CO_2 by utilizing high-chemical-potential compounds generated in intermediary metabolism.

(2) Enzymes which obtain the energy for CO_2 fixation and reduction from the radiant energy of the sun.

Both plants and animals contain enzymes of the first category. We discussed certain aspects of this mode of CO_2 fixation in the previous chapter when we described the importance of CO_2-fixing reactions in the metabolism of invertebrate facultative anaerobes, and we shall not consider this topic further.

In plant cells, by far the greater share of CO_2 fixation occurs via reactions of the second class, those which receive the necessary energy from the high energy compounds produced photosynthetically. Light-driven CO_2 fixation is of fundamental importance on two counts. Firstly, these reactions represent the primary source of carbon and energy for all biological systems; they are the locus where light energy is transduced into chemical energy for the support of life processes. Secondly, and of primary interest in the context of this volume, the enzyme which bears the responsibility for CO_2 fixation exhibits a remarkably *low* affinity for CO_2; its apparent K_m for CO_2 is approximately 50 times greater than ambient CO_2 concentrations. The seemingly inherent "inefficiency" of this enzyme, ribulose diphosphate carboxylase (RuDP carboxylase), has led to the elaboration of important adaptations for bolstering plant capacities to extract CO_2 from a substrate-poor environment. To appreciate these adaptive mechanisms fully, we must first review certain of the basic mechanics of the photosynthetic process.

THE CALVIN CYCLE AND THE PRIMARY PATH OF CARBON IN PHOTOSYNTHESIS

The mechanism of light-driven CO_2 fixation remained a mystery until the late 1940s, when radioisotope methodology first became available to biology. In a series of brilliant time-course experiments with *Chlorella* cells, Calvin and co-workers monitored the appearance of radioactivity in different carbohydrate compounds (Figure 3–3). They identified 3-phosphoglycerate (3PGA) as the first *stable* photosynthetic product. This compound is, of course, an intermediate in the glycolytic series and, therefore, it was logical to view the photosynthetic sequence as at least a partially reversed glycolytic pathway. Both the time course of C^{14} labelling and the distribution of C^{14} in each of the carbons of the key intermediates (Figure 3–3) are consistent with the following linear reaction scheme:

$$\begin{matrix} CO_2 \\ + \\ \text{acceptor} \end{matrix} \rightarrow \text{3PGA} \rightarrow \text{1,3 DPG} \rightarrow \begin{matrix} \text{G3P} \\ + \\ \text{DHAP} \end{matrix} \rightarrow \text{FDP} \rightarrow \text{F6P} \rightarrow \text{G6P} \rightarrow \text{glucose}$$

This reaction scheme is theoretically satisfying in that it represents a modified utilization by photosynthetic organisms of a primitive, pre-existing pathway (pp. 20–26). Very early on, however, it be-

$CH_2OPO_3^{=}$
$CO_2 \rightarrow C{=}O$
HCOH
HCOH
$CH_2OPO_3^{=}$

RuDP

$+ H_2O \longrightarrow$

$CH_2OPO_3^{=}$
HCOH
COO^-

+

COO^-
HCOH
$CH_2OPO_3^{=}$

3PGA

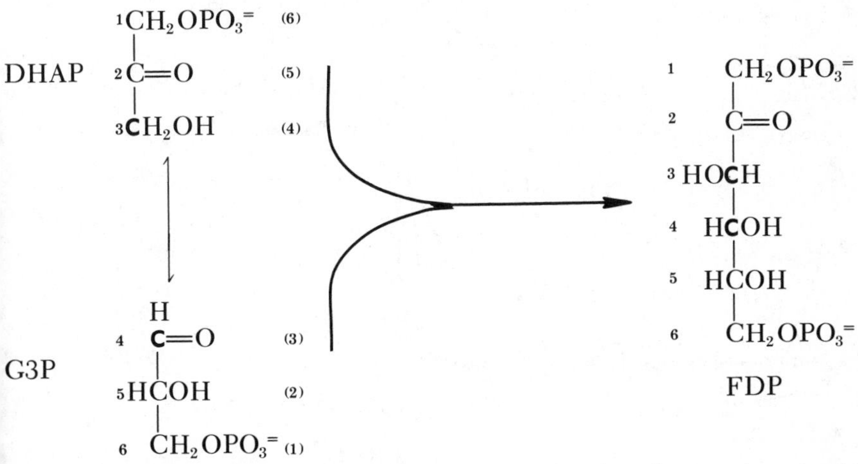

Figure 3-3 Scheme of CO_2 fixation by RuDP carboxylase. By this scheme, carbons 3 and 4 of hexose should be equally labelled following exposure to $C^{14}O_2$, a prediction of the Calvin cycle that allows experimental verification.

came evident that in addition to modifying the glycolytic reactions for "uphill" function, plants possess a modified pentose phosphate pathway which provides a cyclic and catalytic route for CO_2 fixation; by this mechanism, the initial CO_2 acceptor, ribulose diphosphate (RuDP), is regenerated with each "turn" of the cycle (Figure 3–4). We have already discussed the functional advantages of cyclic and catalytic pathways when we described the Krebs cycle. Similar advantages are gained by the Calvin cycle. Thus, six turns of the cycle provide a net yield of one hexose molecule *without a net utilization of the intermediates of the cycle.* Were the above, linear pathway to be the channel of CO_2 fixation, a continuous external supply of CO_2 acceptor would be required.

Plants which utilize the CO_2 fixation scheme outlined in Figure 3–4 are called "C-3 plants" because the primary product of the CO_2 fixation reaction is a 3-carbon phosphorylated acid, 3PGA.

THERMODYNAMIC PROPERTIES OF THE CALVIN CYCLE REACTIONS

Since the Calvin cycle was erected on pre-existing metabolic sequences, and since some of the Calvin cycle reactions occur in

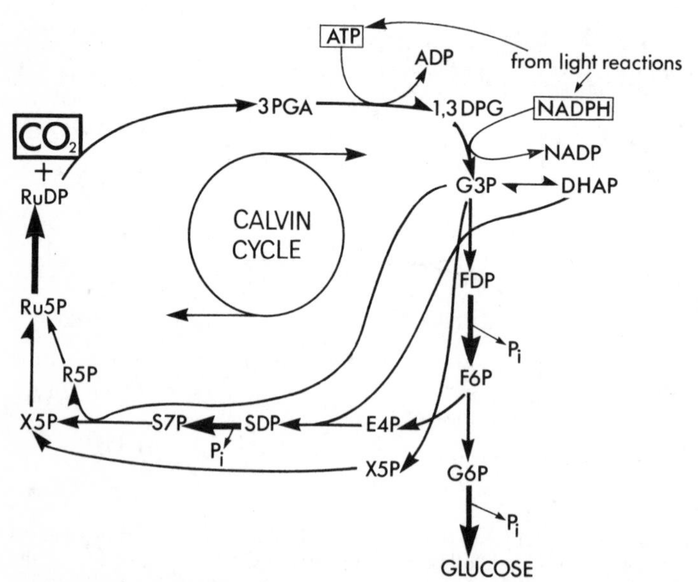

Figure 3–4 Diagrammatic representation of the Calvin cycle of CO_2 fixation. Four of the reactions of the cycle are thermodynamically irreversible and are "downhill." Phosphoglycerate kinase, on the other hand, represents a distinct thermodynamic barrier to net CO_2 fixation by this scheme. Other reactions of the cycle are readily reversible.

the opposite direction from the reactions utilized as "raw material" for elaborating the cycle, we might expect certain thermodynamic barriers to the production of hexose. Before considering the major thermodynamic "challenge" to be overcome during the development of the Calvin cycle reactions, it is important to realize that four of the cycle's most important reactions tend to poise the flow of the cycle towards hexose synthesis. These four reactions are:

1. FDPase: FDP $\rightarrow$ F6P + P_i.
2. G6Pase: G6P $\rightarrow$ Glucose + P_i.
3. SDPase: SDP $\rightarrow$ S7P + P_i.
4. RuDP Kinase: Ru5P + ATP $\rightarrow$ RuDP + ADP.

Each of these reactions proceeds with a sizeable decrease in free energy, and thus each is essentially pre-adapted for key roles in the net flow of carbon towards hexose.

In contrast to the foregoing reactions, the phosphoglycerate kinase (PGK) reaction represents a potent thermodynamic barrier to the photosynthetic pathway. This reaction,

$$\text{1,3 DPG} + \text{ADP} \leftrightarrows \text{3PGA} + \text{ATP},$$

is strongly poised in the direction of ATP synthesis. The free energy change which occurs during this "substrate-level phosphorylation" of ADP is −4.5 kcal/mole. Thus, for the reaction to operate in the reverse direction, as it must during carbon fixation, special kinetic mechanisms must exist to overcome the thermodynamic unfavorability of 1,3 DPG formation. Two mechanisms seem of special importance:

(1) ATP concentrations are maintained at high levels.

(2) Product (1,3 DPG) concentrations are kept low to "pull" the reaction towards 1,3 DPG and ADP production. (One might also predict that the phosphoglycerate kinase of chloroplasts would have low affinity for 1,3 DPG, relative to 3PGA, but we know of no test of this possibility.)

The primary source of large quantities of ATP to help poise the PGK reaction in the thermodynamically "uphill" direction is photosynthetic phosphorylation (Figure 3–5).

The second kinetic pull is accomplished by an active NADPH-linked triose phosphate dehydrogenase (TDH). This enzyme catalyzes the conversion of 1,3 DPG to glyceraldehyde-3-phosphate (G3P):

$$\text{1,3 DPG} + \text{NADPH} \rightarrow \text{G3P} + \text{NAD} + P_i.$$

The free energy change for this reaction (−1.5 kcal/mole) and its high activity insure that 1,3 DPG levels are kept low.

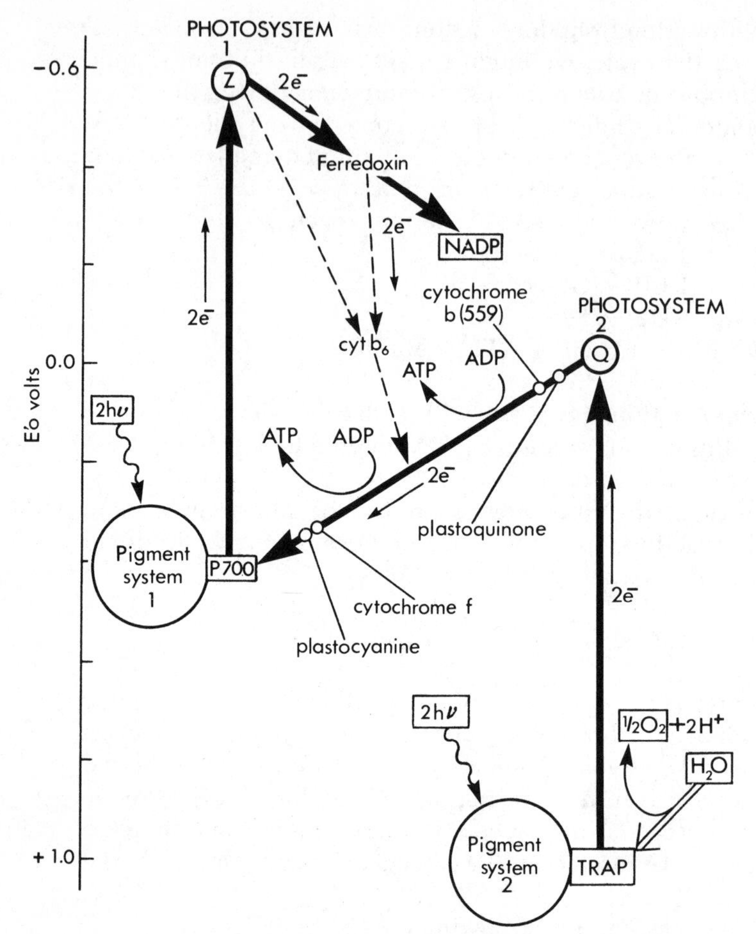

Figure 3–5 Series arrangement of photosystem I and photosystem II in chloroplasts of higher plants. Note the sites of production of ATP (by cyclic and non-cyclic photophosphorylation) and the production of NADPH. Photosystem II is absent in many C-4 plant species; in these, light-dependent production of NADPH is reduced.

The chloroplast TDH, in contrast to "typical" animal TDHs, is NADPH linked. Use of this cofactor, rather than NADH, "makes sense" physiologically: since NADP is the electron and proton acceptor during photosynthetic electron transfer (Figure 3–5), the use of an NADPH-linked TDH closely couples the activity of this enzyme to light-dependent NADPH production. When the plant is producing large quantities of NADPH, it simultaneously will be actively driving the TDH reaction towards G3P and thus "pulling" the thermodynamically unfavorable phosphoglycerate kinase reaction "uphill." This particular adaptation thus represents a situation in which *the kinetic properties of one enzyme can offset the un-*

favorable thermodynamic properties of an adjacent enzyme in the pathway.

Parenthetically, it should be added that plants also contain a "typical" NAD-linked TDH which probably functions in glycolysis and which occurs in the cytosol. Even the chloroplast TDH may utilize NADH as a cofactor under certain conditions: in the dark, the NADPH-dependent form of the enzyme is converted into an NADH-requiring form. This interconversion is reversible, and the enzyme becomes NADPH-dependent when light returns to the system. The so-called "light activation" of this transition is not direct. The actual mediator of this interconversion is NADPH. Thus, as light-dependent synthesis of NADPH occurs, the reduced cofactor promotes a change in the enzyme which leads to the enzyme's gaining a capacity to use NADPH.

ENZYME-CO_2 AFFINITIES AND THE RATE-LIMITING STEP IN THE CALVIN CYCLE

A priori, one might predict that any of the Calvin cycle reactions characterized by large changes in free energy could serve as rate-determining valves for the entire cycle. However, in common with other metabolic control schemes, control of this cycle is vested in the enzyme responsible for the first reaction of the pathway. This enzyme, RuDP carboxylase, catalyzes the actual "fixation" of carbon dioxide:

$$CO_2 + \text{RuDP} \rightarrow (\text{C-6 intermediate}) \rightarrow \text{3PGA}.$$

What distinguishes this enzyme from many other regulatory enzymes is its high concentration in the cell: RuDP carboxylase may constitute up to 50 per cent of the total protein of green plant cells. In sharp contrast, PFK, the major control site in glycolysis, is known to be the enzyme present in least amounts in the case of many glycolytic systems.

Can we account for the need for such large concentrations of RuDP carboxylase? We have already alluded to the most apparent basis for this state of affairs: whereas the CO_2 content of atmospherically equilibrating plant leaves is approximately 7 μM, the K_m of RuDP carboxylase is approximately 450 μM. Few enzymes display such a great difference between E-S affinity and physiological substrate concentrations. Thus RuDP carboxylase operation under conditions so far below saturation may necessitate its maintenance at unusually high concentrations.

It might be worth digressing at this point to consider why plants have not evolved RuDP carboxylases which are more effective at complexing substrate (CO_2). While we can only speculate on this matter, there are certain features of the substrate which would seem to make it less than an ideal candidate for efficient binding. Firstly, it is small and lacks a large amount of stereochemical structure. Consequently, the molecule has few groups which can interact with the enzyme surface. Secondly, the compound is uncharged. And, since charge is an important feature of E-S interactions, CO_2 is again limited in the extent to which it can be attracted and attached to the enzyme surface. Indeed, in other carboxylation reactions (pyruvate carboxylase, acetylCoA carboxylase), HCO_3^- (not CO_2) is the catalytically active and preferred substrate, the charge on the molecule presumably providing the required additional source of stabilization energy between enzyme and substrate.

REGULATION OF RuDP CARBOXYLASE

Keeping the kinetic limitations of the enzyme in mind, let us examine the known mechanisms by which RuDP carboxylase is regulated. These include:

(1) Light activation of the enzyme via direct effect of radiant energy on the enzyme protein.

(2) Secondary light activation of the enzyme via the interaction of RuDP carboxylase with a low molecular weight protein, which itself is converted to an activating state by light.

(3) F6P activation which, in turn, results from the stimulation of FDPase activity by reduced ferredoxin.

This latter scheme is perhaps the most important mechanism for activating RuDP carboxylase, for it further integrates the events of photophosphorylation (Figure 3–6) with the events of carbon fixation. Briefly, this regulatory scheme begins at the level of light-dependent reduction of ferredoxin. In the reduced state, ferredoxin potently activates chloroplast FDPase. The nearly 10-fold activation of this enzyme leads to reduced FDP levels and elevated F6P concentrations. The latter metabolite activates RuDP carboxylase by increasing its V_{max} slightly and by decreasing its K_m of CO_2 approximately 6-fold. Given the condition of limited CO_2 availability, the latter effect may be an especially important stimulus for CO_2 fixation.

In the dark, the above sequence of events is reversed. As ferredoxin is oxidized, the activity of FDPase is reduced, and the rising FDP levels then serve as a metabolic signal for reducing RuDP carboxylase activity. Thus, we find that CO_2 fixation can be closely coordinated with the supplies of ATP and reducing power which are generated by the photophosphorylation reactions (Figure 3–6).

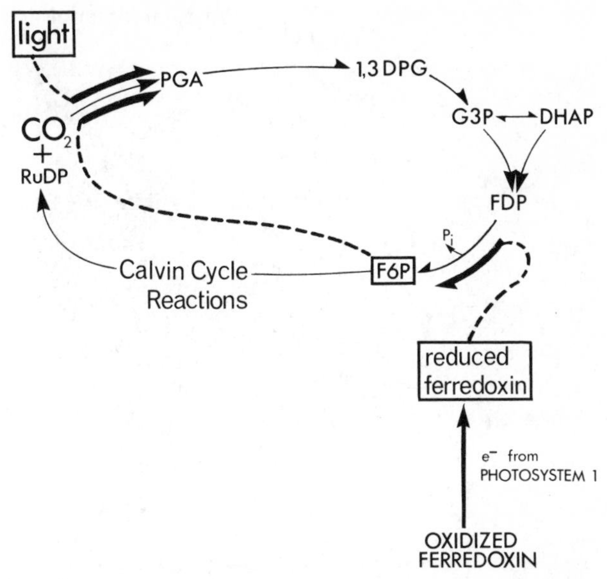

Figure 3–6 Currently known control circuits in the Calvin cycle. Activations are shown by thick arrows.

LIMITATIONS OF THE CO_2 FIXING SYSTEMS IN C-3 PLANTS

The above mechanism for CO_2 fixation is characteristic of the class of plants we earlier termed "C-3" plants. These organisms must accommodate an extensive amount of ventilation of their leaves in order to draw CO_2 from the environment. For terrestrial plants, the concomitant of large scale CO_2 inflow is massive loss of water vapor. For plants living in dry and/or hot regions, we might expect to find adaptations which enable the plants to avoid large water losses while yet accumulating enough CO_2 to support the reactions of carbon fixation and, hence, growth. In fact, there is another class of plants, the "C-4" plants, which have succeeded in developing mechanisms for concentrating carbon dioxide without the associated loss of large amounts of tissue water.

HIGH CO_2-AFFINITY SYSTEMS AS CO_2 "TRAPPING" MECHANISMS IN C-4 PLANTS

Two important characteristics are displayed by plant species which utilize CO_2 concentrating mechanisms for minimizing trans-

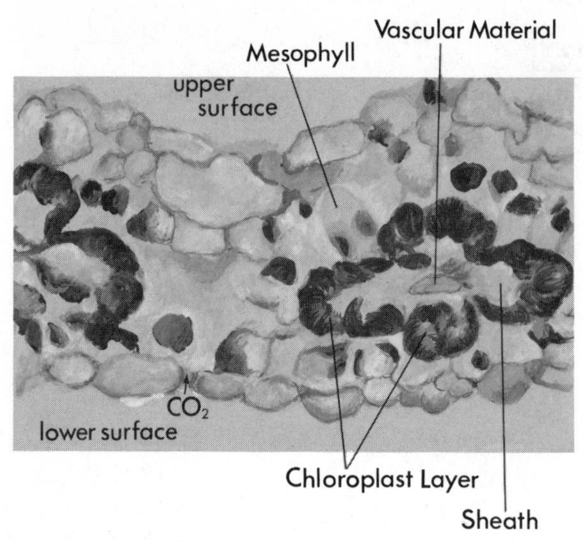

Figure 3–7 Kranz type of leaf anatomy. In these leaves, the bundle sheath is not exposed directly to atmospheric CO_2. Hence, CO_2 for the Calvin cycle in the bundle sheath chloroplasts must be transported from the external environment, through the mesophyll, and into the bundle sheath.

piratory water loss. One trait is biochemical: in all cases, these species synthesize C-4 dicarboxylic acids as initial photosynthetic products (rather than 3PGA). This synthesis is achieved by high CO_2 affinity enzymes which are totally independent of the light reactions.

The second identifying characteristic is morphological: all plants with the C-4 pathway have a "Kranz" type of leaf anatomy (Figure 3–7). The vascular bundles are surrounded by (1) an inner "bundle sheath" layer that is very rich in specialized chloroplasts, and (2) an outer mesophyll layer of cells that are less rich in chloroplasts. It has recently become evident that this characteristic leaf anatomy serves to compartmentalize biochemical events.

Because of the morphology of these leaves, only the mesophyll cells are exposed to atmospheric CO_2. CO_2 entering the mesophyll cells is "trapped" by a high affinity PEP carboxylase, which catalyzes the reaction

$$CO_2 + PEP \xrightarrow{Mg^{++}} OXA + P_i.$$

The phosphate acceptor is H_2O, and the reaction occurs with a large loss of free energy. For this reason, and because of a low K_m for CO_2

(about 7 μM), the carboxylation reaction is essentially irreversible under physiological conditions. For a given exposure to atmospheric CO_2 (i.e., for a given period during which the stomata are open to allow entry of atmospheric CO_2 and the transpiratory loss of H_2O), this system would far outcompete the RuDP carboxylase system for CO_2.

It is therefore not surprising that the enzyme was selected in those species placing an unusual premium on water. The system is made all the more efficient by the presence of a highly active MDH which efficiently converts the newly formed OXA to malate. The equilibrium for this reaction, here as elsewhere, is far in the direction of malate, and the reaction presumably plays the same function here as in invertebrates: it maintains low OXA levels and hence further prevents any reversal of PEP carboxylation.

Interestingly enough, PEP carboxylase is under negative feedback modulation by malate in particular, and, secondly, by aspartate (which can be formed from OXA by transamination). Hence, it is clear that in the C-4 plant, PEP carboxylase is an important (and efficient) locus of control of CO_2 fixation (Figure 3–8). Malate appears to be the primary product of CO_2 fixation, and its further metabolism is of great importance.

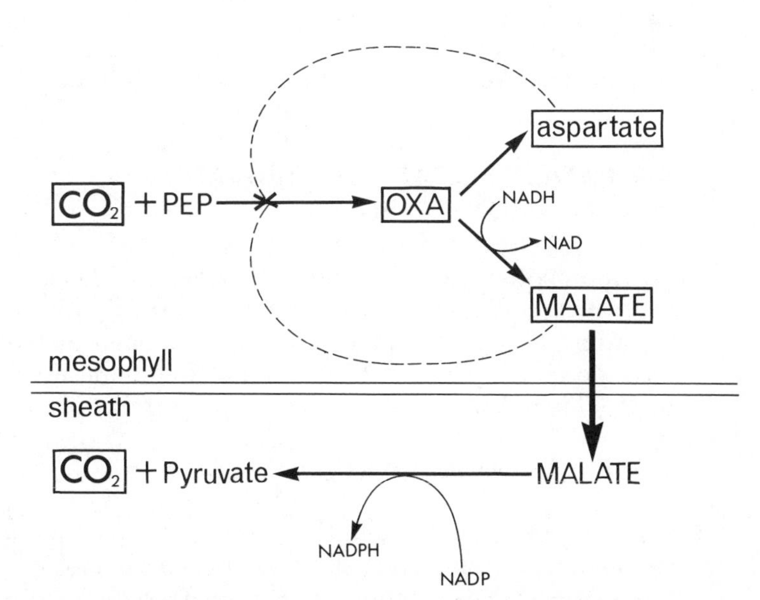

Figure 3–8 Negative feedback regulation of PEP carboxylase, an enzyme which in C-4 plants is strategically positioned for control of CO_2 fixation.

MALATE AS A CO_2 TRANSPORT MECHANISM

Malate, formed by the above reactions in the mesophyll cells, may have a number of metabolic fates. Chief among these is diffusion or transport into the bundle sheath cells. Here it serves as a substrate for "malic enzyme," which in the presence of a divalent cation catalyzes the reaction (Figure 3–8):

$$\text{Malate} + \text{NADP} \rightarrow \text{pyruvate} + CO_2 + \text{NADPH}$$

It is particularly important to note that the malic enzyme shows a strong localization in the bundle sheath. Here it is properly positioned for the delivery of (1) CO_2 and (2) NADPH to the Calvin cycle reactions. *The malic enzyme reaction is the only source of CO_2 for these cells, since they are well protected from the outside atmosphere.* CO_2 concentrations build up as a result of this reaction to estimated intracellular levels in the bundle sheath of 1 mM, levels high enough to saturate the RuDP carboxylase of the Calvin cycle. The malic enzyme reaction is "poised" to the right, presumably because it has a relatively higher affinity for malate than for pyruvate or CO_2. Moreover, the NADPH generated by this reaction is utilized by the TDH step in the Calvin cycle; as if to be assured that this NADPH is utilized for this purpose, the chloroplasts of bundle sheath cells have a reduced capacity for light-dependent generation of NADPH because of the absence of photosystem II (Figure 3–5). Thus, TDH must draw half of its required NADPH from the malic enzyme reaction and so contributes to "pulling" the malic enzyme reaction in the decarboxylating direction.

THE METABOLIC FATE OF PYRUVATE IS RECONVERSION TO PEP

An important question remains concerning the fate of the pyruvate which is formed by malic enzyme in the bundle sheath cells. According to recent evidence, the predominant metabolic fate of pyruvate is reconversion to PEP following its return to the mesophyll. This reconversion is catalyzed by pyruvate, P_i dikinase

$$\text{pyruvate} + \text{ATP} + P_i \rightarrow \text{PEP} + \text{AMP} + PP_i$$

This enzyme occurs only in C-4 plants and apparently functions to complete a kind of carbon shuttle between the mesophyll and the bundle sheath. The reaction proceeds to the right primarily because these cells also possess high activities of pyrophosphatase and adenylate kinase; these act upon AMP and PP_i and in effect maintain them at low concentrations, thus contributing to "pulling" the pyruvate, P_i dikinase reaction to the right. The function of pyruvate, P_i dikinase is clearly to regenerate PEP for the PEP carboxylase reaction.

In a sense, then, the net function of this series of reactions is cyclic and catalytic (Figure 3–9): There is neither a net fixation of CO_2 nor the net utilization of cycle intermediates. These reactions

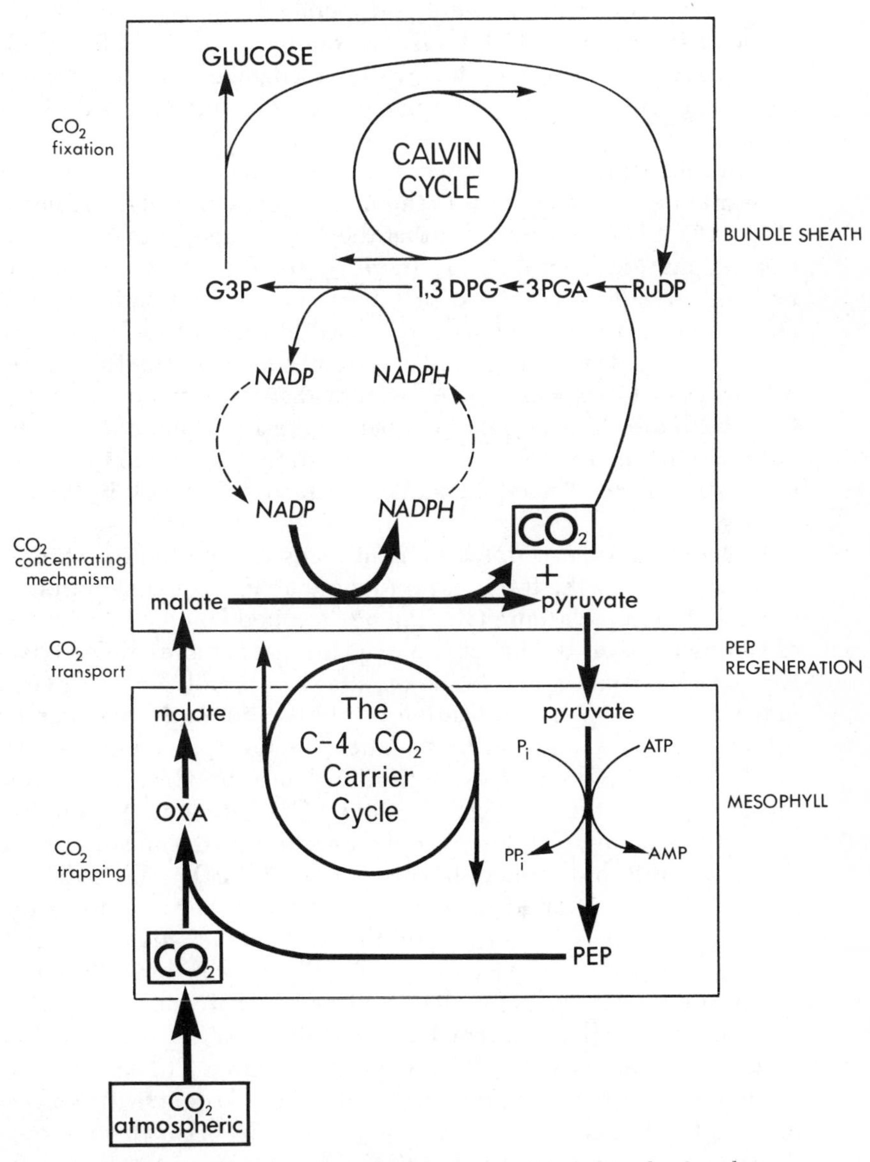

Figure 3–9 The effective path of carbon in CO_2 fixation by C-4 plants.

must be viewed as a means of (1) trapping CO_2 in the mesophyll, which is in contact with the atmospheric CO_2 source; (2) transporting the "fixed-CO_2," as malate, into the bundle sheath; and (3) depositing and concentrating the CO_2 at the site of the Calvin cycle by decarboxylation of malate.

THE SELECTIVE ADVANTAGES OF THE C-4 PATHWAY

There may be a number of peripheral advantages to utilization of the C-4 pathway of CO_2 transport coupled with the Calvin cycle for CO_2 fixation. However, it is widely recognized that the primary advantages relate to (1) minimizing water loss and (2) maximizing growth rates.

In regard to water balance, the arguments are clear and there appear to be few exceptions to them. The geographical distribution of C-4 plant species indicates that they originated primarily in the tropics, and they certainly are more successful than C-3 plants in hot desert regions. Because of high temperatures, both environments put a premium on water. As we have indicated, PEP carboxylase is a very high-affinity system compared to RuDP carboxylase; in C-4 species, it is evident that for a given amount of CO_2 fixed, the stomata need be open to the atmosphere for shorter time periods than in C-3 species. That is, for a given amount of CO_2 trapped, *less water need be transpired in the C-4 than in the C-3 plant.*

The other side of this argument leads us into the question of maximizing growth, for it is evident that for a given period of transpiration, much more CO_2 can be "trapped" by the mechanism of C-4 plants than by that of C-3 plants. What is more, the mechanism of CO_2 concentration in the bundle sheath probably achieves a much more efficient CO_2 fixation by the Calvin cycle than can be achieved in C-3 plants, for in the latter, the low CO_2 affinity system must compete for low atmospheric CO_2 concentrations. Hence, one must assume that this enzyme is seldom CO_2 saturated. In contrast, in the C-4 species, CO_2 levels in the bundle sheath can surpass the K_m by about 2-fold; hence, in these species, RuDP carboxylase is probably fully saturated during peak photosynthetic periods. For these reasons, it is not surprising that C-4 species include highly productive species (such as sugar cane) and can typically attain photosynthetic rates about 2 to 3-fold higher than can C-3 species.

Parenthetically, it might be added that the C-4 plant species typically do not perform well at low temperatures. Their origin was in high temperature environments and they still seem to thrive best there today. For example, Bjorkman and his coworkers have shown that maximum photosynthetic rates in C-4 plants of Death Valley, California occur at temperatures as high as 50°C. Important differences in thermal response of key enzymes such as RuDP and PEP carboxylases have undoubtedly arisen as a result of evolutionary adaptations of plants to such extreme thermal regimes.

SPATIAL VS. TEMPORAL SEPARATION OF LOW AND HIGH CO_2-AFFINITY SYSTEMS

The C-4 plants *spatially* separate the low and high CO_2-affinity systems: thus, PEP carboxylase is located predominantly in the mesophyll while malic enzyme is coupled to RuDP carboxylase function in the bundle sheath. Another group of plants (mainly succulents) utilize both high and low CO_2-affinity systems within the same cell and tissue, but they separate the two functions *temporally.* The phrase that is used in the literature to describe this phenomenon is crassulacean acid metabolism, or CAM for short. CAM refers to a massive CO_2 fixation into C-4 acid compounds during the dark and the subsequent decarboxylation and utilization of this "stored" form of CO_2 by the Calvin cycle in the light. In these plants, malate concentrations fluctuate widely on a day-to-night basis (Figure 3–10). The enzyme reactions involved in this metabolism are identical to those described for other C-4 plants.

It is well known that CAM activity (i.e., night CO_2 fixation and diurnal fluctuations in malate concentrations) is most pronounced when night temperatures are low and day temperatures are high. In part, the effect of temperature on this system is correlated with stomatal resistance to CO_2 transfer. Thus, under some circumstances, greater stomatal opening occurs in the dark at low temperatures. But other processes must be superimposed upon this coupling between low temperature and stomatal opening, because fluctuations in CO_2 fixation occur even if the epidermal barrier of CO_2 is removed.

That at least a part of the net CO_2 fixation in the dark must depend upon the thermal properties of the enzymes involved can be shown by comparing the effect of temperature on PEP carboxylase (which, of course, is the chief determinant of the rate of OXA and hence of malate production) and on malic enzyme (which determines the rate of malate disappearance during the day). The thermal optimum for PEP carboxylase occurs at about 35°C, while that for malic enzyme is not reached by 55°C! If one plots the rate of these two enzymes functioning simultaneously, it is clear that malate accumulation would be favored at lower (night) temperatures while malate decarboxylation (i.e., CO_2 production for the Calvin cycle) would be favored at higher (daytime) temperatures (Figure 3–10). That is, the thermal characteristics of these two key enzymes are closely integrated with the diurnal thermal changes that the succulent encounters in nature. It is not known whether the

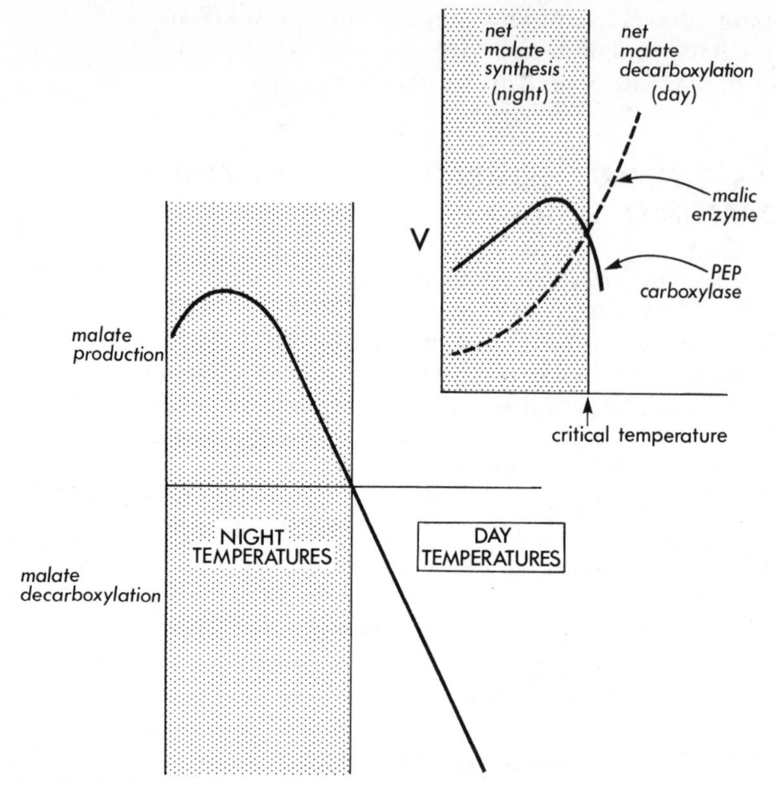

Figure 3–10 The effect of temperature on net malate production and net malate decarboxylation (left panel) in succulent plants is a consequence of the thermal properties of PEP carboxylase and malic enzyme (upper right). (Data from Brandon, P. C. (1967). *Plant Physiol. 42*, 977.)

"compromise" temperature for these two enzymes will vary between different species, depending upon the light and temperature characteristics of their environment, but this is a good question for new students in environmental biochemistry.

SUGGESTED READING

Books and Proceedings

Photosynthesis and Photorespiration (1971). (Edited by Hatch, M. D., C. B. Osmond, and R. O. Slayter), Wiley-Interscience Publ., New York. 560 pp.

Reviews and Articles

Björkman, O., R. W. Pearcy, A. T. Harrison, and H. Mooney (1972). Photosynthetic adaptation to high temperatures: a field study in Death Valley, California. *Science 175*, 786–789.

Broun, G., E. Selegny, C. Tran Minh, and D. Thomas. (1970). Facilitated transport of CO_2 across a membrane bearing carbonic anhydrase. *FEBS Letters 7*, 223–226.

Maren, T. H. (1967). Carbonic anhydrase in the animal kingdom. *Fed. Proceedings 26*, 1097–1103.

Preiss, J., and T. Kosuge (1970). Regulation of enzymic activity in photosynthetic systems. *Ann. Review Plant Physiol. 21*, 433–466.

CHAPTER 4

WATER AND SOLUTE PROBLEMS

THE "FITNESS" OF WATER

No chemical parameter of the environment has a more universal influence on living systems than water. As so elegantly stated by L. J. Henderson in his 1913 classic *The Fitness of the Environment*, the properties of water help render the earthly environment a "fit" place for life to originate and evolve. The solvent capacity, the surface tension, the specific heat, the expansion on freezing, the weak ionization, and the latent heat properties of water all render this ubiquitous substance especially well suited for its roles in biological systems.

Whereas the inherent physical and chemical properties of water suit this molecule so admirably for the genesis and elaboration of living forms, problems associated with the control of water and solute concentrations and movements are some of the most severe ever encountered by living organisms. These problems fall into two broad categories:

(1) Maintenance of low concentrations of the large numbers of different kinds of solute molecules in the cell.

(2) Control of the exchange of solutes and water between the cell and its environment.

The first of these, the maintenance of the solvent capacity of water, is universal in its implications. The second basically reduces itself to the requirements of salinity adaptation, i.e., the requirement for balancing both the concentration and composition of the intracellular fluids with the outside medium. We shall discuss each of these general problems in turn.

METABOLIC COMPLEXITY AND THE PRESERVATION OF INTRACELLULAR SOLVENT CAPACITY

Evolution of the vast array of metabolic reactions characteristic of living systems—reactions which, by and large, must occur in an aqueous phase—has been marked by the elaboration of thousands of distinct types of enzymes. In a given cell, as many as two or three thousand different enzyme species may be present. Associated with each type of enzyme are one or more substrate molecules plus, in many cases, one or more organic or inorganic cofactors. In total, therefore, the individual cell must contain thousands of different *kinds* of molecules, of which there may be a few to many millions of copies of each (Table 4–1).

The evolutionary elaboration of a complex metabolic machine has thus created the serious problem of "packaging" many millions of molecules into each cell without surpassing the solvent capacity of the intracellular water. The adaptations which permit such a wide array of metabolic reactions to occur in the cell's aqueous phase, without saturating intracellular water, are reflected in the present architecture of the cell and, indeed, in the chemical structures of many intermediary metabolites.

ENZYME "STRUCTURAL" ADAPTATIONS TO A LIMITED INTRACELLULAR WATER CONTENT

In the case of enzymes, one solution to the potential difficulties posed by a limited intracellular solvent capacity is quite simple and

TABLE 4–1 APPROXIMATE CHEMICAL COMPOSITION OF A RAPIDLY DIVIDING *ESCHERICHIA COLI* CELL. FROM WATSON (1970).

Component	Average Molecular Weight	Approximate Number Per Cell	Number of Different Kinds
H_2O	18	4×10^{10}	1
Inorganic ions (Na^+, K^+, Mg^{2+}, Ca^{2+}, Fe^{2+}, Cl^-, PO_4^{3-}, CO_3^{2-}, etc.)	40	2.5×10^8	20
Carbohydrates and precursors	150	2×10^8	200
Amino acids and precursors	120	3×10^7	100
Nucleotides and precursors	300	1.2×10^7	200
Lipids and precursors	750	2.5×10^7	50
Other small molecules (heme, quinones, breakdown products of food molecules, etc.)	150	1.5×10^7	200
Proteins	40,000	10^6	2000 to 3000
Nucleic acids			
DNA	2.5×10^9	4	1
RNA			
16s rRNA	500,000	3×10^4	1 (?)
23s rRNA	1,000,000	3×10^4	1 (?)
sRNA	25,000	4×10^5	40
mRNA	1,000,000	10^3	1000

effective: many, perhaps most, enzymes exist not as free solutes but rather as bound entities, more or less tightly associated with subcellular structures. In particular, membrane association is of major significance as a means for keeping the active sites of enzymes exposed to the aqueous cytosol in which substrates and products are found, without the enzymes themselves having to be free solutes.

The fraction of enzymes which are effectively removed from aqueous solution, save for their active sites, via compartmentalization is undoubtedly very large. For example, a study of enzyme localization in the eucaryote unicell, *Euglena*, revealed that *none* of the enzymes studied could be detected as free solutes in the cytosol. The fact that many enzymes are still spoken of as "soluble" may, in reality, simply mean that the procedures used to study the enzymes were not sufficiently gentle to prevent stripping the enzyme from the intracellular structure(s) to which they are normally bound.

The solvent conservation role of enzyme binding to subcellular structures is, of course, but one advantage of this particular mode of biochemical design. The fixture of enzymes into organized structures such as membranes has the additional advantage of juxtaposing enzymes which are involved in the sequential steps of a particular pathway. The mitochondrially bound respiratory enzymes serve as a classic example of this organizational design.

ENZYME-SUBSTRATE AFFINITIES AS ADAPTATIONS TO A LIMITED INTRACELLULAR WATER SUPPLY

The above solution to the solvent capacity problem, which characterizes many enzymes and probably most other macromolecules, is obviously inapplicable in the case of low molecular weight substrate molecules. The flux of enormous numbers of substrate molecules on-and-off the active sites of enzymes precludes the former molecules from being bound to subcellular structures. Substrates and cofactors must remain in true solution for enzymatic activities to occur at rates consistent with life. The problem thus becomes one of maintaining all substrate molecules in concentration ranges which are at once high enough to permit satisfactory rates of metabolic activity, and low enough to prevent (1) *solubility problems*, and (2) the occurrence of *uncontrolled reactions* which are governed only by mass action rules, rather than by tightly regulated enzymic activities.

The regulation of substrate concentrations within narrow limits is a major consequence of the evolution of enzyme regulatory properties. We have already pointed out (Chapter 1) the advantages of maintaining substrate concentrations in the same range as the affin-

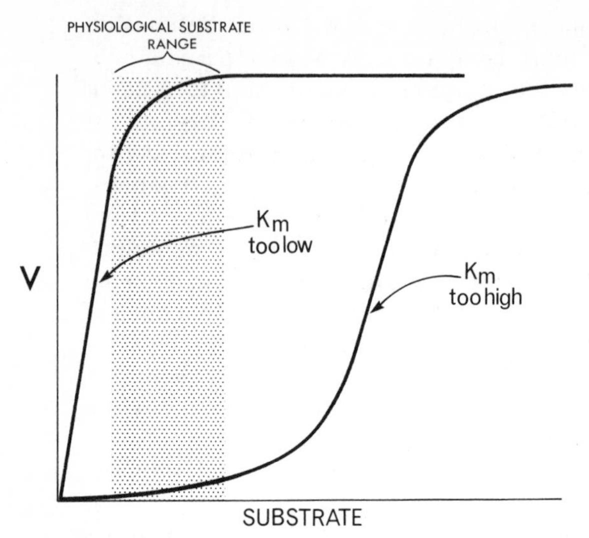

Figure 4–1 The consequences of K_m values deviating from physiological concentration ranges of substrate.

ity parameters (K_m or $S_{0.5}$) of their enzymes: this allows the enzyme to respond to small changes in substrate concentration. Viewed from the angle of maintaining substrate concentrations at constantly low levels, we can appreciate that this similarity between substrate levels and affinity parameters prevents maladaptive increases in the concentrations of free substrate molecules.

If the affinity is either too high or too low (Figure 4–1), substrate concentrations could rise to physiologically dangerous levels. In the first instance where the affinity for substrate is too low, high reaction rates would necessitate high concentrations of substrate. In the second case, where affinity for substrate is too high, the enzyme could be saturated at relatively low substrate levels, and further increases in substrate concentration would not be accompanied by an increased rate of product formation. In such a situation substrate concentrations could rise without limit. Clearly, too great a departure of K_m or $S_{0.5}$ values from the optimal ranges of substrate concentrations has been strongly selected against during the evolution of enzymic regulatory functions.

STABILITY OF SUBSTRATE CONCENTRATIONS DURING LARGE FLUCTUATIONS IN METABOLIC ACTIVITY

A clear demonstration that enzyme regulatory properties are geared to prevent large scale build-ups of substrate molecules (and consequent problems of solubility and uncontrolled reactions) comes from studies of short-term changes in substrate concentrations during transitions from relatively low metabolic rates to exceedingly high rates. Recall, for example, that during the initiation of flight in the blowfly (p. 68), the flux of carbon through the glycolytic path increases by a factor of 100. Yet within the first minute of flight the concentrations of most glycolytic intermediates return to control values. This conservation of low substrate levels is achieved by means of a coordinated activation of all the enzymes which are rate-limiting in the glycolytic sequence. Closely coordinated activation of potential "bottleneck" enzymes prevents the rise of substrate (product) concentrations at any point in the series, while simultaneously allowing a 100-fold increase in flux through the pathway.

COORDINATE REPRESSION-DEREPRESSION OF ENZYME SYNTHESIS

The close integration of enzymic activities which occurs during such sudden changes in metabolic rates is precisely mirrored in longer-term processes which involve the induction of enzyme synthesis. For example, when the qualitative nature of the nutrient supply is changed, an organism may synthesize a new array of enzymes to utilize the newly available substrates. When a sequence of metabolite transformations is involved, the concentrations of all the required enzymes are increased in a uniform manner. This coordinated synthesis thereby prevents serious metabolic imbalances which could arise if one enzyme of a pathway were to be present in either too high or too low a titre to keep substrate (product) concentrations within an optimal range.

SIMILARITIES IN ENZYME-SUBSTRATE AFFINITIES AMONG ENZYMES IN A SEQUENCE

A final manner in which enzyme-substrate affinities have become adapted to keep substrate levels within a certain narrow range involves similarities in K_m or $S_{0.5}$ values among the enzymes of a given sequence. It is usual to find that most enzymes of a pathway have affinity parameter values of the same order of magnitude. In consequence, the *product* of one reaction in the pathway can be fed to the next enzyme, as *substrate*, without harmfully large fluctuations in its concentration occurring.

Summarizing, we can state that enzyme-substrate affinity parameters play a dual role in controlling the state of the cellular chemistry. On the one hand, they closely modulate the activities of the diverse metabolic sequences. In addition, and of equal importance, they provide a necessary and potent constraint to the rise of substrate concentrations. In this latter role, enzyme-substrate affinity parameters can be viewed as a vital mechanism for (1) keeping the solute concentrations *within* the solvent capacity of the cell water, and (2) preventing undesirable and uncontrolled side reactions.

THE ROLE OF SUBSTRATE MODIFICATIONS IN MAINTAINING SOLVENT CAPACITY

In addition to these enzymic mechanisms for controlling the concentrations of intracellular metabolites, the evolution of metabolism has also been characterized by important changes in the chemical structures of substrate molecules, such that the concentrations of substrate permitting a given rate of a particular reaction are greatly reduced. One can calculate, for example, that the important metabolic transformation

$$\text{OXA} + \text{acetate} \rightarrow \text{citrate}$$

would be a physiological impossibility were acetate residues to be supplied as free acetate, because the equilibrium of the reaction is strongly poised towards the cleavage of citrate. For a significant amount of citrate to be produced by this reaction, the concentrations of acetate would have to be enormously larger than could be tolerated by the cell. To effect a high rate of citrate synthesis, organisms take advantage of an evolutionary and chemical "modification" of acetate; acetate is fed into the citrate synthase reaction as a "high energy" Coenzyme A derivative:

$$\text{acetylCoA} + \text{OXA} \rightarrow \text{citrate} + \text{CoASH}$$

The equilibrium for this reaction strongly favors citrate synthesis because cleavage of the thiolester bond in acetylCoA proceeds with a large free energy drop (p. 23). As the CoA derivative, acetate is supplied for citrate synthesis in a *"high chemical potential" form, and therefore only low chemical concentrations need be maintained in the cell.* The evolution of so-called "high energy" or "activated" derivatives of key substrate molecules can therefore be viewed as an adaptation permitting necessary metabolic reactions with unfavorable equilibrium constants to occur at high rates without there being a need for physiologically dangerous—or impossible—concentrations of substrates.

PROBLEMS IN SALINITY ADAPTATION

The adaptive mechanisms which permit enzymes and organic substrates to lessen their demands on the solvent capacity of the cell are, in general, inapplicable in the case of inorganic ions. Although large fractions of the total amount of inorganic ions are bound to cellular constituents, such as the binding of divalent cations to nucleic acids, membrane phospholipids, enzymes, and organic metabolites, some fractions exist as free solutes. Because inorganic ions are the most prevalent solute molecules in the organism (Table 4–1), and because the ionic content and composition of the external environment is often grossly different from the ionic state of the cell, we expect and, indeed, find that adaptations to the ionic environment are extensive and highly demanding of energy.

Problems of ionic adaptation stem from two sources. There is first the basic osmotic problem concerning the *overall ionic concentrations* of the cell and its environment. The concentrations of inorganic ions in the external environment range from zero, in the case of most terrestrial forms, to levels vastly exceeding those within the cells (e.g., for many marine fishes). These more or less sharp ionic gradients between the organism and its environment create classic problems in the maintenance of disequilibrium. The energy costs involved in keeping millions of ions per cell from attaining equilibrium with the external environment are high and would reach enormous levels were it not for the fact that most of the body surfaces of organisms are relatively impermeable to ions.

The second basic problem in ionic adaptation derives from the fact that *qualitative* differences exist in the ionic composition of the cell and the external environment. Thus, the maintenance of ionic gradients involves a selective component. Not only must the cell be able to keep large numbers of ions from attaining equilibrium with the environment, but in addition it must possess regulatory systems with precise abilities to *recognize* different ions and to treat these ions differentially. This consideration leads to the expectation that mechanisms responsible for ionic regulation may share certain similarities with the enzymic mechanisms involved in establishing substrate concentrations, for the latter mechanisms exhibit the precise specificity necessary for close control of molecular concentration and composition.

STRATEGIES AND DEGREES OF IONIC REGULATION: AN OVERVIEW

In theory, at least, there are four potential strategies of ionic adaptation. Firstly, and most simply, the organism may regulate neither the total concentration nor the qualitative composition of the inorganic ions in its fluids. Considering the dependence of many biochemical and physiological functions on the proper ionic conditions, it is not surprising that this first pattern of ionic adaptation is not found in Nature.

Secondly, it is conceivable that an organism could regulate its total ionic content, i.e., its osmolarity, without regulating the qualitative composition—the relative concentrations—of different ionic species. For the reason given above, this pattern also is not observed.

Thirdly, an organism might regulate qualitatively, but not quantitatively. In this pattern of ionic adaptation, the osmolarity of the organism's fluids would equal that of the surrounding water, but the qualitative compositions of the body or cell fluids would differ from those of the environment. In this adaptational scheme the organism would need to expend energy solely to maintain concentration gradients of specific ions; no energy would have to be spent maintaining osmotic differences. This type of ionic adaptation is very common among marine microorganisms, marine invertebrates, and primitive vertebrates like the hagfish.

In the fourth pattern of ionic adaptation, organisms exhibit the abilities to maintain both the osmotic content and the qualitative ionic composition of their fluids within narrow limits and relatively independent of the external environment. This type of ionic adaptation is obviously quite costly, energetically. However, it confers on the organisms all of the advantages which accrue from stabilizing the chemical state of the internal milieu. Most of the more advanced invertebrates and vertebrates display this dual ability to control the ionic status of their cells and body fluids.

In considering the overall problems faced in ionic regulation, it is important to stress the different situations which pertain in unicells and metazoans. The unicell is in direct contact with the environment and, therefore, it must perform its regulatory functions at the cellular-environmental interface. In contrast, most cells of a metazoan are not in direct contact with the external milieu. Usually the blood circulation is separated from the environment by at least one epithelial layer. Thus we expect and, indeed, find that metazoans display ionic regulatory mechanisms at two interfaces: between the blood and the cell and between the cell and the environment. That is, metazoans encounter ionic regulation problems for intra- and extracellular fluids. Before we consider how these distinct problems are "solved," let us examine the manners in which microorganisms adapt to situations where the external salinity impinges directly on the biochemical machinery of the cell.

IONIC ADAPTATION IN MICROORGANISMS: BIOCHEMICAL FUNCTION IN THE ABSENCE OF OSMOTIC REGULATION

Bacteria are osmotic conformers *par excellence.* Over an extremely wide range of salinities, the osmotic content of the cytosol equals that of the external milieu. Because of this great variation in the salt concentration of the intracellular environment in which macromolecules must function, we would expect to find major structural differences among the macromolecules of differently salt-adapted bacteria. In working with such "normal" organisms as *E. coli* and its mammalian hosts, biochemists frequently use high salt concentrations to precipitate proteins, disrupt membranes, dissociate ribosomes, and so forth. In many cases the salt concentrations needed to effect these techniques with "normal" organisms are the *required* salt concentrations for growth and metabolism of halophilic ("salt-loving") microbes. Thus, not only are many of the proteins of halophiles able to function under high salt conditions, but, indeed, they can function only under these circumstances.

Halophilic Bacteria: General Ecological Attributes. The halophilic bacteria are among the most salt-tolerant of organisms. They are capable of exploiting environments which are inimical to most living forms. In such habitats as salt lakes (the Great Salt Lake and the Dead Sea), salty sea muds and sediments, and coastal "salt pans," these organisms survive under conditions where both the external and intracellular environments can be nearly saturated with inorganic salts (Table 4–2). Some halophiles exhibit salt requirements for growth in the range from 3 to 5 M. The inability of the metabolic machinery of non-halophiles to function under these conditions raises two important questions: (1) Do halophiles successfully conduct the same array of metabolic functions which are

TABLE 4–2 INTRACELLULAR CONCENTRATIONS OF SALTS IN VARIOUS BACTERIA. FROM BROWN (1964).

Organism	Concentration (M) in Growth Medium		Intracellular Concentration*		$\frac{\text{Cell Na}^+}{\text{Medium Na}^+}$	$\frac{\text{Cell K}^+}{\text{Medium K}^+}$
	NaCl	KCl	Na^+	K^+		
Nonhalophiles						
Staphylococcus aureus	0.15	0.025	98	680	0.7	27
Salmonella oranienburg	0.15	0.025	131	239	0.9	10
Moderate halophiles						
Micrococcus halodenitrificans	1.0	0.004	311	474	0.3	120
Vibrio costicolus	1.0	0.004	684	221	0.7	55
Extreme halophiles						
Sarcina morrhuae	4.0	0.032	3,170	2,030	0.8	64
Halobacterium salinarium	4.0	0.032	1,370	4,570	0.3	140

*As milliequivalents per kilogram of cell water.

known to occur in "normal" organisms? (2) What biochemical adaptations permit the macromolecules of halophiles to function under conditions of high salt content?

Halophiles are Metabolically Similar to "Normal" Microorganisms. The success with which halophiles have adapted to life under extremes of salinity is witnessed by the fact that they have the same metabolic apparatus, functionally speaking, as their non-halophilic counterparts. The same basic metabolic pathways of ATP generation, the same mechanisms of gene replication, the same mechanisms of protein synthesis occur in halophiles and non-halophiles. These findings indicate that *for such functional similarities to exist, enormous underlying biochemical differences must distinguish halophiles from non-halophiles.* Thus, as we stressed in Chapter 1, apparent similarities in physiological function often exist only because the biochemical machinery responsible for these functions is particularly tailored for function under the exact environmental conditions experienced by the organisms.

The Acidic Nature of Ribosomal Proteins of Halophilic Bacteria. The first clue as to how the proteins of halophiles differ from the homologous proteins of non-halophilic organisms came from studies of the ribosomes of differently salt-adapted bacteria. In terms of "gross anatomy," the ribosomes of halophiles appear similar to those of non-halophiles. Thus, *Halobacterium* and *E. coli* have ribosomes with (i) the same size of subunits, (ii) the same ratio of ribosomal RNA/ribosomal protein, and (iii) the same number of Mg^{++} ions per RNA phosphate group.

The first difference noted between halophilic and non-halophilic ribosomes is in the salt-dependence of ribosomal structure. The halophilic ribosomes exhibit extremely high salt dependencies (Figure 4–2). The 30s and 50s subunits are stable at "merely" 1M concentration of KCl or NaCl, but the formation of the 70s particle requires 3 to 4 M KCl! Under these conditions the ribosomes of non-halophilic bacteria would lose their structure. The halophilic ribosomes, in contrast, dissolve when placed in 5 to 10 mM KCl, conditions which favor the stability of non-halophilic ribosomes.

The molecular basis of these differing salt dependencies is to be found in the amino acid compositions of the ribosomal proteins of the differently salt-adapted microbes. In "normal" organisms, both procaryotic and eucaryotic, the ribosomal proteins are strongly basic. The halophilic microbes, in comparison, possess extraordinarily large amounts of acidic amino acids in their ribosomal proteins. This latter observation explains why the halophilic ribosomes are stable only in high-salt-content media: the negatively charged carboxyl groups repel each other, and will cause the proteins to disaggregate and dissolve, unless they are neutralized by

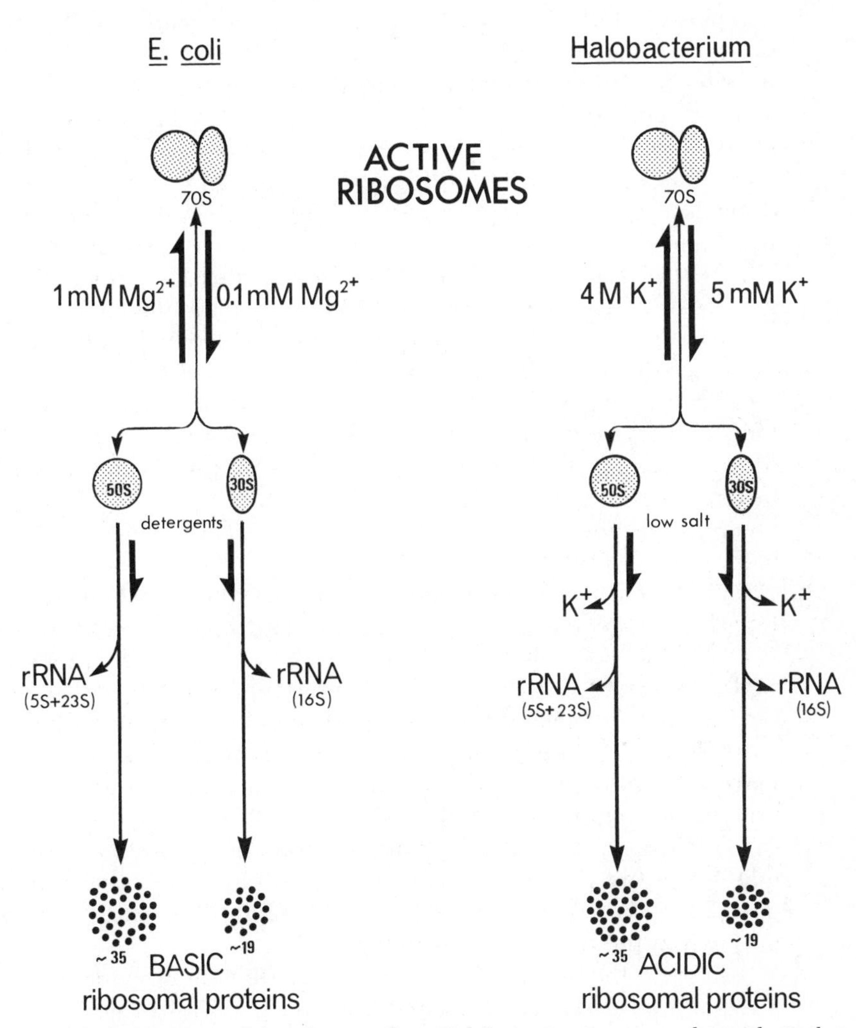

Figure 4–2 The structure of the ribosome from *Halobacterium* is apparently similar to that of *E. coli*. The active ribosome is usually called the 70s ribosome, since 70s (Svedbergs) is a measure (the sedimentation constant) of how fast the active ribosome sediments in a centrifugal field. The *E. coli* 70s ribosome dissociates at low Mg^{++} concentrations to 50s and 30s ribosomal subunits. The process is reversible, and the subunits reassociate in the presence of 1 mM Mg^{++}. The 70s ribosome of *Halobacterium* dissociates to 50s and 30s subunits in the absence of high K^+, but reassociation occurs in the presence of 4 M K^+. The smaller subunits in both species can be further dissociated into constituent proteins and rRNA fractions. In *E. coli* this is done with detergents; in *Halobacterium*, low salt (Na^+ or K^+) leads to dissociation of the 50s and 30s subunits. In *E. coli*, the ribosomal proteins released are notably basic; in *Halobacterium*, they are strongly acidic.

monovalent cations. Thus, *in situ,* the acidic amino acid residues probably are fully titrated by monovalent cations such as K^+.

Acidic Nature of Proteins of the Cell Walls and Cell Membranes of Halophiles. The differences in amino acid composition of ribosomal proteins of halophilic and non-halophilic microbes are also noted in other proteins. Particularly dramatic effects were observed in the cases of cell wall and cell membrane proteins. It has long been known that the cell "envelope" (composed of the cell wall and cell membrane) of halophiles will lyse when the salt content of the medium is reduced to approximately 5 per cent (NaCl). Contrary to what one might predict, cell envelope lysis is not due to osmotic effects. High concentrations of nonionic solutes such as sucrose do not prevent cell envelope lysis. Puzzling over the question of what causes the lysis, Gibbons and his coworkers concluded over 10 years ago that the cell wall and cell membranes are held together by weak secondary interactions and that the structures remain intact so long as their negative charges are screened by high concentrations of Na^+. This postulate in effect predicted a preponderance of acidic proteins in the cell envelope.

Subsequent studies of isolated cell envelopes of extreme halophiles established:

(1) that isolated envelopes mimic the lysis of whole cells in that they dissolve extremely rapidly in distilled water, even at 0° C;

(2) that high concentrations of Na^+, K^+, or H^+ (i.e., low pH) will preserve isolated cell wall and membrane structures;

(3) that increasing the acidity of membrane proteins by succinylation of ϵ-amino groups of lysine in a marine bacterium converts these into more "halophilic-type" membranes, requiring high salt concentrations for stability (Figure 4–3);

(4) that all the carboxyl groups of membrane proteins are normally *exposed* (and hence available for titrating) but that ϵ-amino groups are *buried* in the membrane structure, becoming exposed only when the membrane is dissociated;

(5) that the phospholipids of halophilic membranes contain unusual hydrocarbon chains and are strongly acidic; and

(6) that the proteins of cell wall and cell membranes, on the basis of amino acid composition and electrophoretic mobility, are in fact highly acidic.

Such observations led to a broad acceptance of Gibbons' original speculation that (1) the structural integrity of the cell envelope depends upon the presence of high cation concentrations and (2) the cations stabilize membrane structure by neutralizing the abundant negative charges that are found on the component, acidic proteins. Indeed, this theory would predict that the more acidic the membrane proteins, the higher the salt requirement of the microorganism; the less acidic, the lower the salt requirement. This pre-

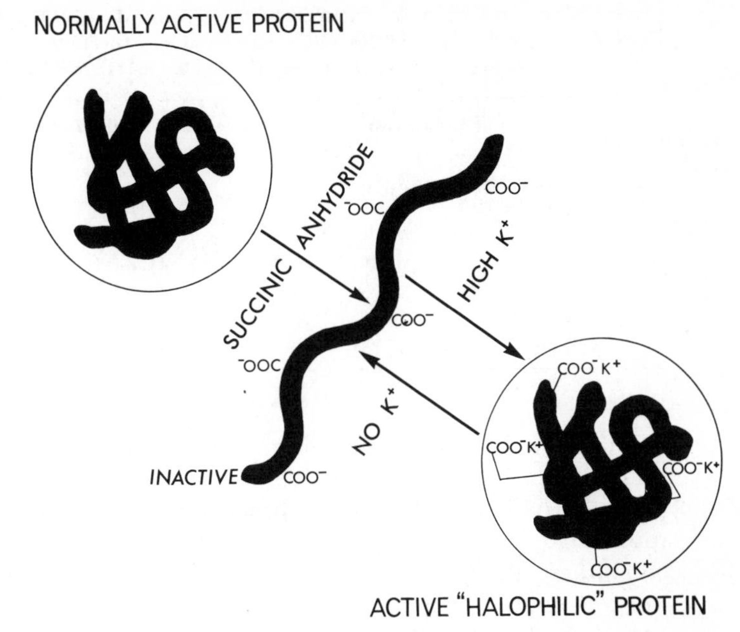

Figure 4–3 Diagrammatic representation of the influence of succinic anhydride on the stability of membrane proteins from marine microorganisms. Succinylation in effect converts the positive charge of the ϵ-amino group of any lysine residue to a negative charge:

$$\underset{\text{(Succinic anhydride)}}{\begin{array}{c} H_2C-C(=O) \\ | \quad\quad \backslash \\ \quad\quad\quad O \\ | \quad\quad / \\ H_2C-C(=O) \end{array}} + H_3N^+-(CH_2)_4-\underset{\substack{| \\ C=O \\ | \\ NH \\ \downarrow \\ \text{chain backbone}}}{\overset{\substack{\uparrow \\ C=O \\ | \\ NH \\ |}}{CH}} \longrightarrow {}^-O-\overset{\substack{O \\ \|}}{C}-CH_2-CH_2-\overset{\substack{O \\ \|}}{C}-\overset{H}{N}-(CH_2)_4-\underset{\substack{\downarrow \\ \text{chain backbone}}}{\overset{\uparrow}{C}}-H$$

Under these conditions, the membrane protein of a marine pseudomonad takes on halophyllic properties and requires much higher salt concentrations for stability.

diction was realized (Table 4–3), and the theory has served as the point of departure for much of the recent research in this area, including studies of other proteins in halophilic bacteria.

Acidic Nature of the "Bulk" Proteins of Halobacteria. The question still remains whether the acidity of membrane and ribosomal proteins is confined to these particular subcellular structures, or whether it is a fundamental characteristic of most of the proteins of the halophilic cell. At least a partial answer to this question became available in 1970 in studies comparing the amino acid com-

TABLE 4–3 THE RELATIVE PROPORTIONS OF BASIC/ACIDIC AMINO ACIDS IN CELL MEMBRANES FROM MICROORGANISMS DIFFERING GREATLY IN THEIR SALT REQUIREMENTS. FROM BAYLEY (1973).

Kind of Organism	Ratio of Basic/Acidic Amino Acids
most non-halophilic bacteria	0.9–1.1
marine pseudomonads	0.5–0.6
extremely halophilic bacteria	0.2–0.3

position of the "bulk" proteins of halophilic bacteria with their non-halophilic counterparts. The term "bulk proteins" is an operational one and refers to the protein precipitable with acid after removal of the cell walls and membranes. This protein fraction constitutes about 80% of the total protein of the cell and includes most "soluble" enzymes. As expected, these proteins contain unusually high concentrations of the acidic amino acids. The percentage excess (on a molar basis) of acidic over basic amino acids is about 10:1, a value in fact higher than that observed for the acidic proteins of ribosomes and membranes. Thus, it appears that most, if not all, proteins of extremely halophilic bacteria require high salt concentrations for essentially the same reason: to neutralize excess negative charges, which in the absence of cations presumably lead to electrostatic repulsion and to the distortion of normal structure and function. A final matter to consider, then, is the effect of cations on the functions of halophilic proteins.

Functional Properties of Halophilic Enzymes. Since the work of Baxter and Gibbons in the mid-1950s it has been clear that enzymes from halobacteria are more salt-tolerant and, indeed, more salt-requiring than the homologous enzymes from non-halophiles. Over twenty enzymes from halophiles have now been described, and from these studies two general conclusions emerge:

(1) The more halophilic the organism, the greater is the salt requirement for any given enzyme.

(2) Enzyme-ligand affinities are especially sensitive to salt concentration. V_{max} values are often less sensitive.

These two generalizations are well illustrated by data gathered on two important regulatory enzymes: NADP-isocitrate dehydrogenase (IDH) and aspartate transcarbamylase (ATCase).

Halophilic NADP-Isocitrate Dehydrogenase. In the presence of a divalent cation, IDH catalyzes the reaction:

$$\text{isocitrate} + \text{NADP} \rightarrow CO_2 + \alpha\text{-KGA} + \text{NADPH}$$

In common with other enzymes from halophiles, IDH from halobacteria exhibits a distinct requirement for Na^+ or K^+, being equally active with either. V_{max} activity is greatest at 1 M salt. However, while salt-dependent, the V_{max} of the reaction varies less than two-fold between salt concentrations of 0.5 and 5.0 M.

In contrast to the slight effect of salt concentration on V_{max}, the apparent enzyme-substrate affinity changes markedly with changing salt concentration. The apparent K_m of isocitrate exhibits a minimal value between 1 M and 4 M salt concentration. As the salt concentration falls below 1 M, the apparent K_m of isocitrate increases more than five-fold. *In vivo,* this sharp rise in the K_m of the substrate would lead to a dramatic inhibition of the reaction velocity (Figure 4–4), as well as a probable loss of control over catalysis. NADP-IDH is product-inhibited by α-KGA and NADPH, and feedback activated by ADP; all of these potential modulators affect the apparent enzyme-substrate affinity. Hence, any salt-dependent interference in enzyme-substrate affinity would strongly affect enzyme regulation as well. Direct measurement of this kind of phenomenon comes from studies of ATCase.

ATCase—The Model Regulatory Enzyme. The first step in pyrimidine biosynthesis,

$$\text{aspartate} + \text{carbamyl-P} \rightarrow \text{N-carbamyl aspartate}$$

is catalyzed by ATCase. The enzyme is subject to feedback inhibition by CTP, the ultimate end product of the pathway; since it is a biosynthetic step, it is not surprising to find that ATP is a potent deinhibitor of the enzyme. As a result of a number of penetrating investigations by Gerhardt, Pardee, Schachman and others, ATCase has come to be the most thoroughly understood of known regulatory enzymes. As such, it has served as a model for most other studies. We therefore shall digress briefly into a discussion of *E. coli* ATCase, before returning to the homologous enzyme in halobacteria.

In *E. coli,* ATCase consists of both catalytic and regulatory subunits. The normal, active enzyme is an oligomer consisting of six catalytic subunits (each of 33,000 MW) and six regulatory subunits (each of 17,000 MW). From the finding of a 3-fold and 2-fold symmetry in the crystal structure, it is now clear that the ATCase molecule (Figure 4–5) consists of two C-protomers (each with three C-subunits) and three R-protomers (each with two R-subunits). Aspartate saturation curves for the enzyme are clearly sigmoidal in the absence or presence of the feedback inhibitor, but in the presence of CTP there is a marked decrease in the apparent enzyme-aspartate affinity. In contrast, the V_{max} does not change (Figure 4–5). The inhibition does not involve CTP-aspartate competition

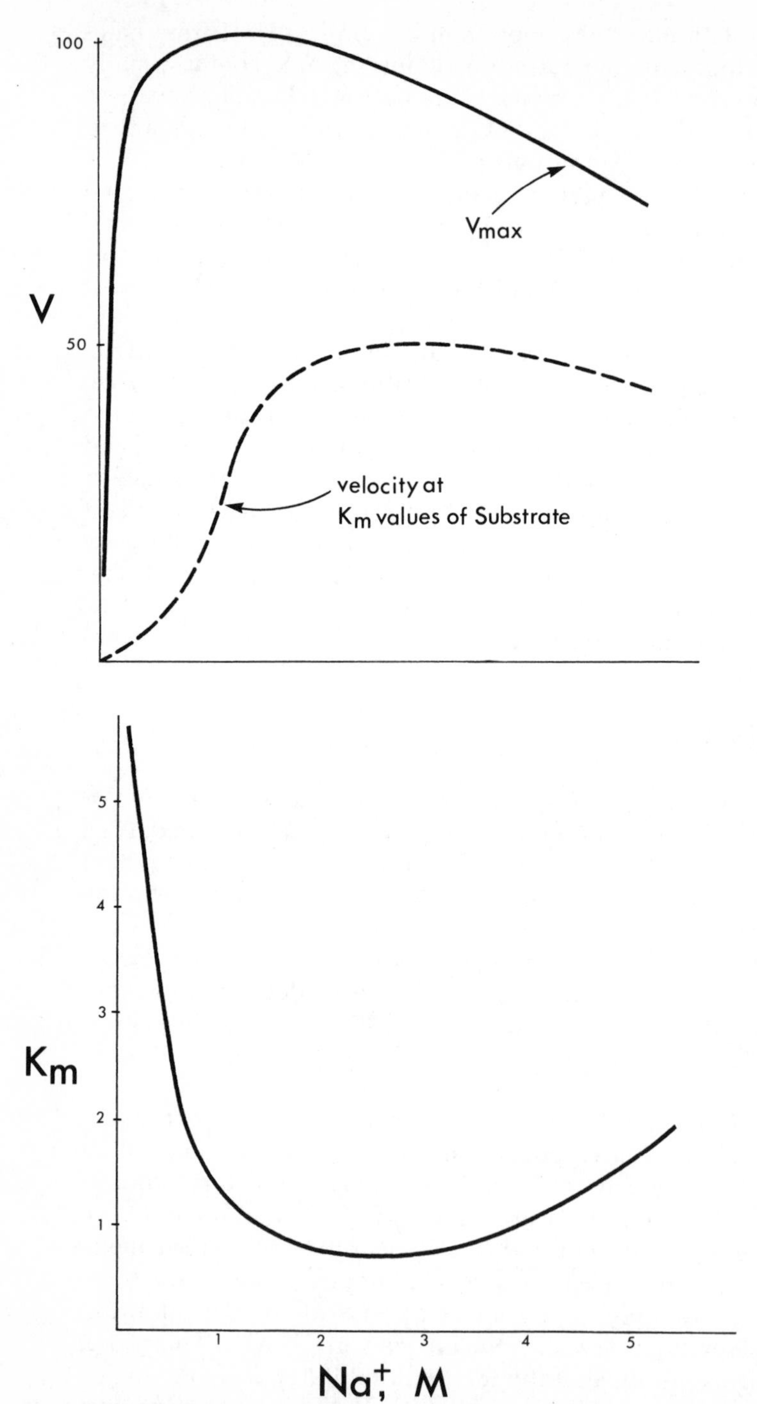

Figure 4–4 The influence of high salt on the relative reaction rate of NADP-IDH from a halophilic bacterium under saturating conditions of substrate (V_{max}) and at K_m concentrations of substrate (upper panel). The high Na^+ dependence of the enzyme at low isocitrate concentrations is due to the extreme effect of Na^+ on the K_m for isocitrate (lower panel). Data from Aitken et al. (1970). *Biochem. J. 116,* 125.

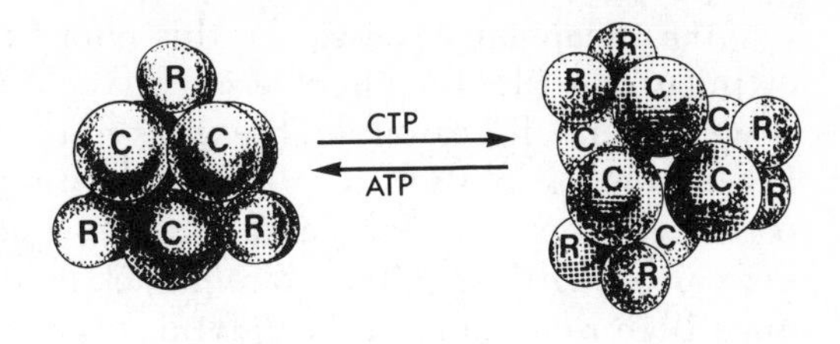

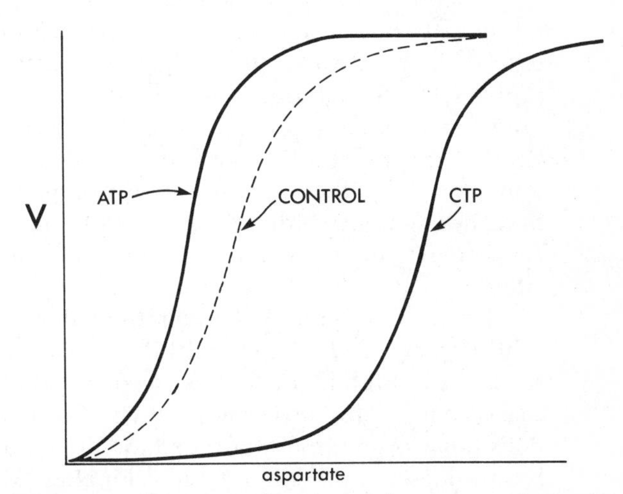

Figure 4–5 Upper panel: postulated structure of *E. coli* ATCase, showing two catalytic protomers (each with three C-subunits) and three regulatory protomers (each with two R-subunits). The enzyme is shown in the inhibited conformation, favored by CTP, and the active conformation, favored by ATP. The domain of bonding between R and C chains changes upon binding CTP. Other bonding domains (C:C and R:R) do not change. Lower panel: aspartate saturation curves for ATCase, in the presence of the negative modulator, CTP, and of the positive modulator, ATP.

for the catalytic site; rather, CTP binds only to the regulatory subunit (i.e., to the allosteric site). The subsequent inhibition of catalysis is thought to be "transmitted" by conformational change in the catalytic subunit. The catalytic subunit apparently occurs in at least two conformations, one fully active and one inhibited or inactive. The binding of CTP by a regulatory subunit apparently "locks" the C-subunit in its inactive conformation; this conformational transition can then be physically transmitted to an adjacent catalytic subunit, locking the latter into its catalytically inactive form.

CTP inhibition can be reversed by ATP. ATP binds to the regulatory subunit at the same site as does CTP. ATP is not an intrinsic activator; it acts primarily by inhibiting the binding of CTP.

The Halophilic ATCase. In this context, it is particularly interesting to consider the effect of salt on the ATCase from halobacteria. If we assume, for example, the marked increase in acidity that typifies other halophilic proteins, what structural consequences will occur? Are both catalytic and regulatory protomers present? Is the enzyme equally sensitive to feedback modulation? Does salt influence both regulatory and catalytic subunits in a similar manner?

Although definitive answers for all of these questions cannot be given, some highly intriguing data are available. Thus, ATCase from halobacteria requires 3 to 5 M salt for activity. Substrate saturation of the enzyme does not follow sigmoidal kinetics, as in the case of *E. coli.* Since the S-shaped saturation curve of non-halophilic ATCase is thought to involve positive cooperative interactions between the two catalytic protomers, these data suggest either (1) that the halophilic enzyme possesses only one catalytic protomer or (2) that the increased acidity of the C-protomers has altered their interactions so drastically that the binding of substrate at one catalytic site does not at all influence the substrate affinity of other sites.

Just as in *E. coli,* ATCase in the halophilic bacteria is potently inhibited by CTP. The inhibition is totally dependent upon the presence of high K^+ or Na^+, as, of course, is the maximum activity of the enzyme. This observation certainly suggests that the enzyme does possess an important regulatory function. However, CTP inhibition does not follow sigmoidal kinetics (Figure 4–6), again suggesting either (1) that there is only one regulatory protomer or (2) that the regulatory protomers also do not influence each other in the binding of the regulatory metabolite, CTP.

It will be evident that the number of studies yet to be done on this enzyme vastly exceeds those that are already completed. Hence, it is difficult at this time to assess fully what is occurring during the evolutionary elaboration of halophilic regulatory enzymes. One conclusion, however, can be made with confidence: from a functional point of view, the effects of salt upon enzyme-substrate and enzyme-modulator affinities are far more critical than are the effects upon the V_{max}. Under physiological conditions, as we have already stressed, substrate concentrations are seldom high enough to saturate enzymes. Since reaction rates *in vivo* are strongly determined by E-S affinities, any effects of salt on the K_m or $S_{0.5}$ are instantly reflected in the metabolism of the cell. That E-S affinities appear to be highest at optimal salinities for growth probably is not a fortuitous outcome, but rather is an outcome of evolutionary forces "tailoring" these enzymes for halophilic function.

Halophilic Proteins Also May Occur in Certain Metazoan Cells. Parenthetically, it should be mentioned that "halophilic proteins" also may occur in metazoan cells. For example, Conte and his co-

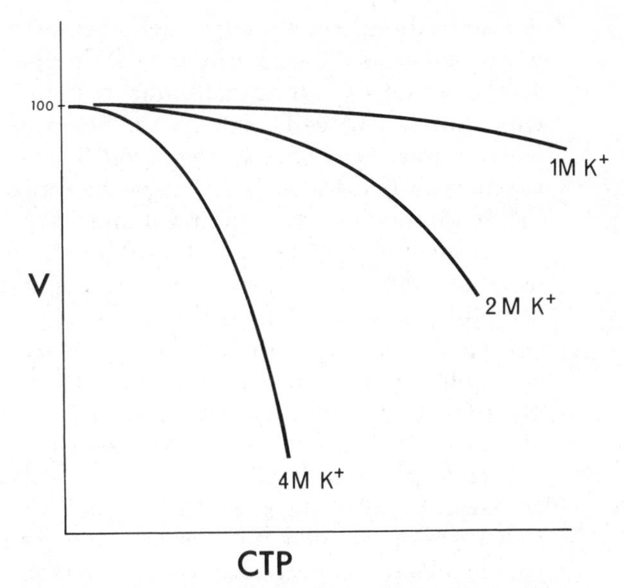

Figure 4–6 The absolute dependence of CTP feedback inhibition upon high salt in the case of ATCase from *Halobacterium.* This dependence of a regulatory property of the enzyme upon high salt levels is greater than the V_{max} dependence on salt. (Data courtesy of Dr. J. G. Kaplan, Univ. of Ottawa.)

workers have shown that the brine shrimp possesses a "salt gland" for Na^+ regulation in this little halophilic crustacean. A postulate being tested is that adaptation to the nearly salt-saturated intracellular conditions has involved the elaboration of acidic proteins in the "salt gland," as in the halobacteria. Moreover, the activity of the cation transport mechanism in this tissue depends upon unusually wide Na^+ and K^+ concentrations. By and large, however, the strategy of halophilic enzymes is not commonly utilized by metazoans. Even in the brine shrimp, other strategies must be brought into play if the organism is to maintain itself.

A SECOND STRATEGY OF SALINITY ADAPTATION: MAINTAINING INTRACELLULAR OSMOLARITY

Many marine and euryhaline invertebrates, capable of tolerating broad ranges of salinity, maintain their blood somewhat hyperosmotic to the medium, particularly when the latter is more dilute than sea water. The degrees of blood hypertonicity vary greatly in different groups. Whereas these organisms succeed in maintaining their blood hypertonic to the medium in all cases, a decrease in outside salinity is invariably accompanied by a decrease in the blood tonicity. Such changes in blood concentration, which occur as the concentration of the medium changes, must be accompanied by appropriate adjustments in the internal osmotic concentration of cells; otherwise, *the cells would take up water from, or release water into, the blood.* That is, the cells would either swell or shrink. In fact, they do neither and the question, then, is: "How do they maintain their volume?"

Recent studies indicate that although initial changes in cellular osmolality are due to changes in levels of inorganic ions, most of the osmotic adjustments of tissue cells are due to variations in the concentrations of intracellular amino acids. In crustaceans, the major contribution to changes in total intracellular osmolality is made by glycine, alanine, proline, and glutamate, although the relative importance of each of these components varies among species. In the echinoderm *Asterias* sp., glycine and taurine are the only amino acids involved in the regulation of cellular osmolality. In the annelid *Arenicola,* glycine, alanine, and glutamate play the major part in adaptation of cells to osmotic change, whereas proline levels do not vary. As a result of such changes, cell volume (i.e., intracellular osmolarity) is closely regulated in many marine invertebrates, even in the face of substantial osmotic changes in the blood and adjacent sea water. What are the signals for these changes in intracellular amino acid concentrations? What are the mechanisms which bring them about?

The Key Role of the GDH Reaction. It is now widely believed that the enzyme glutamate dehydrogenase (GDH) is closely involved in initial phases of cell-volume regulation. The reaction catalyzed by GDH,

$$\alpha\text{-KGA} + NH_4^+ \xrightarrow[\text{NADH} \rightarrow \text{NAD}]{} \text{glutamate}$$

is strongly activated by inorganic ions. It is of fundamental importance in the biosynthesis of amino acids in all species since it is the primary, if not only, pathway for the formation of α-amino groups directly from NH_4^+. Moreover, transamination of α-keto acids with glutamate as the amino-group donor represents the major pathway for the introduction of α-amino groups in the biosynthesis of most other amino acids, *including glycine, alanine, and proline.* The equilibrium position of the GDH reaction favors glutamate formation. Hence, as demonstrated by Schoffeniels and his coworkers, a transient increase in intracellular levels of ions (reflecting the change in blood concentrations) would be expected to bring about a GDH activation with a consequent increase in the level of free glutamate in the cell as well as in the levels of alanine, glycine, and proline. In consequence, a series of highly specific regulatory "messages" are transmitted through the cell, as shown in Figure 4–7.

In Figure 4–7, the primary event is ion activation of GDH; both cations and anions are known to activate GDH, but by different mechanisms. Glutamate formed by the reaction then serves as an amino donor for both alanine and glycine (in this way favoring increased synthesis of the latter two amino acids at a time when increased ion levels in the blood must be osmotically balanced by increased amino acid levels within the cell). The two amino acids, alanine and glycine, as well as serine, feedback inhibit glutamine

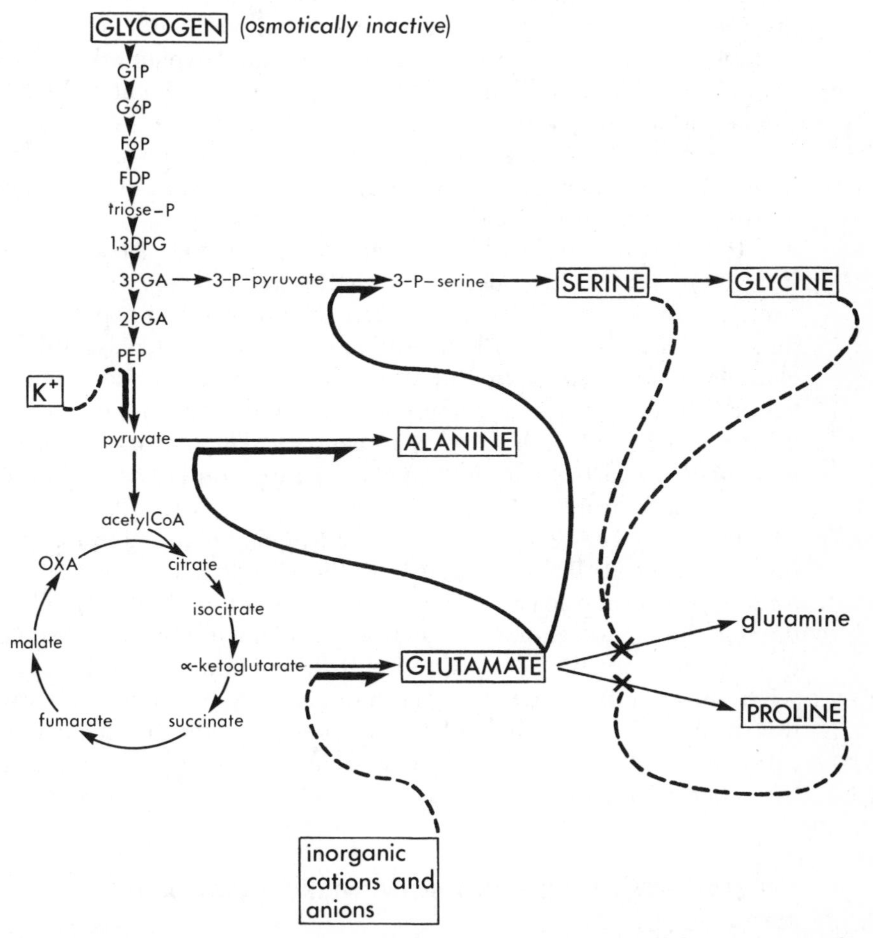

Figure 4–7 The pivotal role of glutamate dehydrogenase in intracellular osmotic regulation in marine and euryhaline invertebrates. The initial signal eliciting the increase in synthesis of glutamate, alanine, glycine, and proline is a temporary change in salt concentration which reflects changes in the blood osmolarity.

synthetase, a major pathway for the further utilization of glutamate, and in this way allow an additional increase in glutamate concentration, which is then available for additional alanine and glycine synthesis. Interactions such as these *lead to an exponential increase in the levels of all four amino acids*—glutamate, alanine, serine, and glycine (Figure 4–7); the initial stimulus triggering the control "cascade" may be something as simple as a change in Na^+ or Cl^- concentrations, encountered first in the outside environment, and then reflected in the blood and finally within the cell. A system of this type is autocatalytic and automatic: a change in the outside salinity is reflected very quickly by an appropriate change in the intracellular concentration of amino acids in order to maintain osmotic (and, therefore, cell volume) balance.

In addition, glutamate serves as an immediate precursor of proline; hence, involvement of proline in cell volume regulation in certain marine invertebrates could be readily explained by this control scheme. Should any of the enzymes in the direct pathways of alanine and glycine synthesis be affected by ions in the same manner, control of cell osmotic pressure by these amino acids would be even further facilitated. The enzyme pyruvate kinase (PK), generating pyruvate from PEP, may be one such enzyme.

In intertidal molluscs, for example, PK shows a distinct requirement for K^+, as does the homologous enzyme from other sources. Since pyruvate is an immediate precursor of alanine, the degree of activation of PK by intracellular K^+ might also reflect changes in ionic concentration of the blood and of the outside sea water. More significantly, alanine is a competitive feedback inhibitor of PK, and in this way *it serves as a sensitive signal for its own further synthesis* from pyruvate. Proline similarly controls its own synthesis from glutamate.

These mechanisms, then, would appear to be adequate for regulating intracellular osmolarity in those organisms whose blood tonicity reflects, at least to some extent, the external salinity. They do not, however, explain how the organism is able to regulate closely the inorganic composition of its blood and other fluids even when the overall blood tonicity is changing. The biochemical key to this problem, as we shall elaborate below, is the occurrence of cation pumps which actively transport specific ions into, or out of, the blood of aquatic organisms.

THE STRATEGY OF ISOLATING BOTH INTRACELLULAR AND EXTRACELLULAR FLUIDS

In contrast to organisms such as the euryhaline invertebrates, most vertebrates closely regulate both the ionic and osmotic composition of their fluids. In all cases, this strategy *isolates* both extracellular and intracellular fluids from the direct effects of salinity changes in the external environment. This is achieved by both passive and active processes. The passive component is a body surface essentially impermeable to water and to ions. The active, controlled component of this strategy harnesses the energy of ATP hydrolysis to the "uphill" transport of ions into or out of the body. Energy-requiring cation pumps are strategically placed in tissues that are geared for the regulation of large ionic exchanges between the organism and its environment. These ionic regulatory organs conduct the dual functions of controlling both (1) the total ionic concentration (the osmoregulatory function) and (2) the qualitative composition of the extracellular and intracellular fluids (the ion regulation function). In all metazoans, these two functions depend upon trans-epithelial (i.e., *trans-cellular*) transport of solute (usually Na^+), a situation substantially more complex than that occurring in unicellular organisms (where transport occurs across a single membrane). Because water movement across such epithelia always accompanies solute movement, we must briefly consider the relationship between the two transport processes.

Coupling of Water Transport to Active Solute Transport. It turns out that the mechanism coupling water to solute transport is intimately related to the ultrastructural geometry of the transporting epithelium; this geometry typically is of two kinds. The major difference between the two kinds of epithelia is the polarity of a series of "dead-end channels" which supply a route by which both water and solutes are moved. Virtually all fluid-transporting epithelia are characterized by some structures—lateral spaces, basal infoldings, or intracellular canaliculi—that conform to this geometry of long, narrow dead-end channels (e.g., gall-bladder, vertebrate intestine, renal proximal and distal tubule, ciliary body, salivary gland striated duct, liver, pancreas, stomach, avian salt gland, the gills of aquatic animals, and so forth).

The way in which this shared geometrical ground-plan couples water transport to solute transport is illustrated in the top of Figure 4–8. As solute is dumped into the channel and diffuses down its concentration gradient towards the open mouth of the channel, water enters osmotically across the channel walls, progressively reducing the osmolarity. In the steady state the channel contents will be hypertonic, a standing osmotic gradient will be maintained along the length of the channel by active solute transport, and a fluid of constant osmolarity—isotonic or hypertonic, depending on factors such as the geometry and water permeability—will continually emerge from the mouth of this standing-gradient flow system. The progressive approach to osmotic equilibrium along the channel accounts for the maintenance of water flow, and there is no neces-

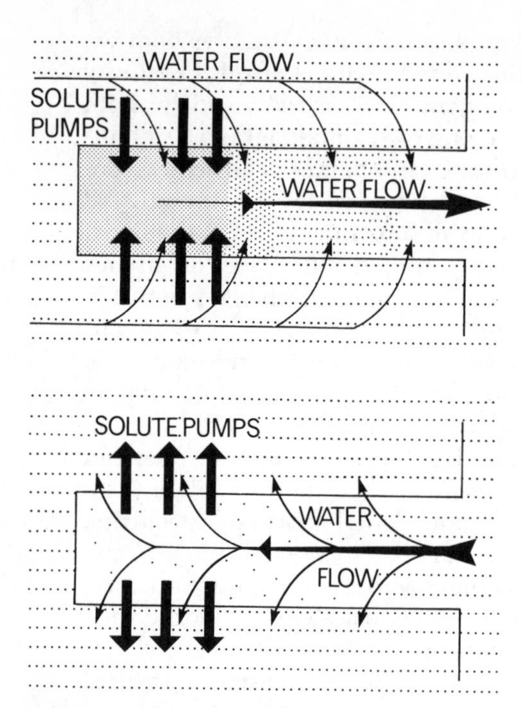

Figure 4–8 Comparison of "forwards" and "backwards" operation of a standing-gradient flow system, which consists of a long narrow channel closed at one end (e.g., a basal infolding or lateral intercellular space). The density of dots indicates the solute concentration. Forwards operation (top): Solute is actively transported into the channel across its walls, making the channel fluid hypertonic. As solute diffuses down its concentration gradient towards the open mouth, more and more water enters the channel across its walls owing to the osmotic gradient. In the steady state a standing osmotic gradient will be maintained in the channel by active solute transport, with the osmolarity decreasing progressively from the closed end to the open end; and a fluid of fixed osmolarity (isotonic or hypertonic, depending upon the values of such parameters as radius, length, and water permeability) will constantly emerge from the mouth. Backwards operation (bottom): Solute is actively transported out of the channel across its walls, making the channel fluid hypotonic. As solute diffuses down its concentration gradient towards the closed end, more and more water leaves the channel across its walls owing to the osmotic gradient. In the steady state a standing osmotic gradient will be maintained in the channel by active solute transport, with the osmolarity decreasing progressively from the open end to the closed end; and a fluid of fixed osmolarity (isotonic or hypertonic, depending upon the parameters of the system) will constantly enter the channel mouth and be secreted across its walls. Solute pumps are depicted only at the bottom of the channels for illustrative purposes but may have different distributions along the channel. (After Diamond (1971).)

sity for osmotic gradients between the final secretion or absorbate and the bathing solution.

The transported fluid is most nearly isotonic when all the solute is dumped in at the closed end of the channel, and becomes progressively hypertonic (the magnitude of the hypertonicity depending on other parameters such as the water permeability) *as the solute pump is spread over an increasing fraction of the channel length,* because the distance to the channel mouth is then shorter and there is less opportunity for osmotic equilibration. This relation may explain why mitochondria often are concentrated towards the closed end of the channels, since the distribution of the energy sources parallels the distribution of the pumps. The above reconstruction applies to epithelia where the channels are polarized for fluid transport *from the closed to the open end,* a situation occurring in many epithelia, including gall-bladder, vertebrate intestine, and kidney. In some other epithelia, such as the avian salt gland, the anatomy is "backwards"—i.e., fluid transport is in the direction from *the open to the closed end of the channel.* As illustrated in the lower half of Figure 4–8, solute is probably being pumped out of rather than into the channels of these "backwards" epithelia: the channel goes hypotonic, water is absorbed out of the channel and flows into its open mouth, and standing gradients are established such that the channel osmolarity progressively decreases from the open to the closed end.

In either system, the final osmolarity of the fluid being transported across the epithelium will depend upon such parameters as

(1) the length of the channel;

(2) the plasma concentration of the solute being pumped;

(3) the number (and positioning) of the active pumping sites along the channel; and

(4) the water permeability of the channel.

From a functional, biochemical point of view, by far the most important component of this system is the active cation pump, since it supplies the "driving" power for the overall process.

The Fundamental Role of the Na^+ Pump. In a large number of different species and different tissues, the primary active pumping job is carried out by the so-called Na^+ pump; usually other cations and many anions (although not necessarily all anions, since some organisms possess Cl^- pumps) come to be redistributed largely according to concentration and electrical gradients. Although the pump has not been isolated in "pure" form (a task that many biochemists are trying to achieve even as this essay is going to press), many of its kinetic properties are known from studies of Na^+ movements in animals under different experimental conditions. From such studies, it appears that Na^+ active transport is usually "coupled" with an equivalent but oppositely directed movement of a counter-ion, such as K^+, NH_4^+, or possibly H^+.

The Na^+ Pump = The Na^+ K^+ ATPase. It has now been conclusively established that the Na^+ K^+ stimulated ATPase is responsible for Na^+ movements and that this enzyme is vectorial, or directional, in its action. The most unequivocal evidence, stemming from studies of Na^+ K^+ ATPases in mammalian erythrocytes, can be summarized as follows: when red blood cells are exposed to distilled water under controlled conditions, they swell and their membranes become very "leaky." As a result, they lose their hemoglobin and cytoplasmic proteins as well as their internal electrolytes. Such "ghosts" can now be "loaded" with a variety of compounds, for when an isotonic medium is added back to the ghosts, they shrink back to their normal size and their membranes return to their usual relative impermeability. In this manner, erythrocyte ghosts with internal ATP and various concentrations of internal Na^+ and K^+ can be reconstituted (Figure 4–9). Then it is a simple matter to test the effect of various internal and external concentrations of Na^+ and K^+ on the rate of enzymatic hydrolysis of ATP.

In such experiments, the movements of Na^+ and K^+ are in the anticipated directions during hydrolysis of ATP. Thus, when Na^+ is present in high concentration on the inside and the K^+ is high on the outside, maximal ATP hydrolysis is evoked; *as ATP is hydrolyzed,* Na^+ *moves out of the cell and* K^+ *moves into it.* Moreover, it is found that these preparations can only utilize *internal ATP*; external ATP is not attacked at all. The actual substrate is a Mg-ATP complex, which accounts for the absolute dependence of the enzyme for Mg^{2+}. Recent studies indicate that the specificity for ATP is exceedingly high, the dissociation constants for ITP, CTP, AMP and ADP being 1500, 300, 300, and 10 times greater than for ATP. External K^+ can be replaced by other monovalent cations, but internal Na^+ (the chief inorganic "substrate") is an absolute requirement for the ATPase activity. The apparent affinity for Na^+ is reduced by K^+ and vice versa because each competes for the other's binding sites.

As Na^+ K^+ ATPase is positioned on membrane boundaries—and may in fact be a component of membrane structure—it is not surprising that the enzyme appears to have an absolute phospholipid requirement. Binding of phospholipid in a specific ratio to the protein leads to a conversion of the enzyme from an inactive to a fully active form. The activator species of phospholipid may depend upon the organism and tissue under study; in the enzyme from the salt gland of marine birds, it appears to be a sulfatide (Figure 4–10).

Although it is not established experimentally, we are assuming that the Na^+ pump in all species is at least in part composed of the Na^+ K^+ ATPase system. Specific properties, such as the *vectorial component, the affinity constants, or the counter-ion requirements,* may differ between species, but overall stoichiometry, overall binding sequence of Na^+ and other substrates, in short, the general nature of the active sites on the protein molecule—these we currently assume are fundamentally similar wherever the Na^+ K^+

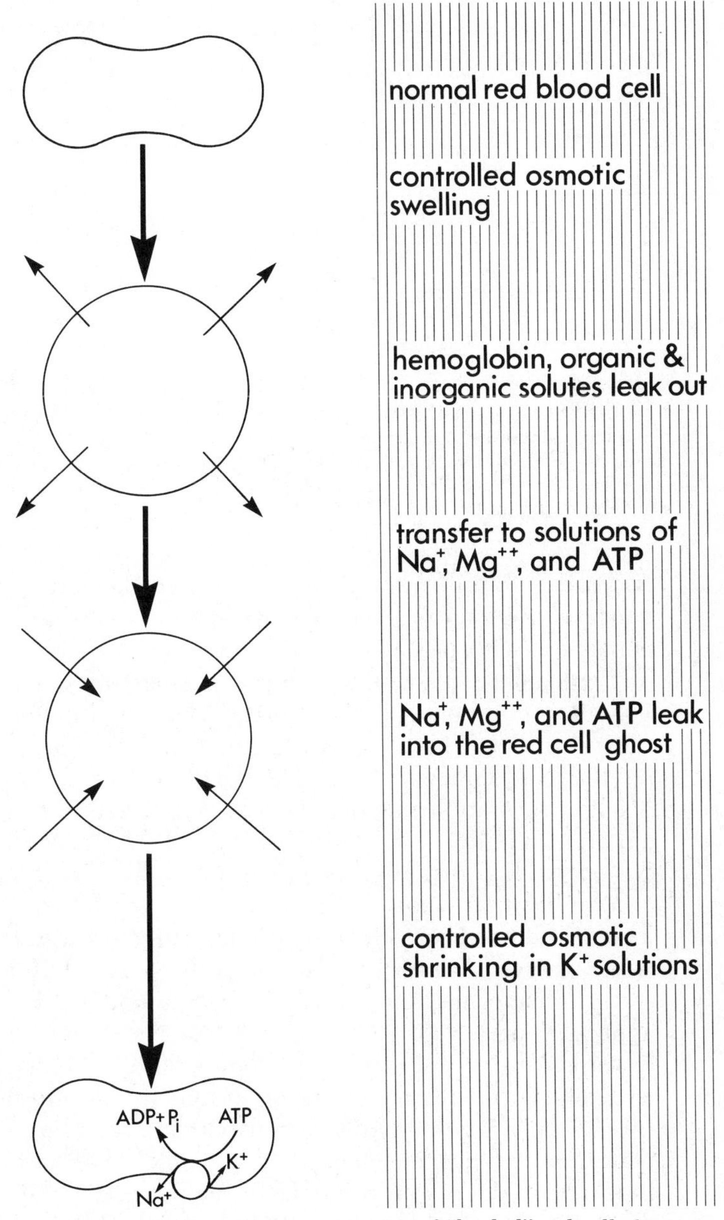

Figure 4–9 Steps in the preparation of "loaded" red cell ghosts.

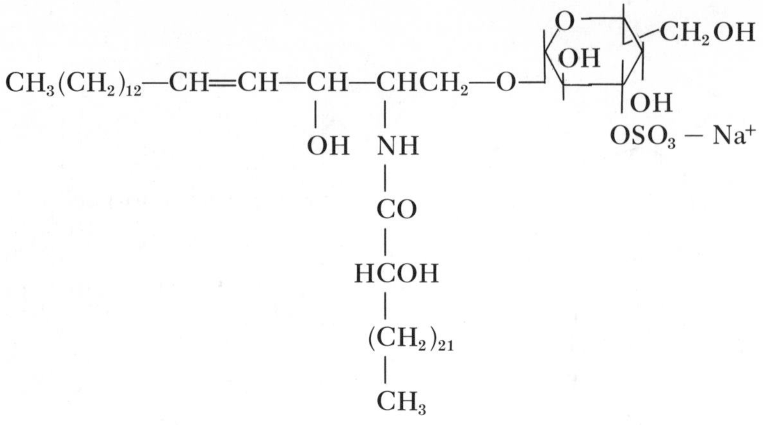

$$CH_3(CH_2)_{12}—CH{=}CH—\underset{\displaystyle OH}{\underset{|}{CH}}—\underset{\displaystyle \underset{\displaystyle \underset{\displaystyle \underset{\displaystyle (CH_2)_{21}—CH_3}{|}}{HCOH}}{CO}}{\underset{|}{NH}}CH_2—O—\text{galactose}(CH_2OH,\ OH,\ OH,\ OSO_3^-\ Na^+)$$

Figure 4–10 The major molecular species of sulfatides of the salt gland of salt-loaded ducks, containing sphingosine, 2-hydroxytetracosanoic acid, and galactose esterified in position 3 with sulfuric acid. The glycosidic bond of human brain sulfatide is of β-configuration. This is the major glycolipid of the salt gland, and its concentration parallels that of Na^+ K^+ ATPase during salt adaptation.

ATPase is found in the animal kingdom. Current evidence on the conservative nature of active sites in the evolution of proteins is consistent with this assumption.

"Halophilic" Properties of Na^+ Transport in Marine Invertebrates. From the point of view of our present discussion, however, what is most significant is the dependence of the Na^+ transporting system on the Na^+ concentration. This dependence follows Michaelis-Menten saturation kinetics; hence, it is possible to quantitate the apparent affinity of the system for its chief inorganic "substrate," Na^+. It so happens that the apparent K_m for Na^+ depends in *most critical fashion upon the species origin of the transport system.* Thus, in crustacean species normally inhabiting full-strength sea water, the apparent K_m for Na^+ is in the order of 20 mM. In fresh-water crustaceans, the K_m is reduced to about 1/100 this value—about 0.2 mM Na^+. (Interestingly, the Na^+ concentrations of sea water and fresh water differ by about 100-fold!) Euryhaline, brackish water forms display transport systems with intermediate affinities for Na^+. The active Na^+ transport inward allows these organisms to maintain their bloods hyperosmotic in all three (freshwater, brackish, and sea water) media. To achieve this, the Na^+ transport system apparently has been *tailored in each case for controlled function in the salt concentration range in which the system normally functions.*

There is a clear necessity for a close correlation between the $K_{m(Na^+)}$ and the Na^+ levels in which the transport enzyme functions. The high $K_{m(Na^+)}$ displayed by the transport systems of open-ocean invertebrates is simply unsuited for function in brackish and freshwater environments. As shown in the saturation curves in Figure 4–11, if a low-affinity transport system were placed in environments of low Na^+ concentration, *both the absolute transport rate and the responsiveness to normal changes in Na^+ concentration would be drastically low.* Put another way, large changes in Na^+ concentration would be required for small changes in Na^+ transport rate. This is an impossible situation if efficient control of the system is to be attained, since large Na^+ concentrations do not occur in fresh water. For this reason, if none other, *the low affinity Na^+ transport systems of marine invertebrates are linked to the high salinity environment in which they live.* In this sense, these are halophilic enzymes.

By the same token, the high-affinity Na^+ transport system in brackish and freshwater forms *is equally dependent upon low Na^+ concentration ranges for controlled catalytic function.* In this case,

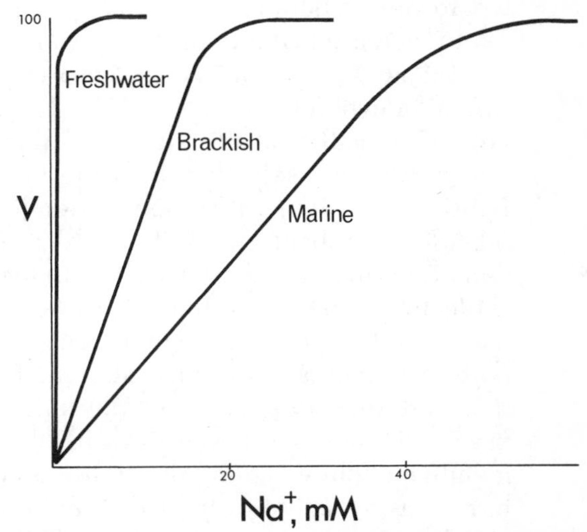

Figure 4–11 Hypothetical Na^+ saturation curves for Na^+ K^+ transport systems from related freshwater, brackish water, and marine invertebrates. The apparent $K_{m(Na^+)}$ in freshwater forms is in the 0.2 mM range; in the brackish water forms it is about 9 mM; in the open ocean species, it is 20 mM. It is evident that the transport system from open ocean species is quite unsuited for function in low salinity environments. As shown, this form is highly unresponsive to small changes in Na^+ concentration, as would be observed in freshwater environments. Increasing Na^+ from 0 to 2.5 mM leads only to a 5% saturation of the marine form of the enzyme and about 10% saturation of the brackish water transport system. The same change in Na^+ concentration fully activates the enzyme from freshwater forms. For these reasons, the three kinds of transport systems are linked to environments of different salinities. For further discussion, see Lockwood (1967).

the system would be continuously saturated in an environment of high Na^+, and consequently would be entirely unresponsive to changes in Na^+ concentration.

These large differences in the enzyme-cation affinity between various species of invertebrates undoubtedly reflect underlying differences in structure of the transport system. Lockwood and Croghan, in their studies of a Baltic isopod species, a relic of the last glacial age, allow some estimate of the length of time which is required for these kinds of evolutionary changes to occur. One race of this species lives throughout the Baltic in saline waters; other races inhabit a number of freshwater lakes. The apparent $K_{m(Na^+)}$ for the saline race is about 9 mM Na^+, a value typical of transport systems in euryhaline crustaceans. The apparent $K_{m(Na^+)}$ for the system from a freshwater race in Lake Malaren is about 3 mM Na^+. A point which Lockwood and Croghan stress is that this lake, once a part of the Baltic Sea, was isolated and became a freshwater lake only about 700 years ago. That is, 700 years of evolution of this particular species in an environment of gradually declining salinity has led to the elaboration of a Na^+ transport system with at least a 3-fold increased affinity for the cation.

Salinity Adaptation Through Elaboration of Na^+ K^+ ATPase Variants. The high evolutionary rates of Na^+ K^+ ATPase indicate a high degree of versatility at this site, as well as an enzyme system that is under extreme selective pressure. From extensive studies by Bonting and others, it is now generally appreciated that Na^+ K^+ ATPase is probably ubiquitously distributed in the animal kingdom. It occurs in highest activities in those tissues whose major job is electrolyte transport, but it is present in lower activities in most tissues of the body. Although the enzyme usually specifically transports Na^+ and a counter-ion such as K^+ wherever it occurs, this transport subserves different physiological functions in different tissues. In nervous tissue, it is involved in repolarization of the membrane following impulse transmission. In the kidney, it occurs in increasing activities towards the distal end of the loop of Henle and functions in the reabsorption of Na^+ from the ultrafiltrate; this process generates the "power" required for operation of the counter-current multiplier mechanism of concentrating the urine. In the intestine it functions in the transport of Na^+ across the intestinal wall. In the cochlea, the organ which transduces acoustical signals into nervous excitation, the enzyme is responsible for maintenance of large concentration differences between one chamber containing endolymph (an extracellular fluid with 12 mM Na^+) and two surrounding chambers containing perilymph (extracellular fluid containing 150 mM Na^+). (See Bonting, 1970, for further discussion of ATPase functions in different tissues of the mammal.)

In all these cases, *a primary signal for activation of the Na^+ K^+*

ATPase activity appears to be the concentration of Na^+. For this reason, and because the Na^+ concentration ranges for these various functions are specific to each tissue, one would expect that the apparent Na^+ affinity of the enzyme in different tissues would also be carefully modulated. This indeed appears to be the case. Thus, in the guinea pig, for example, the K_m for Na^+ for intestinal ATPase is about 1.0 mM; under comparable conditions the K_m for Na^+ for kidney ATPase can be as high as 35 mM, while the K_m value for cochlear Na^+ K^+ ATPase is about 4.5 mM (Table 4–4). The Na^+ concentration range in the kidney tubular microenvironment (particularly in the region of the papilla) is much higher than in the

TABLE 4–4 The variation in the apparent $K_{m(Na^+)}$ for Na^+ K^+ ATPase from various tissues and various species. Unless specified below, the K_m for Na^+ was determined in the presence of 20 mM K^+. Note the effect of increasing the K^+ concentration on the K_m for Na^+ (guinea pig heart and kidney). In the case of human red blood cells, the K_m for Na^+ is relatively insensitive to K^+ concentrations below 80 mM K^+, but it is dramatically increased at above 80 mM K^+. The thermal properties of these ATPases depend upon the origin of the enzyme. For the ATPases from ectothermic organisms (octopus, squid, etc.), the Q_{10} is less than 2.0, while for the ATPases from the endothermic species, the Q_{10} is as high as 4.0 (guinea pig heart Na^+ K^+ ATPase).

Species and Tissue	$K_{m(Na^+)}$ in mM
guinea pig	
intestine	1.0
stria vascularis	4.5
heart at 5 mM K^+	10.0
at 130 mM K^+	30.0
kidney	15.0
at 80 mM K^+	35.0
rabbit	
kidney	25.0
pancreas	10.0
rat	
kidney	8.0
liver	6.0
brain	10.0
human	
red blood cells	24.0
gull	
salt gland	12.5
goose	
salt gland	8.0
electric eel	
electric organ	25.0
octopus	
gill	100.0

endolymph, which in turn contains higher Na^+ concentrations than are likely to occur in intestinal juices. For reasons already considered (p. 125), controlled Na^+ K^+ ATPase function would be most efficient (sensitive) when the apparent K_m for Na^+ approximates the effective Na^+ concentration range *in vivo*. Since Na^+ levels vary between tissues, as do the concentrations of other monovalent cations which competitively interfere with the binding of Na^+, *strong selective pressures have favored the development of tissue-specific "functional variants" of Na^+ K^+ ATPase.* The structural bases of these variants are not known; the evidence for their existence is entirely kinetic. On the basis of the known general structure of ATPases, we would predict that changes in the enzyme protein or in the associated lipid moiety required for enzymic activity could provide a mechanism for generating functional variants.

Just as functionally distinct Na^+ K^+ ATPases appear to be elaborated for particular functions within different organs of the same individual, the necessity could readily arise for more than one functional form of Na^+ K^+ ATPase *in the same organ under different environmental conditions.* The gill of catadromous and anadromous fishes, which successfully migrate across the freshwater-salt water barrier, presents us with an excellent case in point.

Freshwater Adaptation in Fishes. In freshwater fishes, the major problems of ion and osmoregulation are (1) maintaining ion concentrations in the blood at values much higher than the outside fresh water, and (2) preventing an excess osmotic flow of water into the cells and tissues of the body. These problems are common to any and all freshwater animals. Theoretically, the organism could cope by adjustments of ion and water flows through the gill, the gut, the kidney, and the skin. Of these, the skin plays a negligible role since it is continuously impermeable to both water and ions. In fresh water, the gut is not much involved either, as the organism usually reduces the water load which is taken in by drinking. This leaves the organism with the gills and kidneys as the major effector organs; solution of osmotic and ion problems involves the active *uptake* of Na^+ ions at the gill surface and the production of a dilute and copious urine at the kidneys (Figure 4–12). Of these, the active uptake of Na^+ at the gills is quantitatively the more important process.

The gill transport system in freshwater-adapted fish displays two essential properties which, as we shall see, are not present during salt adaptation:

(1) The salt pump is *polarized* in function so that the direction of net Na^+ movement is *inwards.*

(2) It must be a high-affinity system (with a low apparent K_m for Na^+) because the Na^+ concentrations in the outside fresh water can be very low (below 0.2 mM). In the case of trout, for example,

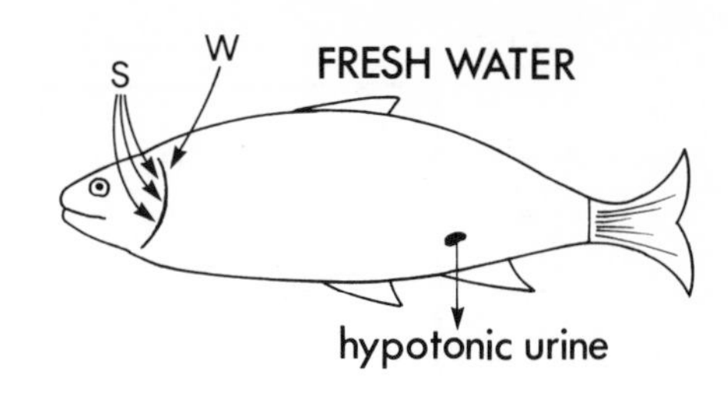

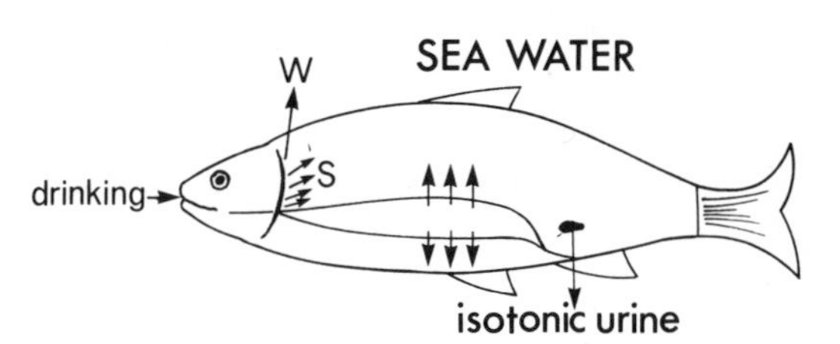

Figure 4–12 Schematic representation of the role of various effector organs in the maintenance of salt (S) and water (W) balance in freshwater and sea water teleosts.

the $K_{m(Na^+)}$, estimated in the living animal, is about 0.4 mM – representing a high Na^+-affinity system.

Recent studies of freshwater (FW) and sea water (SW) adapted trout by Kirschner and his coworkers indicate some of the specific biochemical changes which occur in the gill during FW and SW adaptation. In the gill of FW adapted trout, these workers discovered two kinds of Na^+-linked ATPases. One is the classical Na^+ K^+ ATPase with properties resembling those of mammalian enzymes. Thus, it is completely inhibited by ouabain, which binds at the K^+ site, and is maximally activated by 20 mM K^+ in the presence of 100 mM Na^+. The activity of this transport enzyme is so low that its primary function appears to be only the regulation of intracellular Na^+/K^+ ratios.

A second, much more active ATPase in FW adapted gills is described by Kirschner as a *Na^+ ATPase.* It does not display a requirement for K^+, as far as these workers could tell, nor for any other monovalent cation. However, the enzyme could be utilizing H^+ of the buffer medium, since counter-ion specificities of salt-linked ATPases are not absolute (p. 124). In any event, the loss of K^+ requirement probably is adaptive, since it will be evident (Figure 4–13) that linking the Na^+ input with K^+ output would potentiate an already large (and unfavorable) K^+ gradient outwards. Probably for this reason, K^+ is not used as the counter-ion in freshwater adapted fishes. For some time it was believed that the K^+ requirement was replaced by NH_4^+. Since freshwater teleosts are ammonotelic, this mechanism supplies an easy route for the excretion of the major end product of nitrogen metabolism – the NH_4^+ ion. But it will be readily apparent that the demands of nitrogen metabolism and Na^+ metabolism may not always be in balance, as would be

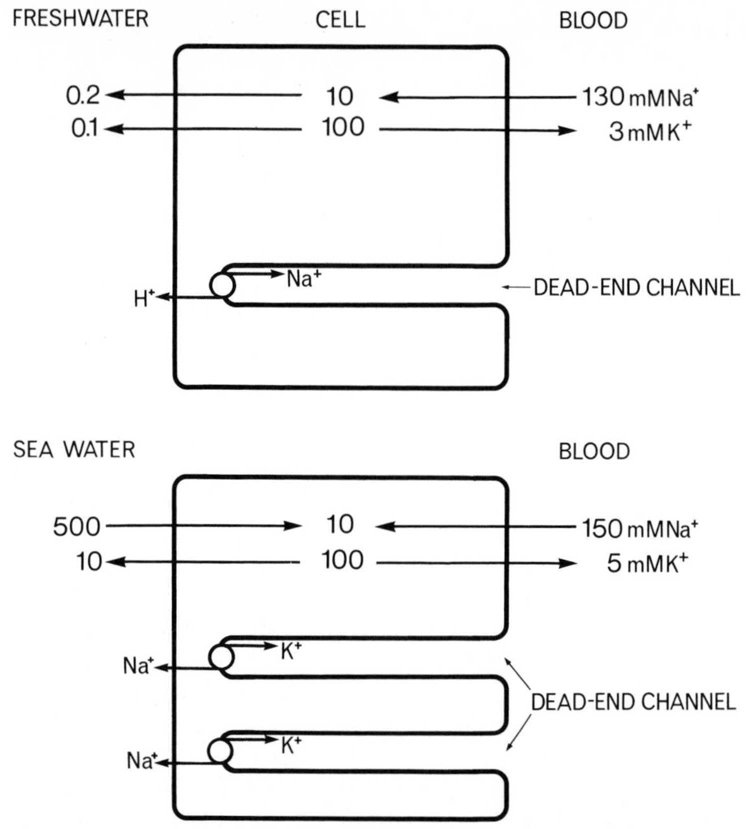

Figure 4–13 Functional model of the chloride cell in the gills of freshwater and sea water adapted fishes. The model is to be looked upon strictly as a summary of the physiological and biochemical information that is available on the transport of monovalent cations. The anion "pump" which presumably accounts for Cl^- and HCO_3^- movements is not shown.

required if there occurred an obligatory link between Na^+ and NH_4^+ transport at the gill. Indeed, it is well documented that Na^+ transport inwards can be varied by nearly a factor of 10 without change in NH_4^+ transport outwards; under these conditions, H^+ is used as the counter-ion for the Na^+ transported inwards. These physiological data can be readily accommodated by Kirschner's biochemical data if the Na^+ ATPase were utilizing H^+ as its counter-ion.

Unequivocal evidence on the localization of the Na^+ pump enzymes unfortunately is unavailable. Philpotts and his coworkers suggest that the active Na^+ pump is localized on the "dead-end" standing gradient channels of the gill, pumping Na^+ into the channel. Whatever the precise mechanism for Na^+ movement into the gill chloride cell, its inward transport is coupled directly or indirectly to an outward transport of H^+. The source of this H^+ is a question which we must briefly consider.

The origin of H^+ for this process is the carbonic anhydrase catalyzed reaction,

$$CO_2 + H_2O \rightarrow H^+ + HCO_3^-.$$

On this, the evidence is fairly clear and there is widespread agreement among the workers in the field. The bicarbonate in turn is thought to exchange with chloride ion of the outside medium. The chloride uptake may be an active process, possibly coupled to bicarbonate transport by a specific anion-linked ATPase. Anion-linked ATPases have been identified in frog gastric mucosa (which carries out an active secretion of H^+ and Cl^- ions into the stomach), in the gills and stomach of *Necturus,* and in the proximal tubules of the mammalian kidney. If present also in the gills of freshwater fishes, such an enzyme could contribute to the controlled movements of ions between the outside and the inside of the cell. At the moment, this possibility has not been ascertained in fishes; it clearly is an important question to be settled with all possible haste. Whatever these latter mechanisms, it is clear that HCO_3^- is excreted at the gill in exchange for Cl^- uptake at the same time as H^+ is excreted in exchange for the Na^+ transported inwards. This coupling of Na^+ and Cl^- uptake with H^+ and with HCO_3^- release admirably integrates several functions of the gill at once: (1) electrolyte balance, (2) acid-base balance, and (3) release of respiratory CO_2 as HCO_3^-.

Salt Water Adaptation in Fishes. Upon encountering brackish water, and then full strength sea water, an organism like the salmon or the eel is faced with quite different problems of ion regulation. Here, the animal must deal with (1) an osmotic loss of water at the gills, the consequences of which are exaggerated by (2) a continual "downhill" inflow of Na^+ and other ions. Of the various effector organs potentially capable of responding to this new environmental challenge, the increased water intake at the gut (by increased drinking of sea water) balances the osmotic loss of water at the gill. This appears to be the only important contribution of the gut. The kidneys are involved in the adaptive response by reducing the volume of urine eliminated and probably by increasing the Na^+ output of the urine. By far the most important effector organ in quantitative terms is again the branchial gill apparatus (Figure 4–12), which actively pumps out Na^+ in order to maintain blood and cell ion concentrations within physiologically acceptable ranges. Increased plasma Na^+ levels after exposure to sea water definitely constitute the ultimate signal activating salt adaptation mechanisms, although this signal may be mediated hormonally (by cortisol and prolactin).

Ultrastructural Adjustments in the SW Adapted Gill. During salinity adaptation in fishes, important ultrastructural changes occur in the transport cells of the gill. These cells (the chloride cells) display a number of characteristics (Figure 4–14) which are generally accepted as hallmarks of transport cells (and indeed a part of the evidence that these cells are the key transport cells in the gill epithelium arises from their unique ultrastructure). Included in this category are two essential features:

(1) Mitochondria with closely packed cristae are numerous and increase in abundance during salt adaptation.

(2) All transport cells have mechanisms for membrane amplification; in the case of the salt water adapted chloride cell, the surface-amplified membrane is greatly invaginated, forming multiple "dead-end" channels closely associated with the mitochondria. During salinity adaptation in *Fundulus* and in the eel, there occurs a 6- to 7-fold increase in Na^+ K^+ ATPase activity in concert with amplification of this membrane. The pseudobranch (a vestigial gill which has lost the respiratory function but which retains the transport function) represents a much more homogenous population of chloride cells. Here Na^+ K^+ ATPase activity increases 10-fold during salt water adaptation. A similar situation is found in the eel and probably in all catadromous, anadromous, and euryhaline fishes.

Na^+ K^+ ATPase Function in the SW Adapted Gill. The above studies establish that salinity adaptation involves an extensive activation of biosynthetic machinery, probably involving both transcriptional and translational processes. One of the products of the

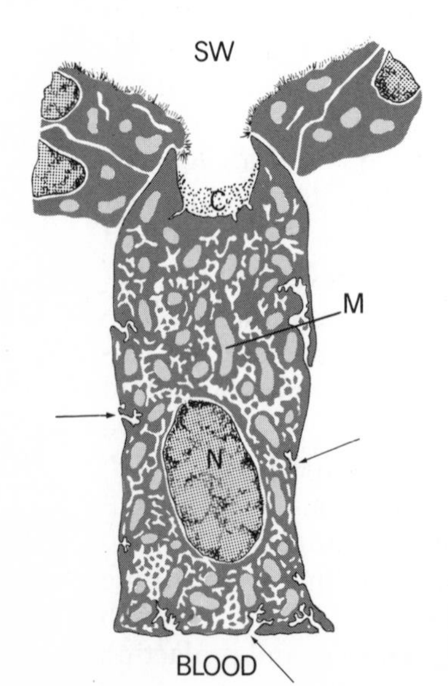

Figure 4–14 Diagrammatic representation of a typical salt water adapted chloride cell. The apical crypt (C) of the cell communicates directly with the external sea water. Arrows indicate connections between the intracellular tubular system, which forms many standing gradient channels within the cell, and the plasma membrane of the basal and lateral surfaces. Note the abundance of mitochondria (M). The nucleus is indicated by the letter N. This schematic is based on the studies of Philpott and his coworkers.

biosyntheses occurring at this time is a Na^+ K^+ ATPase variant which apparently differs kinetically from the predominant form present in freshwater adapted fish gills. Its appearance is *preceded* by the rapid turnover of an mRNA which presumably specifies some or all of the polypeptide components of this ATPase. This "new" transport system in the salt water adapted gill displays the following properties:

(1) It is polarized functionally so that Na^+ is now being pumped *outwards* rather than inwards in order to maintain Na^+ levels in the blood at lower concentrations than they are found in the sea water. For this purpose, a Na^+ K^+ ATPase presumably "replaces" the Na^+ ATPase in the wall of the standing gradient channels, for the latter enzyme is not detectable in high activities in SW adapted trout.

(2) The Na^+ K^+ ATPase in salt water adapted fish gills displays an absolute specificity for Na^+ and, at least under physiological conditions, likewise has an absolute requirement for K^+. Altering the external K^+ levels dramatically alters the outward pumping of Na^+. The exchange ratio is 1 Na^+:1 K^+ and the optimal activity occurs at a ratio of 1:1. Thus, this ATPase is a classical Na^+ *and* K^+ linked ATPase, whereas in the freshwater adapted gill, the K^+ requirement is either replaced by H^+, or lost entirely.

(3) In addition, the Na^+ K^+ ATPase in salt water adapted gill must *retain* an acceptably high affinity for K^+ in the presence of high concentrations of Na^+, for the counter-ion is being "presented" to the ATPase in the presence of high (sea water) Na^+ levels. This outside Na^+ decreases the apparent affinity for K^+, which explains why altering the external K^+ levels alters the Na^+ pumping-out rate. Such a situation has selected for an ATPase with a high affinity for K^+ to reduce sensitivity to competitive Na^+ inhibition, a feature of the enzyme which has only been casually examined by the workers in this field.

From these considerations it is evident that the Na^+ transport system on the effective outer barrier of the SW adapted gill is qualitatively different from that observed in the freshwater adapted one. Direct evidence for this contention has come largely from kinetic studies of isolated enzyme preparations. Whether or not the entire ATPase molecule elaborated during salinity adaptation is different from the analogous activity in the freshwater adapted gill is not known. However, Conte has shown that at least several new protein components in the gill are synthesized during this time, and it is probable that some of these are directly connected to salt transport functions. However, an equally attractive postulate suggests that the kinetic properties of the cation pump are altered in SW adaptation simply by changing its phospholipid component.

Summary of Adaptive Events During Freshwater–Sea Water Transition in Fishes. To recapitulate, it appears that in fishes, freshwater or salt water adaptation depends upon the type of salt-linked ATPase present in the gill. In freshwater adapted gills, the ion pump seems to be linked to a "*Na^+ ATPase*," which probably is positioned for net Na^+ pumping in an *inward* direction. Unlike most salt-linked ATPases, this ATPase appears to have no K^+ requirement; rather, this type of pump couples the transport of Na^+ inwards with H^+ movement outwards. H^+ is generated by the carbonic anhydrase reaction, which also converts CO_2 to HCO_3^-. HCO_3^- transport outwards is coupled with Cl^- uptake.

In the salt water adapted state, both the *direction* and *degree* of pumping change: now, the dominant pumping activity is linked to a Na^+ and K^+ ATPase, which is increased dramatically in total activity and which displays absolute dependence upon both Na^+ and K^+, at least under physiological conditions (Table 4–5). This adjustment is reasonable. If the salt-water ATPase system maintained a H^+ coupling, it would pump H^+ into the cell. Of course, this is the reverse of what should happen. Also, intracellular K^+ concentrations are about 10-fold higher than sea water K^+ levels. Hence, linking the salt pump to external K^+ serves to reverse the "downhill" flow of intracellular K^+ to the outside.

Central to this view of salt adaptation in these organisms is the basic assumption that kinetically different forms of salt-linked ATPases are required to account for the different *directions and degrees of pumping* and for the different *ion specificities* that are maintained *in vivo*. In some instances, one would predict that both forms of these enzymes are present at all times, and may be modulated in relation to the environmental situation. Thus, in rainbow trout perfused gill preparations, the Na^+ concentration in the perfusate determines whether Na^+ is actively transported inwards or outwards across the gill surface. This situation could be nicely accommodated by a model which assumes the *continuous presence* of both ATPases; the balance of their oppositely poised but unidirectional functions would presumably be determined by their relative affinity constants and by the relative concentrations of Na^+, K^+, and H^+. A similar situation may occur in species such as *Fundulus*,

TABLE 4–5 COMPARISON OF THE Na^+ ACTIVE TRANSPORT SYSTEM IN FW AND SW ADAPTED FISH

	FW	SW
Direction of Na^+ transport	in	out
Apparent Na^+ affinity	high	low
Counter-ion	H^+	K^+
Pump localization	dead end channels	dead end channels

which is essentially an intertidal form, and may therefore encounter sudden and large fluctuations in outside salinity.

Unfortunately, currently available literature evidence does not allow one to speculate much further on these matters. Most of the published studies of salt-linked ATPases in these species have been carried out under distinctly non-physiological conditions. Concentrations of Na^+ and K^+ do not approach the physiological condition; H^+ and other monovalent ions usually are not tested as potential substrates for ATPases from freshwater adapted fishes, although the physiological evidence is clear that these are the important counter-ions for Na^+. In other organisms, salt-linked ATPases are strongly modulated by the adenylates, by specific phospholipids, and possibly by divalent cations, but these effects have not been systematically tested in fishes. And finally, the temperature of the enzyme assay too often and unfortunately is 37° C. Choice of such non-biological assay conditions may lead to quite distorted conceptions of fundamental regulatory interactions. This area is an obvious one for future research.

Other Examples of Na^+ K^+ ATPase Variants in Salinity Adaptation. Similar requirements for more than one form of salt-linked ATPase probably arise in other species facing different kinds of environmental problems. Certain freshwater lakes and ponds tend to dry up by evaporation during the summer season. As the years wear on, some of these bodies of water acquire very high salinities. Phillips and Meredith examined the problems of ion regulation in mosquito larvae that live in such high-salinity lakes. The situation here as regards Na^+ transport is similar to that in the salmon. Normally, in nature, the larvae apparently pump Na^+ outwards against a concentration gradient. When the larvae are adapted (over about a two week period) to low salinity media (5 mM in Na^+), the direction of pumping is reversed and very effective uptake of Na^+ occurs against a 100-fold concentration gradient. The apparent K_m for this "induced" transport system is about 2 mM Na^+, and the system appears to be located in the anal papillae.

There appears to be a lower limit to the adaptability of the salt water mosquito larvae. Following acclimation of the larvae to 0.05 mM, 0.5 mM, and 5 mM Na^+ (concentrations which are reduced from those seen in nature by 10000, 1000, and 100 fold respectively!), the Na^+ transport system that appears in all three regimes displays a K_m of 2 mM Na^+ and is quite unable to function to any measurable degree in 0.05 mM media. The reason for this is obvious, for at 0.05 mM Na^+ the transport system is only about 1% saturated, and under these conditions, transport rates are simply inadequate to concentrate Na^+ against the 1000-fold concentration gradient that would be required.

These studies attest well to *the importance of having the K_m*

for Na^+ *fall within the* Na^+ *concentration range of the environment.* In this case, if the K_m for Na^+ were lowered to 0.2 mM, the enzyme could function at about 10% of V_{max}; if it were reduced to 0.02 mM Na^+ (i.e., 100-fold lower than actually observed), the enzyme could then become fully saturated and function at maximal rates even at 0.05 mM Na^+ levels of the medium. But the elaboration of Na^+ pumps with such low K_m values (with such high affinities for Na^+) appears to be beyond the regulatory abilities of salt adapted races of the mosquito *Anopheles compestris.* Such properties are not, however, beyond the adaptive powers of related mosquito species selected through evolutionary time for larval existence in normal fresh bodies of water. In these, the apparent K_m for Na^+ is about 0.2 to 0.4 mM Na^+, indicating transport systems with high Na^+ affinities. These function efficiently in fresh water of low salinity and indeed, for the reasons already specified, they are linked to such an environment.

Salinity Adaptation Through Elaboration of Higher Quantities of Na^+ K^+ ATPase. In some instances in nature, salt adaptation involves a requirement *to do more of the same kind of transport work* rather than the requirement (as in the salmon) to do a *different kind of work.* In such instances, there is no necessity to elaborate new variants of the transport system; instead, it is sufficient merely to change the quantity of the enzyme system already present, so that more work can be done per unit time. The nasal salt gland of birds supplies us with an interesting example of this kind of adaptation.

Originally discovered in gulls by Schmidt-Nielsen and his coworkers, the salt gland has subsequently been shown to play important transport roles in the salt metabolism of many marine birds. In all species thus far studied, the basic function of the salt gland is to pump Na^+ (which enters the body by ingestion with food) out of the blood system and into the lumen of the gland. The lumen then leads to the outside via the nostrils. When active, the salt gland excretes droplets of highly concentrated saline (up to 1 M Na^+!) which drip off the end of the beak. The salt gland in this way provides an *extrarenal excretory pathway* which aids the organism in gaining osmotically free water while removing ingested salts. (Comparable extrarenal excretory organs serving strictly analogous functions have arisen frequently and independently in evolution. Included here are the lacrimal "salt" glands of marine reptiles such as sea turtles and sea snakes, the rectal glands of marine elasmobranchs, the coxal glands of blood-sucking parasites, and so forth.)

More recently, it has become apparent that both *the size and the pumping activity* of the salt gland are adapted according to the salt load taken into the body. The adaptation process, stimulated by the salt load and requiring about two weeks, involves the mobilization of two basic processes:

(1) The first is a *general* increase in the growth and differentiation of the tissue so that the salt gland doubles in weight and protein content, with a concomitant amplification and invagination of the surface membrane of the secreting cells. Mitochondria become more abundant during salt adaptation and tend to be localized at the blood (basal) side of secretory cells rather than at the lumen side.

(2) The development of this characteristic ultrastructure is coupled with a *specific* increase in the activity of Na^+ K^+ ATPase, which indeed is paralleled by an appropriate increase in Na^+ secretory capacities. The membrane amplification presumably serves to properly position the salt pumping sites; again, there is good evidence that Na^+ K^+ ATPase is strongly localized in the standing gradient channels of the salt gland.

Careful studies by Holmes and coworkers indicate that both events (1) and (2) are *preceded* by an activation of RNA synthesis in the gland. The best available evidence suggests that the increase in size of the gland and the increase in the specific activity of the transport system are due to *de novo* synthesis and therefore depend upon the initial synthesis of messenger RNA species. As far as the data can indicate, no new components are being built—only more of the kind that are already present in the salt gland. The translation of one part of the genome—that specifying the essential polypeptide subunits of the Na^+ K^+ ATPase complex—appears to be particularly active and leads to a specific increase in both the messenger RNA for this enzyme and ultimately the concentration of the enzyme itself. The glycolipid required for this Na^+ K^+ ATPase increases concomitantly.

Mechanism of Ion Transport in the Avian Salt Gland. There is considerable evidence indicating that the concentration gradient between the blood and the fluid secreted by the salt gland is established at the *effective*, luminal membrane barrier of the secretory cell (Figure 4–15). The relevant data are:

(1) The concentration gradient between the cell and the lumen is greater than between the blood and the cell.

(2) During secretion, the duct of the gland becomes electropositive with respect to the blood by about +50 mV, presumably due to Na^+ transport into the duct.

(3) Ouabain, a specific inhibitor of Na^+ pumping and of the Na^+ K^+ ATPase, when administered into the duct, abolishes both secretion and the potential difference. Hence, *the large quantities of Na^+ K^+ ATPase present in salt glands must be effectively "exposed" to the lumen* (Figure 4–15).

The rate of Na^+ entry into the cell from the blood appears to be dependent upon a Na^+—H^+ "carrier," for Na^+ enters the cell at a much higher rate than would be predicted on the basis of diffusion alone. The source of the H^+ is again thought to be the carbonic

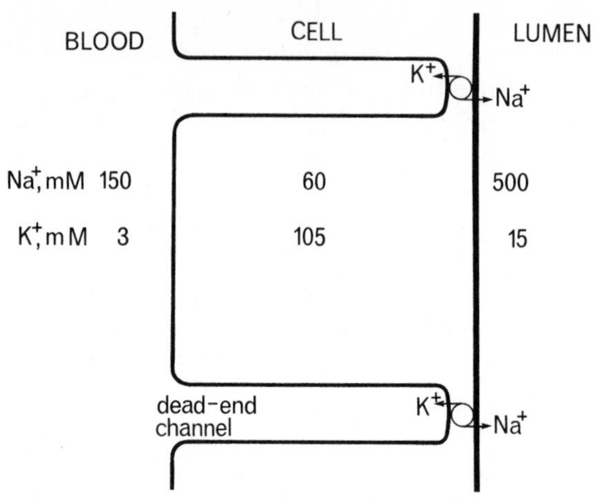

Figure 4–15 Functional model of the salt gland cell in marine birds, showing concentration gradients and the probable roles and positions of dead-end channels and of the Na^+ K^+ ATPase. Whatever the precise localization of the Na^+ K^+ ATPase, it is in effect "exposed" to the lumen since it can be inhibited by injecting ouabain into the lumen. Modified from Peaker (1971).

anhydrase reaction: $CO_2 + H_2O \rightarrow H^+ + HCO_3^-$. The HCO_3^- in turn is thought to exchange with blood Cl^-. Two experimental observations strongly support this interpretation: in the first place, mild acidosis inhibits secretion, as would be expected since the H^+ concentration on the blood side of the cell would be greatly increased (Figure 4–16). Secondly, specific inhibitors of carbonic anhydrase also inhibit secretion.

The secretion rate of the salt gland appears to be under positive feedback regulation. As Na^+ is pumped out across the luminal membrane, ATP hydrolysis leads to higher levels of ADP and P_i; the latter both stimulate cellular respiration, and more CO_2 is formed in consequence. CO_2 is a substrate for carbonic anhydrase; H^+ and HCO_3^-, the two products of the reaction, increase in concentration and in turn accelerate the rate of entry of Na^+ at the blood side of the cell (Figure 4–16). In this way, more blood Na^+ is delivered to the Na^+ K^+ ATPase on the luminal side of the cell. Such *self-potentiation* of the system explains why a maximal stimulus for secretion by the salt gland (an intravenous injection of Na^+) does not lead immediately to a high rate of secretion. Instead secretion rates rise exponentially, reaching a maximum only after about 15 to 20 minutes even though the increased blood flow to the gland peaks out at 5 minutes.

Regulation of Salt Gland Secretion. The ultimate trigger initiating all of the above events is osmotic, for the salt gland of birds only secretes in response to the ingestion of sea water or to the administration of hypertonic solutions. The detectors of excess salt in the body have recently been found to be located in the heart, and the signal to which they respond is the *raised tonicity* of the blood following ingestion of sea water. In addition, however, the secretion by the gland is *elicited and maintained by acetylcholine,* released at the gland by cholinergic nerves. The salt gland, indeed, requires continuous stimulation by acetylcholine, since interruption of nervous transmission while the glands are secreting immediately abolishes secretion. At this physiological level of organization, the evidence for these mechanisms of regulation is unequivocal. What, on the other hand, are the biochemical mechanisms of acetylcholine action upon the transport processes of the salt gland?

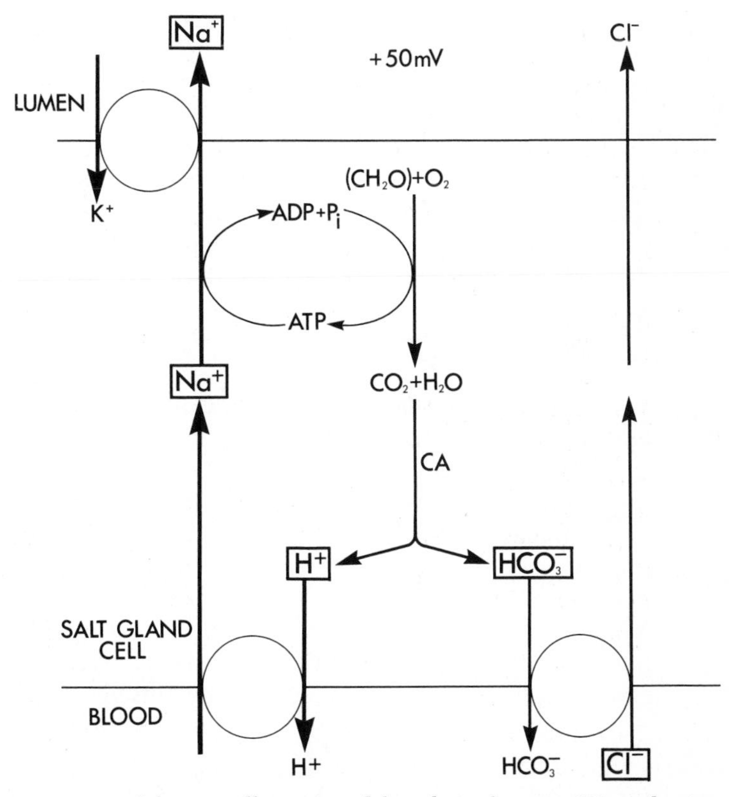

Figure 4–16 Schematic illustration of the relation between H^+ production by carbonic anhydrase (CA) and the movement of Na^+ from the blood into the lumen of the salt gland. Cl^- is thought to exchange with HCO_3^- of the blood; intracellular Cl^- can then move into the lumen down an electrical gradient since the lumen under these conditions is about +50 mV relative to the blood. Upon activation of Na^+ K^+ ATPase, the increased availability of ADP for the electron transfer chain contributes to an activation of cellular respiration and hence to an increased production of molecular CO_2 for the CA reaction.

The simplest theory assumes that acetylcholine acts directly upon Na^+ K^+ ATPase and so brings about an increase in the rate of Na^+ transport into the lumen. There is some strong evidence for this theory. Drug analogues of acetylcholine, for example, competitively interfere with the effects of acetylcholine on Na^+ transport. Also, a molluscan Na^+ pumping system comparable in properties to that of the salt gland is known to be specifically activated by acetylcholine. However, if this were the sole mechanism of acetylcholine action, all of its effects on the salt gland should be blocked by ouabain, which specifically blocks Na^+ K^+ ATPase. Unfortunately, the theory does not survive this most critical test of its completeness. Hence, it is clear that acetylcholine must have additional effects on the avian salt gland.

Many hormones are now known to act by first initiating the intracellular release of a second "messenger," cyclic AMP. The basic action of these hormones is to activate adenyl cyclase (which hydrolyzes ATP to cyclic AMP + PP_i). The cyclic AMP then carries a specific "message" to the intracellular "target." In the case of ion transport tissues, Berridge and his coworkers have demonstrated that the *"target" of cyclic AMP is the cation pumping mechanism, while the "message" most probably is activation of Na^+ K^+ ATPase* (Figure 4–17). A comparable interplay between acetylcholine and cyclic AMP metabolism in the salt gland could well account for the dependence of Na^+ transport upon the continual release of acetylcholine at the gland.

MOLECULAR SITES SENSITIVE TO SELECTIVE PRESSURES DURING SALINITY ADAPTATION

From the above considerations, it is clear that a fundamental biochemical key to electrolyte regulation is active ion transport. Of the various ions actively transported into or out of the body, Na^+ movement by the Na^+ K^+ ATPase-linked pump is established as a major mechanism in ionic and osmotic regulation. This cation pump harnesses the energy of ATP hydrolysis to "drive" Na^+ against substantial concentration gradients. In the absence of additional active mechanisms (such as ion-specific permeability changes or anion pumps), inorganic ions other than Na^+ subsequently redistribute themselves in accordance with electrical and concentration gradients.

In evolutionary terms, the characteristics of the Na^+ pump which seem to be most sensitive to selective pressures are (1) the relative affinities for Na^+, (2) the counter-ion specificity, and (3) the *in vivo* polarity of function. Some problems in salt adaptation (as, for example, those encountered by marine invertebrates)

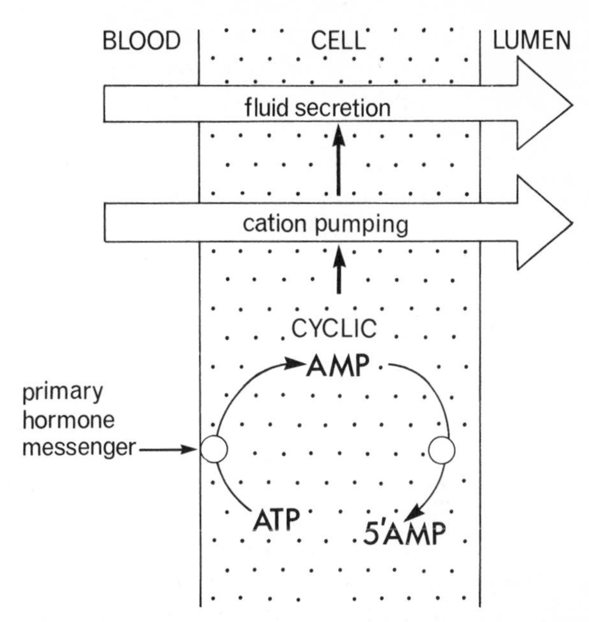

Figure 4–17 The role of cyclic AMP in the control of fluid secretion. The intracellular concentration of cyclic AMP depends on the balance which exists between its synthesis by adenyl cyclase and its destruction by phosphodiesterase. Hormones such as acetylcholine may act by stimulating adenyl cyclase. Modified from Berridge and Prince (1971).

involve the elaboration *through evolution* of Na^+ transport systems with Na^+ affinities approximating the Na^+ concentrations in which each specific pump normally functions (high affinity in fresh water; low affinity in sea water). Other problems of salt adaptation, such as the crossing of freshwater-salt water barriers by fishes, requires the elaboration *during acclimation* of functional variants of salt-linked ATPases. As in evolutionary adaptation, the key characteristics under regulation are (1) the Na^+ affinity, (2) the counter-ion specificity, and (3) the polarity, and these are specific to ATPases in freshwater adapted and sea water adapted organisms.

A third type of salinity challenge can be answered simply by increasing the organism's ability to do more of the same kind of pumping job (rather than to pump at different rates in different directions). The salt gland of marine birds admirably illustrates this strategy. Solution to this problem involves the elaboration *through evolution* of regulatory mechanisms for increasing on demand the pumping capacity of the salt gland, which is achieved by increasing the *amount* of Na^+ K^+ ATPase being synthesized and maintained in the salt gland at any given time. Hence, mechanisms controlling the steady-state concentration of Na^+ K^+ ATPase appear to be another site of evolutionary action.

Finally, we come to the question of signals eliciting adaptive responses. In all of the above examples, the ultimate signal releasing any of the variety of adaptive responses is the salt concentration of the blood and other extracellular fluids of the body. However, it is also probably true that this signal is often mediated by neuroendocrine mechanisms. In the case of the salt gland, the continuous release of acetylcholine at the salt gland by cholinergic nerves is an essential component of adjusting the function of the salt gland to the salt loads being sustained. Although the mode of action of acetylcholine is not yet entirely clear, it is probable that it involves a second "messenger" system, probably the adenyl cyclase mediated release of intracellular cyclic AMP. If, as seems probable, the action of acetylcholine is directed to cyclic AMP formation, via the activation of the enzyme adenyl cyclase, we find an interesting problem. Adenyl cyclase is probably found in all tissues of the body. Yet the action of acetylcholine on this enzyme is noted only in the salt gland. Such tissue specificity suggests that distinct *regulatory* forms of adenyl cyclase may be found in avian tissues. Such regulatory isozymes might share a common catalytic subunit while having different regulatory subunits, each with a particular hormonal specificity, and hence conferring on the given cell type a distinct hormonal responsiveness.

Thus it appears that strong selective forces are applied at four key loci in the ion-regulating system. First, the pumping system itself is "modulated" adaptively in three important ways, so that the pump enzymes have (i) the correct affinities for ions, (ii) the correct counter-ion specificities, and (iii) the correct polarity to conduct the required work. In addition to these essential changes in the pumping mechanism *per se,* selection appears to exert an important shaping influence on the control circuitry which regulates the pumping rate and pumping direction in accord with salinity changes of the environment.

SUGGESTED READING

Books and Proceedings

Water and Solute Transport (1971) *Phil. Trans. Royal Soc. London* B, Vol. 262.

Henderson, L. J. (1913). *The Fitness of the Environment.* Macmillan Co., 317 pp.

Lockwood, A. P. M. (1967). Aspects of the Physiology of Crustacea. W. H. Freeman & Co., San Francisco, 328 pp.

Watson, J. D. (1970). The Molecular Biology of the Gene. Benjamin, New York.

Reviews and Articles

Atkinson, D. E. (1969). Limitation of metabolite concentrations and the conservation of solvent capacity in the living cell. In *Current Topics in Cellular Regulation* (ed. B. L. Horecker and E. B. Stadtman), Academic Press, N. Y., Vol. 1, 29–43.

Bayley, S. T. (1973). Evolution and adaptation of macromolecules by halophilic organisms. I. Halophilic bacteria. In *Biochemical Adaptation* (ed. F. Conte), Univ. of Chicago Press, in press.

Berridge, M. J., and W. T. Prince (1971). The electrical response of isolated salivary glands during stimulation with 5-hydroxy-tryptamine and cyclic AMP. *Phil. Trans. Royal Soc. London* B, *262*, 111–120.
Bonting, S. L. (1970). Na^+ K^+ activated ATPase and cation transport. In *Membranes and Ion Transport* (ed. E. E. Bittar), Wiley-Interscience Publ., Vol. 1, 257–363.
Brown, A. D. (1964). Aspects of bacterial response to the ionic environment. *Bacteriological Reviews, 28,* 296–329.
Conte, F., and T. N. Morita (1968). Immunochemical study of cell differentiation in gill epithelium of euryhaline *Onchorhynchus. Comp. Biochem. Physiol., 24,* 445–454.
Diamond, J. (1971). Water-solute coupling and ion selectivity in epithelia. *Phil. Trans. Royal Soc. London* B, *262,* 141–151.
Epstein, F. H., M. Cynamon, and W. McKay (1971). Endocrine control of Na^+ K^+ ATPase and sea water adaptation in the eel, *Anguilla rostrata. Gen. Comp. Endocrinology, 16,* 323–328.
Ernst, S. A. (1972). Transport adenosine triphosphatase cytochemistry. *J. Histochem. Cytochem., 20,* 23–38.
Ewing, R. D., G. L. Peterson, and F. P. Conte (1972). Larval salt gland of *Artemia salina* nauplii. Effect of inhibitors on survival at various salinities. *J. Comp. Physiol. 80,* 247–254.
Gerhardt, J. C. (1970). A discussion of the regulatory properties of aspartate transcarbamylase from *E. coli.* In *Current Topics in Cellular Regulation* (ed. B. L. Horecker and E. B. Stadtman), Academic Press, N. Y., Vol. 2, 275–325.
Glynn, I. M., J. F. Hoffman, and V. L. Lew (1971). Some "partial reactions" of the sodium pump. *Phil. Trans. Royal Soc. London* B, *262,* 91–102.
Hegyvary, C., and R. L. Post (1971). Binding of ATP to Na^+ K^+ ATPase. *J. Biol. Chem., 246,* 5234–5240.
Karlsson, K., B. E. Samuelsson, and G. O. Steen (1971). Lipid pattern and Na^+ K^+ ATPase activity in the salt gland of the duck before and after adaptation to hypertonic saline. *J. Membrane Biol., 5,* 169–184.
Kempner, E. S., and J. H. Miller (1968). The molecular biology of *Euglena gracilis.* V. Enzyme localization. *Exp. Cell Res. 51,* 150–156.
Kushner, D. J. (1968). Halophilic bacteria. In *Advances in Applied Microbiology, 10,* 73–99.
MacLeod, R. A. (1965). The question of the existence of specific marine bacteria. *Bacteriological Reviews, 29,* 9–23.
Peaker, M. (1971). Avian salt glands. *Phil. Trans. Roy. Soc. London B 262,* 289–300.
Pfeiler, E., and L. B. Kirschner (1972). Studies on gill ATPase of rainbow trout. *Biochim. Biophys. Acta 282,* 301–310.
Phillips, J. E., and J. Meredith (1969). Active Na^+ and Cl^- transport by anal papillae of a salt water mosquito larva. *Nature, 222,* 168–169.
Reistad, R. (1970). On the composition and nature of the bulk protein of extremely halophilic bacteria. *Arch. Mikrobiol., 71,* 353–360.
Ritch, R., and C. W. Philpott (1969). Repeating particles associated with an electrolyte-transport membrane. *Experimental Cell Research, 55,* 17–24.
Schoffeniels, E., and R. Gilles (1970). In *Chemical Zoology* (ed. M. Florkin and B. T. Scheer), Academic Press, New York, Vol. 5, pp. 255–285.
Stewart, D. J., and W. N. Holmes (1970). Relation between ribosomes and functional growth in the avian salt gland. *Amer. J. Physiol., 219,* 1819–1821.

THE DISPOSAL OF NITROGENOUS WASTES AND THE WATER-LAND TRANSITION IN VERTEBRATES

CHAPTER 5

BASIC PATTERNS OF AMMONIA DISPOSAL

The catabolism of nitrogen-containing compounds (amino acids, purines, pyrimidines, etc.) almost invariably leads to the production of ammonia, a highly soluble and potentially toxic waste product. The toxicity of ammonia stems from several effects. Firstly, ammonia is quite strongly basic; the K_a for the reaction

$$NH_3 + H^+ \rightleftharpoons NH_4^+$$

is 5.5×10^{-10} at 25°C. At physiological pH's (pH 7.0 to 7.5), about 99% of the ammonia exists in the protonated form as NH_4^+. The release of excess ammonia in the unprotonated form therefore can markedly raise the pH of the cell fluids. Secondly, ammonia at high concentrations "drives" the glutamate dehydrogenase reaction

$$\alpha\text{-ketoglutarate} + NADH + NH_4^+ \rightarrow \text{glutamate} + NAD$$

in the direction of glutamate, thereby strongly inhibiting amino acid catabolism and cellular energy production. Lastly, ammonia is known to interfere strongly with a number of membrane functions, in particular with the active transport of monovalent cations. For these reasons, strict limitations are imposed on the quantity of ammonia which an organism can safely maintain in its cells and body fluids. In mammals, for example, NH_4^+ concentrations in blood and intracellular fluid are estimated in the 1 to 10 μM range.

The disposal of ammonia presents no major problem to aquatic organisms. In these forms ammonia readily diffuses into the surrounding medium at the respiratory surfaces. However, in terrestrial organisms, whose respiratory surfaces are not bathed with large volumes of water, ammonia disposal is a much more difficult task. As we shall discuss in this and the following chapter, the most important "solution" to the ammonia processing difficulties faced by terrestrial organisms is the conversion of ammonia into relatively non-toxic compounds which can be held within the body with relatively little danger until excretion is possible.

The particular detoxified form of ammonia which is excreted varies quite widely among different invertebrate and vertebrate groups. Many terrestrial invertebrates, as well as birds and reptiles, excrete uric acid (p. 165). Many other species, most notably mammalian forms, produce urea. The choice of excretory product will be seen to have a high dependence on the overall ecology of the species. In particular, the availability of water appears to play a key role in determining which pathway of nitrogenous waste disposal is utilized. Changes in water availability may reshape the pattern of ammonia disposal during the lifetime of the individual as well as over the long spans of evolutionary history of a species. In this chapter we will pursue the manners in which different vertebrate groups have adapted their processes of nitrogenous waste disposal vis-à-vis their particular environmental conditions.

BIOCHEMICAL PATHWAYS OF AMMONIA AND UREA METABOLISM

The basic enzymic pathways of ammonia and urea metabolism in vertebrates are shown in Figure 5–1. The sequence of reactions which lead to urea formation (termed the urea cycle) was first elucidated by Krebs and Henseleit in 1932. In vertebrates these reactions are localized primarily in the liver.

It will be noted that the reactions on the left-hand side of the urea cycle are essentially the steps leading to arginine biosynthesis. As such these reactions are almost universal, occurring in all organisms capable of synthesizing arginine, and, in addition, must be regarded as the primitive foundation on which the urea cycle was later erected. The arginase reaction illustrated on the right-hand side of the "cycle" is found in highest activities in ureotelic (urea excreting) species. In this important reaction, arginine is irreversibly hydrolyzed to urea and ornithine. The generation of the latter compound effectively completes the cycle, and allows the condensation of an additional molecule of carbamyl phosphate to yield citrulline. The urea thus formed is released into the blood and is carried to the kidneys, where it is voided in the urine.

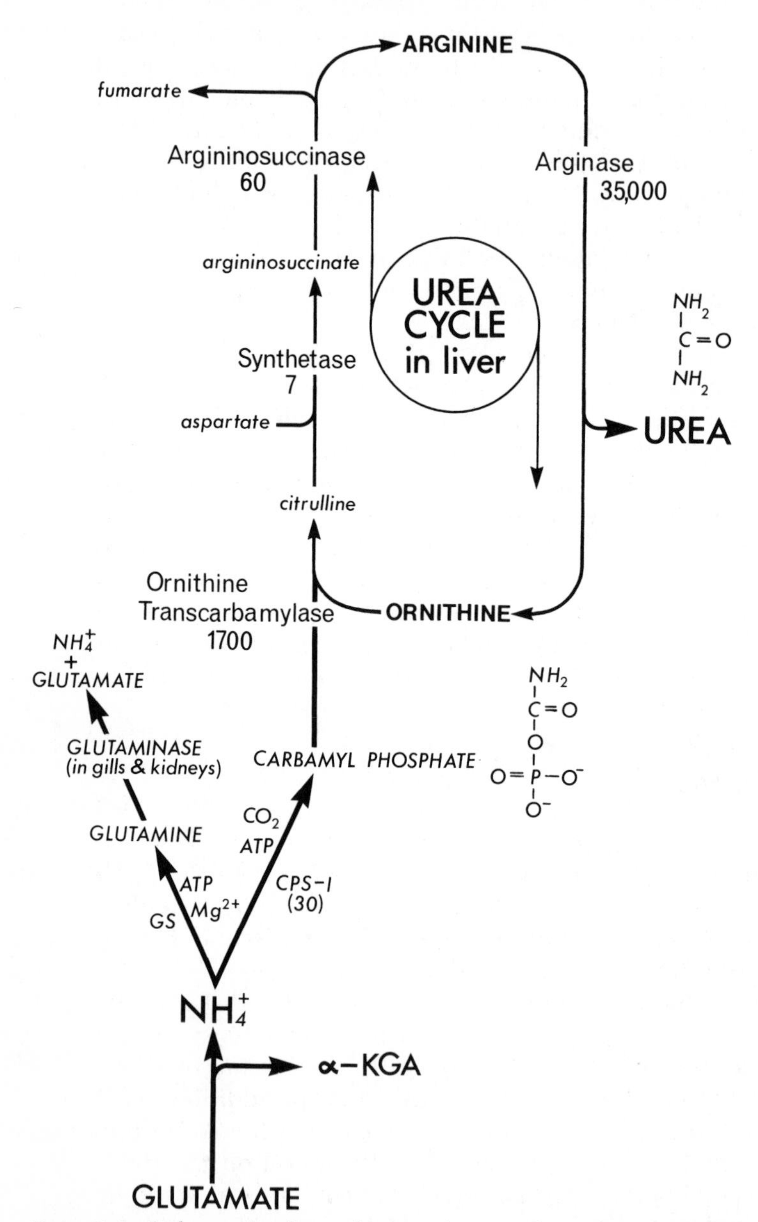

Figure 5-1 Pathways for the production of urea and ammonia in vertebrate liver. Representative activities (in μmoles product/gm tissue/hour), shown for the enzymes of urea synthesis in liver of the African lungfish, *Protopterus*, are taken from Campbell (1972).

The reactions leading to ammonia production and excretion are illustrated at the bottom of Figure 5–1. The reactions which draw metabolically formed ammonia *towards* glutamine synthesis can be regarded as a pathway which *branches away from* arginine synthesis. The glutamine formed in the glutamine synthetase reaction is the major *transport and storage* form of ammonia in ureotelic and ammonotelic (NH_4^+ excreting) animals, and other organisms as well. The glutamine is carried to the excretory sites (the gills in most aquatic vertebrates; the kidneys in mammals and some Amphibia) where it is hydrolyzed by glutaminase to yield glutamate and NH_4^+. The latter is then released to the outside. In many aquatic vertebrates, blood NH_4^+ is itself a major source of the ammonium released at the gills.

ORIGIN AND DISTRIBUTION OF THE UREA CYCLE ENZYMES

The urea cycle undoubtedly arose at an early stage of metazoan evolution. In some terrestrial flatworms, urea is the selected mode of nitrogenous waste excretion, and this may be an adaptation to reduced water supply, as in the case of vertebrates. Among the vertebrates, distinctly different patterns of ammonia disposal are noted in the teleosts (bony fishes) and the cartilaginous elasmobranchs. In the latter, large quantities of urea are synthesized and maintained in the body fluids for osmoregulatory purposes. Teleosts, in contrast, are almost entirely ammonotelic.

Interestingly, the teleost ancestors of terrestrial vertebrates quite likely were ureotelic. The closest living relatives of the ancestors of terrestrial vertebrates (the South American and African lungfishes, and *Latimeria,* the coelocanth) all display a ureotelic mode of ammonia disposal. It therefore seems reasonable to assume that the immediate ancestors of the primordial amphibians possessed the enzymes of the urea cycle before the terrestrial mode of life was adopted.

Following the migration to land, evolutionary divergence took place in the ways of processing nitrogenous wastes in different vertebrate groups. Although the stem reptiles arose with ureotelic habit, most present day reptiles (and birds) display uricotelism (uric acid excretion). However, both marsupial and placental mammals have retained ureotelism.

In all cases, the particular form of nitrogenous waste disposal characteristic of an organism represents a significant biochemical adaptation to the environment of the organism, particularly with regard to the availability of large supplies of water for flooding away excretory products. Perhaps the most vivid illustration of this pattern of evolutionary selection is found in the cases of the lungfishes and certain amphibians which are characterized by both aquatic and terrestrial life stages. In both groups the transition from water to land is accompanied by a shift-over from ammonotelism to ureotelism. Thus, we find an especially dramatic example of one of the major strategies of biochemical adaptation to the environment, namely the production of *new kinds* of enzymes which better supply the requirements for survival in—and, indeed, exploitation of—a new environment situation.

THE WATER-LAND TRANSITION IN VERTEBRATES: EVOLUTIONARY ASPECTS OF THE PROBLEM

The transition from a primarily aquatic to a primarily terrestrial mode of existence was one of the major evolutionary steps in the development of the vertebrates. This evolutionary "quantum jump" is thought to have occurred during the Devonian Era, approximately 300 million years ago. Climatic conditions at this time appear to have been highly unstable. In particular, tremendous fluctuations in rainfall led to times of extreme drought. Aquatic organisms encountering these conditions were thus under intense selective pressure to develop means for surviving under both wet and dry conditions. A primary problem was, of course, the need to avoid toxic increases in ammonia concentration when switching from an aquatic to a terrestrial mode of life.

THE PROCESSING OF NITROGENOUS WASTES IN PRESENT DAY LUNGFISHES AND AMPHIBIA

The patterns of evolutionary change in nitrogen metabolism which must have characterized the original water-to-land transition in the vertebrates can be gleaned from an examination of present day lungfishes and certain amphibian species which display a similar transition during their lifetimes.

Lungfishes are found primarily in the equatorial rain forests of Africa and South America, regions which are prone to extreme fluctuations in seasonal rainfall. In addition, the oxygen content of the swamp and river waters of these rain forests—never very great even during the wet, monsoon season—can fall precipitously. Under these conditions, it is not surprising that both the African and South American lungfish are obligate air breathers. During this time, they also depend upon both ammonia and urea as end products of nitrogen metabolism (Figure 5–2). The bulk of ammonia is probably released at the gills as NH_4^+, which may serve as a

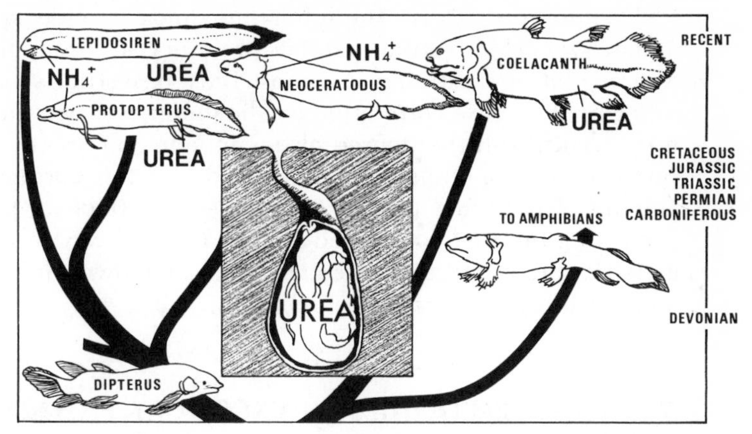

Figure 5–2 Phylogenetic relationships between the living lungfishes and the early amphibians. Two nitrogenous waste products are normally voided —NH_4^+ at the gills and urea at the kidneys. During estivation (shown in the center) the lungfish produces only urea.

counter-ion to Na^+ which is actively taken up from the outside water. The bulk of urea is released at the kidneys.

Much of the habitat in which these lungfish normally live is subject to annual drought, sometimes so severe as to reduce a swamp into a mud flat. Caught under such circumstances, the lungfish burrows into the mud, curls up, secretes a kind of cocoon, drops its metabolism by many fold, and then goes into an extended period of estivation. During this time, NH_4^+ excretion ceases; urea becomes the sole end product of nitrogen metabolism. The lungfish can remain in estivation for years, but usually needs to do so only for months at a time. Upon the return of the rainy season, the organism terminates estivation by reversing the above processes.

In contrast to the African and South American lungfishes, the Australian lungfish cannot estivate. This organism remains ammonotelic throughout its life.

Ecological problems very similar to those faced by the lungfishes are encountered by some amphibians. The South African toad *Xenopus laevis* is normally a fully aquatic amphibian in both larval and adult life stages. As such, one would predict that its nitrogenous wastes would be eliminated largely as ammonia. This, for the most part, is the case. However, *Xenopus* may at times face severe periods of drought, and during these times it is known to estivate in a manner reminiscent of lungfishes. As in the case of the estivating lungfish, *Xenopus* assumes a ureotelic pattern of nitrogen metabolism. Interestingly, even the larval stage of *Xenopus* has the capacity for switching to a ureotelic habit if it is removed from water.

Within the Class Amphibia one finds a wide spectrum of aquatic-versus-terrestrial habitat preferences. From the aquatic *Xenopus* at one extreme to almost fully terrestrial species at the other limit, a close correlation can be found between the degree of terrestriality and the degree of dependence on ureotelism, as opposed to ammonotelism. The best index of this correlation is the activity of carbamyl phosphate synthetase-I (CPS-I), the enzyme catalyzing the first committed step in the pathway to urea: CPS-I activities are highest in terrestrial forms and lowest in larval stages and aquatic species (Figure 5–3).

STRATEGIC SIMILARITIES IN EXCRETORY PATTERNS

The fact that the ammonotelic → ureotelic transition occurs in such varied organisms as the lungfishes and *Xenopus* is instructive on two counts. First, it indicates that this adaptive mechanism is a phylogenetically old one and a very satisfactory "solution" to the problem of ammonia disposal under conditions of limited water supply. Second, it shows that identical biochemical "solutions" can be utilized when the problem is a short-term one confronting the organism during its lifetime and requiring adjustments in the phenotypic expression of its genome, or when it is a long-term problem which is "solved" only over many generations by means of fundamental changes in the genetic apparatus of the organism. What are the molecular mechanisms underlying the ammonotelic-ureotelic transition?

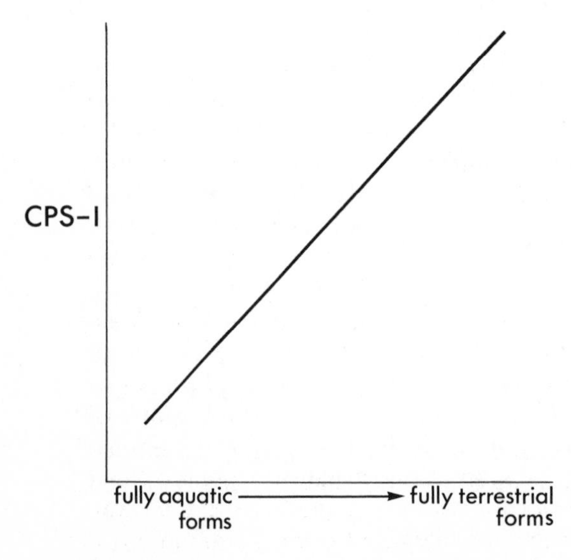

Figure 5–3 Relationship between the amount of CPS-I and the degree of terrestriality of the vertebrate organism.

THE MECHANISMS OF SWITCH-OVER FROM AMMONOTELISM TO UREOTELISM IN AMPHIBIANS AND LUNGFISHES

The basic events involved in the switch-over from ammonia to urea excretion during the water-land transition appear to be fundamentally similar in the various groups—whether the transition is associated with larval metamorphosis in frogs and toads, with a drought encounter on the part of adult *Xenopus*, or with estivation in the lungfish. Specific details may vary from species to species, and some of the generalizations made may not hold in detail for all. Nevertheless, enough information is now available to suggest that the following four basic events probably are involved, at least in part, in the switch-over from NH_4^+ to urea production wherever that switch-over occurs in the vertebrates; predictably, these fall into a distinct hierarchical order:

(1) The initial event appears to be an activation of transcriptional machinery. Although specific signals at the moment are not clear, hormonal control, particularly by the thyroid, is definitely involved, at least in *Rana catesbeiana.* In *Xenopus,* reduced water availability is an important environmental signal, since the same changes in nitrogen metabolism are induced by dehydration and by hypertonicity of the medium. Recent studies by Janssens indicate that increased levels of NH_4^+ in tissues and blood constitute an important metabolic signal for activating ureotelism (Figure 5–4). During early metamorphosis, RNA polymerase activities increase dramatically, followed by the production of various m-RNA species. These presumably are involved in specifying the host of proteins which are being synthesized at this time. Included among these are the urea cycle enzymes.

(2) Large increases in the activities of the urea cycle enzymes constitute the second major event involved in the switch-over (Figure 5–5). These are undoubtedly *de novo* syntheses in all cases, although unequivocal proof of this contention is available for only some species and some of the enzymes. Nevertheless, the time periods involved in the ammonia → urea transition in both the lungfish and amphibian species are easily adequate to allow for changes in enzyme concentrations by alterations in *de novo* synthesis rates. The increase in activities of the urea cycle enzymes at this time is impressively large. A 10- to 20-fold increase, for example, is found in the case of a species of *Rana.* A 6-fold increase in CPS-I activity and a 10- to 20-fold increase in urea levels are observed in *Xenopus* during estivation. A 5- to 10-fold increase in aspartate transaminase activities occurs in estivating lungfish. Estimates of *de novo* synthesis rates of CPS indicate comparable increases during metamorphosis in *Rana.* A number of observations suggest that the production of the urea cycle enzymes concomitant with development of

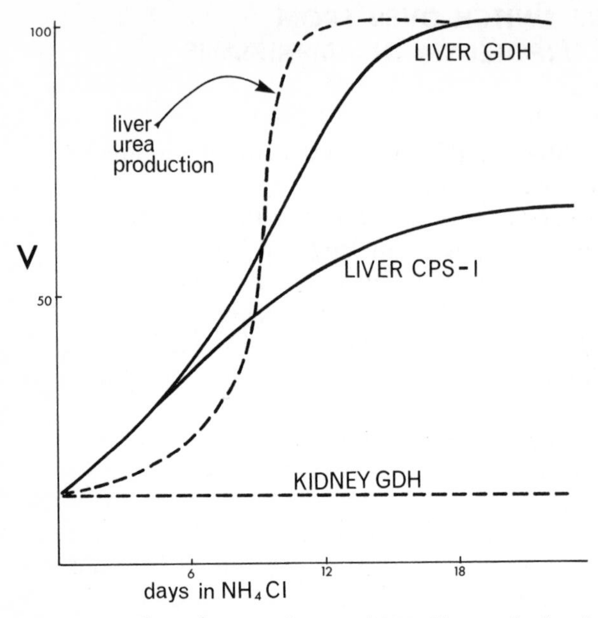

Figure 5–4 The influence of external NH_4Cl upon the levels of glutamate dehydrogenase (GDH), carbamyl phosphate synthetase (CPS-I), and urea biosynthesis in *Xenopus*. Data summarized from Janssens (1972).

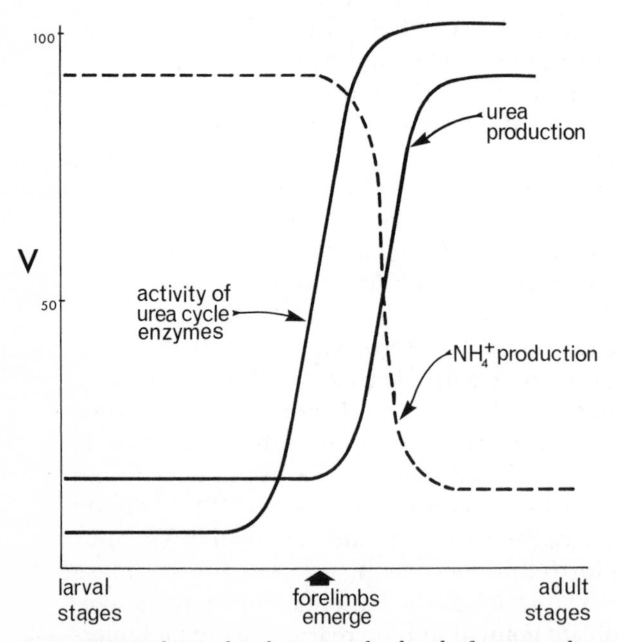

Figure 5–5 Relationship between the level of urea cycle enzymes, urea excretion, and NH_4^+ excretion during the water-land transition occurring in amphibian development.

ureotelism is a single, coordinate event. Thus, (a) the time of RNA polymerase activation, (b) the time of initiation of enzyme synthesis, (c) the time course of synthesis, (d) the approximate percentage increase in the activities of the various enzymes, and (e) the derepression signals for these events (thyroid, NH_4^+ levels, and water availability) are all identical or remarkably similar. *Such behavior would be predicted if all the component enzymes were specified by a closely integrated section of the genome.*

(3) Despite the similarity in response pattern of the enzymes required for urea biosynthesis, there are a number of findings which suggest that a finer level of control is involved than mere synthesis of the enzymes. The best available data concern CPS-I. The biosynthesis of carbamyl phosphate synthetase appears to occur in at least a 2-step process: (a) the initial process appears to be the usual m-RNA dependent synthesis at the ribosomal site. This process is sensitive to the usual inhibitors of both transcription and translation. A second step (b) is an epigenetic one and is thought to involve transfer of inactive precursor "subunits" into the mitochondria, followed by the assembly of these subunits into the active oligomeric enzyme. The second step of this process also may be activated by thyroid hormone. Comparable assembly processes have been described in somewhat lesser detail for glutamate dehydrogenase and may also account for arginase differences in premetamorphic vs. postmetamorphic amphibians. It is indeed probable that some epigenetic control is put upon the assembly of all the other oligomeric enzyme components of the urea cycle. Since epigenetic control of each of the urea cycle enzymes may differ, one would anticipate the observed differences in the percentage increase in activities of different urea cycle enzymes during ureotelic adaptation.

(4) Following the activation of RNA polymerases, translation of the m-RNA into the initial protein subunit products, and the epigenetic assembly of the enzyme subunits into holoenzymes each positioned correctly within the cell, one final level of control is available to the organism: *regulation of the catalytic activities of the new battery of enzymes produced.* As we stressed initially, the pathway to ammonia production quite clearly *branches away* from the pathway leading to urea synthesis. The two pathways compete for a common substrate, NH_4^+, in a most immediate manner, and for glutamate, in a secondary manner. The channelling of glutamate nitrogen between these two pathways is undoubtedly a carefully controlled process. Details of that control are only currently being worked out. However, sufficient information is available at this time to warrant a discussion of the properties of CPS-I, GDH, and glutamine synthetase—three enzymes positioned so strategically at this point in metabolism that control of their function is a most essential component of control of ureotelic function.

THE CHANNELLING OF NITROGEN TOWARDS GLUTAMINE OR TOWARDS UREA

From the discussion thus far, it is evident that in organisms which can produce either ammonia or urea as the ultimate form of nitrogenous waste, a reciprocal relationship is obtained in the relative participation of the urea and the ammonia producing pathways. Consequently, when both CPS-I and glutamine synthetase (GS) are present in the same cell, control circuitry should favor *either* glutamine synthesis *or* carbamyl phosphate synthesis, *but not both simultaneously.* The central role at this branchpoint is played by N-acetylglutamate.

N-ACETYLGLUTAMATE CONTROL OF CPS-I

It is well established that CPS-I has an absolute requirement for the activator, N-acetylglutamate. The latter is formed by acetylation of glutamate as shown in Figure 5–6. Recent studies by Tatibana indicate that N-acetylglutamate levels depend strictly upon the concentration of glutamate (Figure 5–6), providing there is a

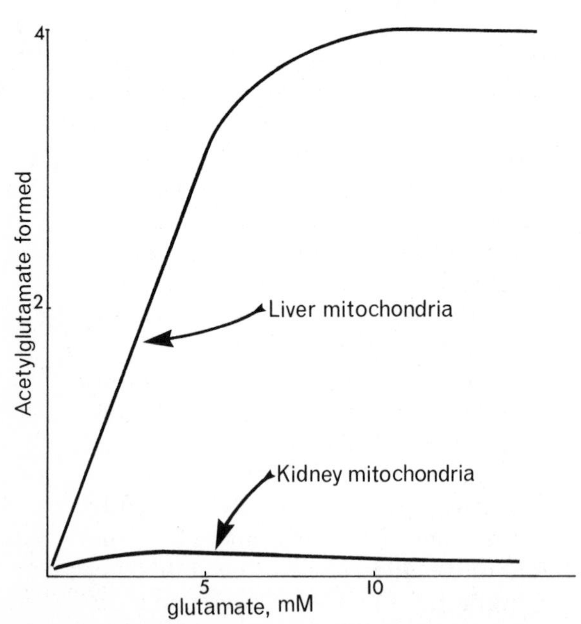

Figure 5–6 Synthesis of N-acetylglutamate from glutamate by liver and kidney mitochondria incubated in the presence of acetylCoA. Data summarized from Shigesada and Tatibana (1971).

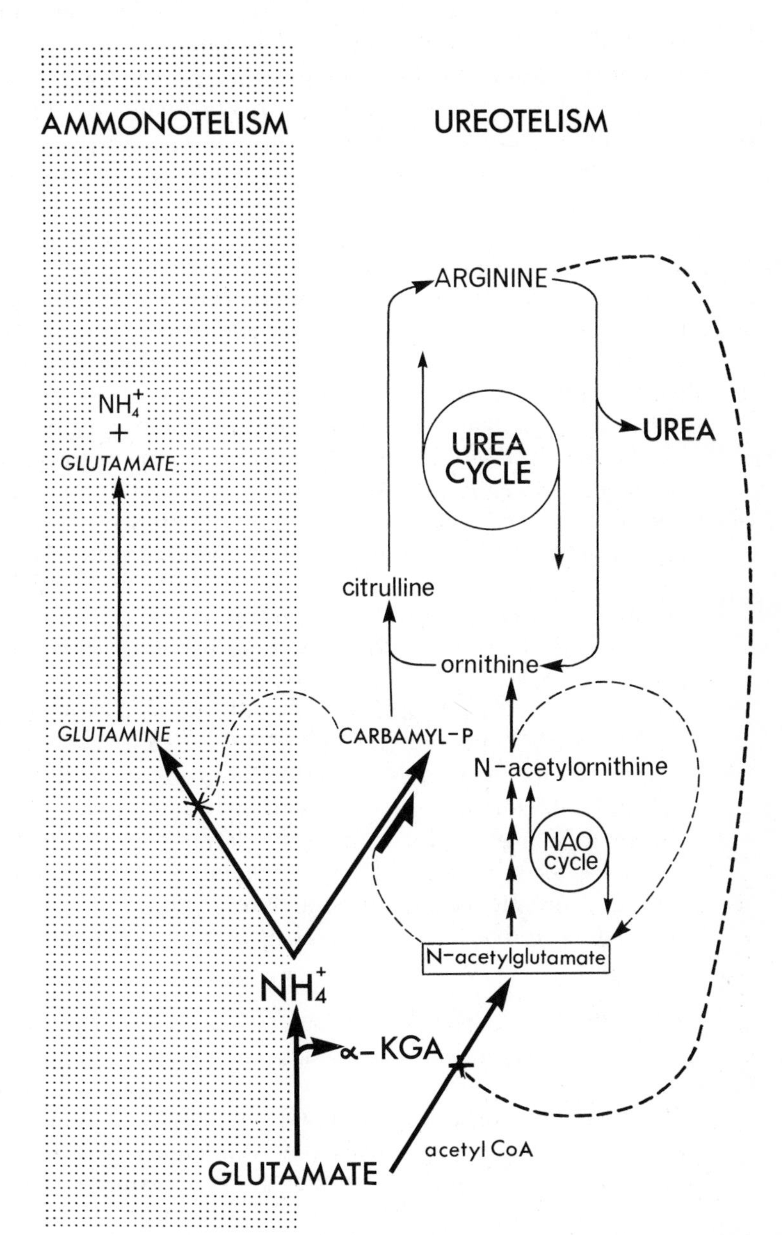

Figure 5–7 Control circuitry involved in the switch-over from ammonotelism to ureotelism during water-land transition in Amphibia or in the lungfishes. Activations indicated by thick arrows; inhibitions, by thick crosses.

source of acetylCoA. The primary functions of N-acetylglutamate cannot be completely specified, but these are known to involve (1) conversion of the enzyme CPS-I to an active conformation, and (2) facilitation of substrate (ATP) binding to the enzyme. Hence, whenever glutamate accumulates, the production of acetylglutamate strongly *facilitates flow of glutamate nitrogen towards carbamyl phosphate* (Figure 5–7).

Under these conditions, a concomitant regulatory interaction can be initiated at the glutamine synthetase (GS) site: as the concentration of carbamyl phosphate rises, *it potently and specifically inhibits glutamine synthetase.* This regulatory effect has the consequence of "sparing" *both glutamate and NH_4^+* (both substrates of the GS reaction) for the CPS pathway. *Such a regulatory process is always autocatalytic*: the carbamyl phosphate inhibition of glutamine synthetase leads to a greater supply of NH_4^+ substrate for carbamyl phosphate synthetase plus a greater supply of glutamate for N-acetylglutamate formation. The latter in turn further accelerates CPS function. As more CP is formed, the process is again potentiated. The pivotal role of the CPS step in regulation of the direction of NH_4^+ metabolism at this point is indicated also by the close correlation observed between its activity and the degree of ureotelism developed in these various species (Figure 5–3).

Parenthetically, it should be noted that another form of CPS, termed CPS-II, occurs in tissues such as mammalian spleen. It functions only in pyrimidine biosynthesis, since these tissues are unable to synthesize urea. Interestingly, CPS-II is not activated by N-acetylglutamate. Such an activation appears to have been specifically selected for the controlled channelling of glutamate nitrogen into the urea cycle, as in amphibian or mammalian liver.

Once formed, the subsequent metabolic fate of N-acetylglutamate in ureotelic organisms is not known with certainty. One possibility suggested by Tatibana and by Campbell is that it is a precursor for ornithine (Figure 5–7):

N-acetylglutamate → N-acetyl-γ-glutamyl-P →
acetylglutamic γ-semialdehyde → N-acetylornithine

The N-acetylornithine is thought to donate its acetyl group to glutamate to yield ornithine plus N-acetylglutamate. This is an attractive postulate, for it supplies a powerful mechanism for "sparking" the urea cycle *by increasing the availability of ornithine at the same time as the availability of carbamyl-P is increasing.* The reaction span is cyclic and catalytic and is often termed the N-acetylornithine cycle (the NAO cycle in Figure 5–6); at steady state it constitutes no net drain of N-acetylglutamate. What is more, a single metabolic signal—acetylglutamate—serves to activate both CPS-I and ornithine synthesis. Although this mechanism has not been completely established, an ornithine "sparking" of the urea cycle can in fact be demonstrated. It is a well established observation that an ornithine load increases urea synthesis and thereby reduces sensitivities to injections of NH_4^+. Similar "protection" against ammonium toxicity is achieved by injecting other urea cycle intermediates such as arginine; *the arginine, cleaved by arginase to urea and ornithine, in this context becomes metabolically equivalent to an ornithine load.* The increased urea synthesis observed following arginine injection therefore appears to be mediated by an ornithine "spark." Should arginase activities be greatly reduced, however, it is evident that such a process could not occur; under these conditions, arginine probably takes on an inhibitory function.

"TURNING OFF" THE UREA CYCLE

The postulated regulatory circuitry at this level in metabolism may itself be under typical negative feedback control (Figure 5–7), for arginine is known specifically to inhibit N-acetylglutamate synthetase, the enzyme catalyzing the acetylation of glutamate. Under normal conditions, of course, arginine levels in the liver are very low because of the exceedingly rapid breakdown to urea and ornithine catalyzed by arginase. In *Xenopus,* the specific activity of arginase is over 1000 times greater than that of CPS-I, and its relative activity is similarly high in other ureotelic species (Figure 5–1). Hence, an accumulation of arginine such as would be required for an arginine feedback loop (Figure 5–7) would be difficult to realize under normal circumstances. Such a feedback inhibition would be physiologically possible and advantageous, however, whenever arginase concentrations were falling owing to reduced rates of arginase synthesis, a situation to be expected only during reversion from ureotelism to ammonotelism (for example, at the termination of estivation in lungfish or in *Xenopus*). Arginine inhibition of N-acetylglutamate synthetase at this time would lead to *dramatic inhibition of CPS-I and hence to an efficient deinhibition of glutamine synthetase.* Under such circumstances, the organism would again be free to synthesize large quantities of glutamine, the compound which carries the amino group "in storage" or releases it as NH_4^+.

REGULATORY FUNCTIONS OF GLUTAMATE DEHYDROGENASE

Workers in this field have stressed the importance of GDH in the control of nitrogen flow through this point in metabolism, since this enzyme is of the allosteric type and is closely regulated by the concentration of the adenylates, GTP, and NAD(H). However, the GDH reaction does not appear to be an efficient site for control of nitrogen metabolism at this point. In the first place, GDH is badly positioned, for it occurs *before* the primary branching point between the two pathways (Figure 5–7). Because of its location, inhibition of GDH would block flow of nitrogen into *both* pathways. Control at this site, therefore, would lack the *specificity* required to account for the observed reciprocal relationship between the pathways leading to glutamine and to urea. This lack of adequate specificity could be circumvented by the elaboration of isozymic forms of GDH, each expressing different regulatory properties. However, the best available evidence suggests that liver GDH occurs as a single electrophoretic species. For these reasons alone, then, we believe that GDH can be ruled out as an important control site for the *channelling of nitrogen* towards ammonia or towards urea. There are other reasons for this conclusion as well; these indicate that GDH is an important control site, but that its major regulatory function is in the *channelling of carbon* into the Krebs cycle. The nub of the argument is presented below:

(a) One of the key functions of GDH is the collection (in the α-amino group of glutamate) of amino groups from various amino acids by the action of transaminases. The combination of the action of these transaminases with that of GDH has the same effect *as if there were individual dehydrogenases for each of the other amino acids*. Consider, for example, the fate of accumulating alanine in the presence of glutamate-alanine transaminase and GDH (Figure 5–8). Alanine reacts with α-ketoglutarate to form glutamate and pyruvate. The resultant glutamate is oxidized by NAD, liberating NH_4^+ and regenerating α-ketoglutarate. *The net effect is the oxidation of alanine by NAD to form pyruvate, NH_4^+, and NADH.*

This elegantly simple system has an obvious advantage. Since the equilibrium constants of the transamination reactions are near unity, their rates will depend upon the rate of removal of products of the reaction. One of these products is usually glutamate, whose further metabolism is tightly controlled by the operation of a tightly controlled GDH. Thus, this system allows the organism to localize control for amino acid catabolism at a single primary locus, which in mammalian liver has a structural counterpart: a multienzyme complex expressing both transaminase and GDH activities.

(b) A second key function of the GDH reaction is the delivery of α-ketoglutarate to the Krebs cycle for further energy metabolism during periods of amino acid catabolism. It is this function in the energy metabolism of the cell which would be best regulated by the energy status of the cell, and this indeed appears to be the case. Although the activity of GDH is altered by a variety of compounds,

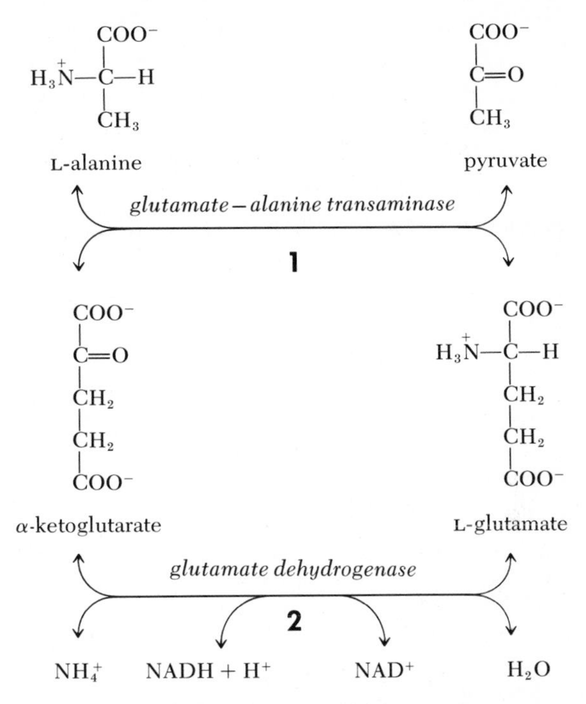

Figure 5–8 The concerted action of glutamate-alanine transaminase and glutamate dehydrogenase achieves the same result as the action of a hypothetical alanine dehydrogenase.

1. The amino group of alanine is transferred to α-ketoglutarate, forming pyruvate and glutamate.
2. The resultant glutamate is oxidized by NAD, with the nitrogen originally in alanine now appearing as NH_4^+. The α-ketoglutarate consumed in the first step is now regenerated and there is no net change in the concentration of this compound. The overall reaction involves only the disappearance of alanine and NAD, with the appearance of pyruvate, NADH, and NH_4^+.

the critical effect appears to be a potent non-competitive inhibition by GTP. This presumably is best positioned to adjust the supply of amino acids to the demands for amino acids in the cell's energy metabolism, for a part of the energy captured during operation of the Krebs cycle appears as GTP during conversion of succinylSCoA to succinate (Figure 5–9). Under energy saturating conditions, GTP will accumulate and lead to an inhibition of its own further production by inhibiting the GDH reaction. Since ATP and NADH concentrations also are high during energy saturating conditions, their concerted inhibition of GDH would simply potentiate the GTP effect on the enzyme. This response of GDH to the "energy charge" of the cell is classical and entirely comparable to many other enzymes which are intimately involved in energy metabolism (see p. 14). Thus in this case, as in many others, a lower energy charge (indicating increasing concentrations of ADP and AMP) leads to GDH activation; in this case, the primary positive modulator is ADP.

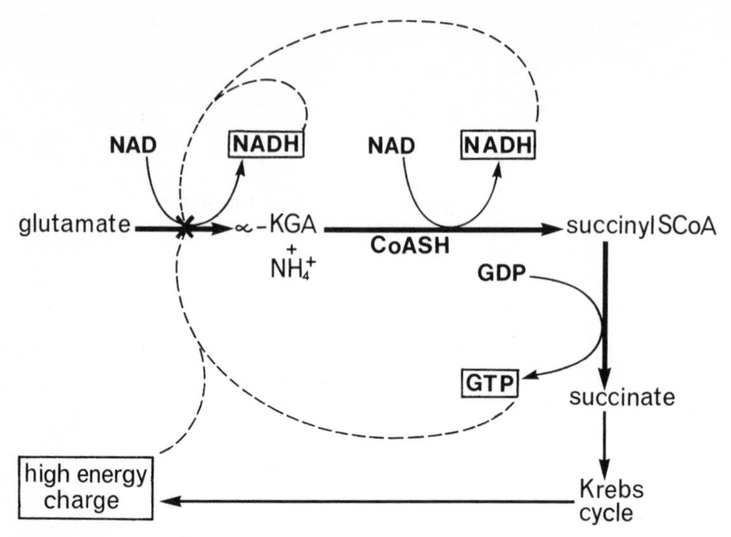

Figure 5–9 Feedback inhibition of glutamate dehydrogenase (GDH) by NADH and by high energy charge.

All of the above regulatory properties of the GDH reaction strongly implicate this enzyme in *energy* metabolism. As we have pointed out, the enzyme is poorly positioned (and lacks adequate specificity) for an important regulatory role in *nitrogen* metabolism. Yet there must be some link between these two roles; otherwise, one can imagine the energy saturated situation where GDH is held in an inactive state by the high energy charge, but its function is required for disposal of nitrogen owing to accumulation of amino acids. In fact, the *accumulation of amino acids is the regulatory link between energy and nitrogen metabolism at this point* (Figure 5–10). Several amino acids, of which leucine is a prominent example, activate the enzyme by binding at sites remote from the nucleotide binding site. The effect of leucine overrides the inhibiting effect of the adenylates and assures increased degradation of amino acids by disposal of their nitrogen whenever they begin to accumulate. This will occur even if the total demand for high energy phosphate (i.e., the total demand for amino acid catabolism) is not very great at the time of amino acid accumulation.

The build-up of the amino acid pool may have two additionally important regulatory effects. First, increased availability of substrate for various transaminases will assure the cell of a continued supply of glutamate. And secondly, the requirement for channelling nitrogen towards the urea cycle is met by alanine and serine inhibition of glutamine synthetase (Figure 5–10).

A final important function of GDH is the maintenance of glutamate reserves. Recently, Janssens has shown that *Xenopus* exposed to 5 mM NH_4^+ responds by increasing steady state levels of GDH and CPS concurrent with a 10-fold increase in urea production (Figure 5–4). In terms of *nitrogen* metabolism, the primary function of the increased GDH activity is to increase the glutamate concentrations and hence the amount of N-acetylglutamate available for CPS-I activation. The student will recognize these events as initial components of the cascade control system shown in Figure 5–7, which lead to an exponential increase in the participation of the urea cycle in nitrogen metabolism.

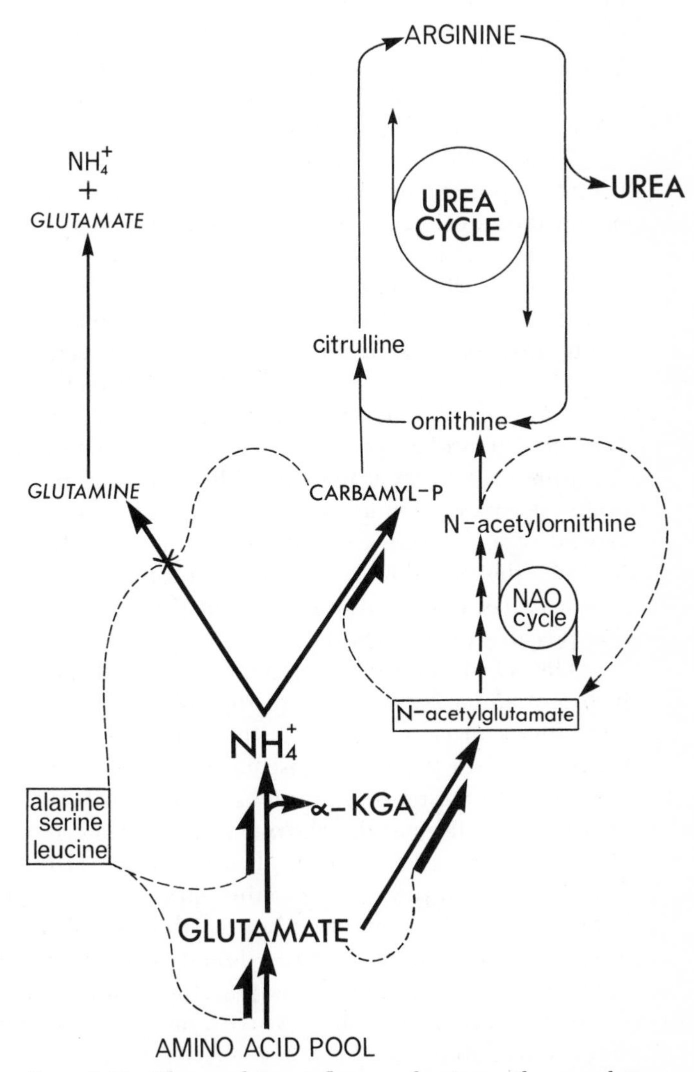

Figure 5–10 The regulatory influence of amino acid accumulation on nitrogen excretory metabolism in vertebrate liver. Activations shown by thick arrows; inhibitions, by thick crosses.

ROLE OF UREA IN OSMOREGULATION IN AMPHIBIANS AND FISHES

Earlier in this chapter, mention was made of the observation that a reduction in available water supplies was an important environmental signal for the transition from ammonotelism to ureotelism. In most amphibians and all the lungfishes, reduction in available water is always associated with exposure to the terrestrial environment. However, there are various species of amphibians that encounter a similar problem of restricted availability of osmotically free water during their excursions into, and exploitation of, saline environments. Included in this latter category are the toad *Bufo viridis* and the crab-eating frog, *Rana cancrivora,* which inhabit mangrove swamps. Both species can withstand salinities as high as 75% of full strength sea water.

In all species examined, excursion into a saline environment initiates an activation of the urea cycle and increased participation of the urea cycle in nitrogen metabolism. However, instead of excreting the urea at the kidneys (which is the usual case in terrestrial amphibians), the marine forms accumulate the urea in their blood and tissues where it contributes to maintenance of the higher osmotic pressure required for survival in salt water.

Mechanisms involved in accumulating the high blood and tissue concentrations of urea are not completely described. Aside from the greater dependence upon the ureotelic habit, salt water amphibians reduce glomerular filtration rate and increase tubular reabsorption of water; presumably, the active secretion of urea typical of terrestrial ureotelic amphibians is "turned off." Indeed, the idea that the *direction* of active urea transport system might be *reversed* has been entertained, but thus far definitive measure of this possibility is unavailable.

The retention of large amounts of urea has necessitated changes in the organism's tolerance to the substance. Whereas at low concentrations urea is not a highly toxic compound, at the 200 mM range found in these various marine amphibians, muscle contraction in the common *Rana pipiens* is 50% inhibited. Muscle contraction in the crab-eating marine frog, *R. cancrivora,* of course, is entirely refractory to such high urea levels. High urea concentrations are known to have profound effects on secondary and tertiary levels of protein structure; hence, the development of enzymes with a high resistance to urea denaturation in marine amphibians appears to be an essential component of their invasion of a part of the marine environment. However, data dealing with this problem are not available.

The classic literature example of the utilization of urea for osmoregulation is the marine elasmobranch. Indeed, it is widely if implicitly assumed that the primary function of the urea cycle in the elasmobranchs is osmoregulatory. The utilization of an end product of nitrogen catabolism rather than amino acids themselves for osmotic purposes (as occurs in marine invertebrates) has an energetic advantage in that the energy derived from carbon catabolism of the amino acids is not lost.

As part of their adaptation to their mode of life, marine elasmobranchs have developed an impressive tolerance for high concentrations of urea in their blood and tissues. Concentrations of urea in these forms routinely surpass 0.5 M levels. Although the problem of tolerance mechanisms to the high urea levels is recognized, it has not been systematically examined.

A number of cartilaginous fishes have successfully survived being geologically trapped in fresh bodies of water. Included in this category are the freshwater sharks of Lake Nicaragua and the freshwater skates of the Amazon. These freshwater elasmobranchs, unlike their marine relatives, apparently are ammonotelic, and maintain plasma urea levels at less than 1/300 of those typically found in marine forms. These low urea levels, clearly of adaptive significance in freshwater elasmobranchs, are achieved by (1) a low rate of urea synthesis, reflected in the low activities of urea cycle enzymes, and (2) an increased rate of urea excretion at the kidneys. Aside from these few observations, little is known of nitrogen metabolism or of its control in these "special case" elasmobranchs.

SUGGESTED READING

Books and Proceedings

Comparative Biochemistry of Nitrogen Metabolism. 2. The Vertebrates (1970). (Ed. J. W. Campbell) Academic Press, New York, pp. 495–916.

Reviews and Articles

Balinsky, J. B. (1970). Nitrogen metabolism in amphibians. In *Comparative Biochemistry of Nitrogen Metabolism* (Ed. J. W. Campbell) Academic Press, N.Y., Vol. 2, pp. 519–637.

Campbell, J. W. (1972). Nitrogen Metabolism. In *Comparative Animal Physiology*, 3rd ed. (Ed. C. L. Prosser), W. B. Saunders Co., Philadelphia, pp. 279–316.

Fahien, L. A., J. H. Lin-Yu, S. E. Smith, and J. M. Happy (1971). Interactions between glutamate dehydrogenase, transaminases, and keto acids. *J. Biol. Chem. 246*, 7241–7249.

Janssens, P. A. (1972). The influence of NH_4^+ on the transition to ureotelism in *Xenopus laevis*. *J. Exp. Zool. 182*, 357–366.

Shigesada, K., and M. Tatibana (1971). Role of acetylglutamate in ureotelism. I. Occurrence and biosynthesis of acetylglutamate in mouse and rat tissues. *J. Biol. Chem. 246*, 5588–5595.

Tate, S. S., F. Leu, and A. Meister (1972). Rat liver glutamine synthetase. Preparation, properties, and mechanism of inhibition by carbamyl phosphate. *J. Biol. Chem. 247*, 5312–5321.

Tiemeier, D. C., and G. Milman (1972). Chinese hamster liver glutamine synthetase. Purification, physical and biochemical properties. *J. Biol. Chem. 247*, 2272–2277.

THE DISPOSAL OF NITROGENOUS WASTES AND THE WATER-LAND TRANSITION IN THE INVERTEBRATES

CHAPTER 6

INTRODUCTION

Much as different anatomical solutions to a common problem of adaptation are in evidence among different groups of organisms (e.g., the employment of endoskeletons by chordate species and exoskeletons by the arthropods), so may the biochemical strategies used in adapting to a given environmental parameter vary widely among different organisms. This is certainly true in the case of adaptation of excretory mechanisms to a limited water supply.

In vertebrate organisms, the assumption of terrestrial existence was made possible in large part by the availability of the urea cycle enzymes for the production of a relatively non-toxic nitrogenous waste product. However, among invertebrate species, the dependence on this mechanism of nitrogenous waste disposal is quite rare. Only a few terrestrial annelids and flatworms are known to be ureotelic. The vast majority of terrestrial invertebrates excrete nitrogenous wastes in the form of purines (Figure 6–1). Aquatic invertebrates, as one would expect, have retained the ammonotelic pattern of nitrogen excretion.

Of the purines used as excretory outlets for nitrogenous wastes, uric acid (Figures 6–1 and 6–2) is particularly important. It is the primary nitrogenous waste of insects, although certain degradation products of uric acid (allantoin, allantoic acid, urea, and glyoxylate) may also be excreted. The Myriapoda (the millipedes and the cen-

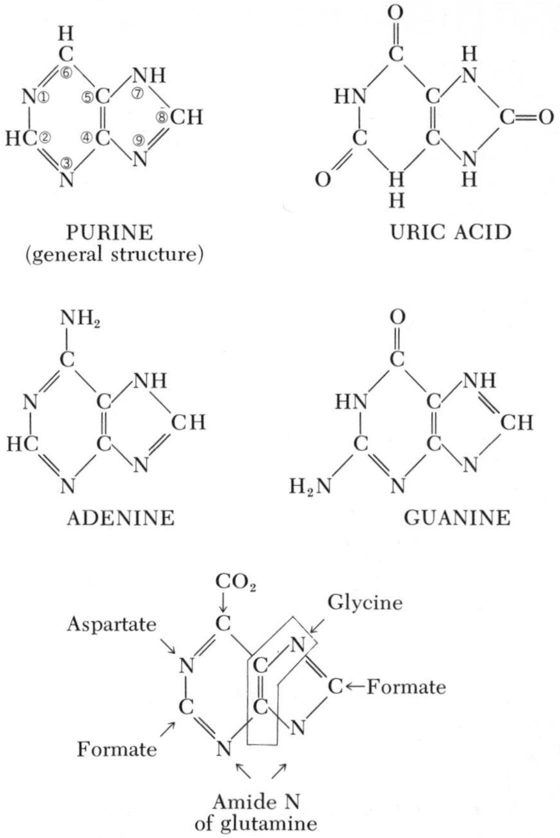

PURINE
(general structure)

URIC ACID

ADENINE

GUANINE

Aspartate
CO_2
Glycine
Formate
C←Formate
Amide N
of glutamine

Figure 6–1 General purine structure is shown at top, left. Specific structures for uric acid, adenine, and guanine are included in the top panel. The origin of the carbon and nitrogen atoms of the purine rings are shown in the lower panel.

tipedes) also excrete uric acid; recently uric acid production was also demonstrated in a terrestrial crustacean. Another purine, guanine (Figure 6–1), is the major excretory product of nitrogen metabolism in arachnids.

In addition to differing from vertebrates in the nitrogenous end product which is excreted, invertebrates also may display unique abilities to find other important physiological uses for so-called nitrogenous "waste" products prior to the latter's actual elimination from the body. This multi-functional aspect of nitrogenous waste excretion is well illustrated in the case of molluscs.

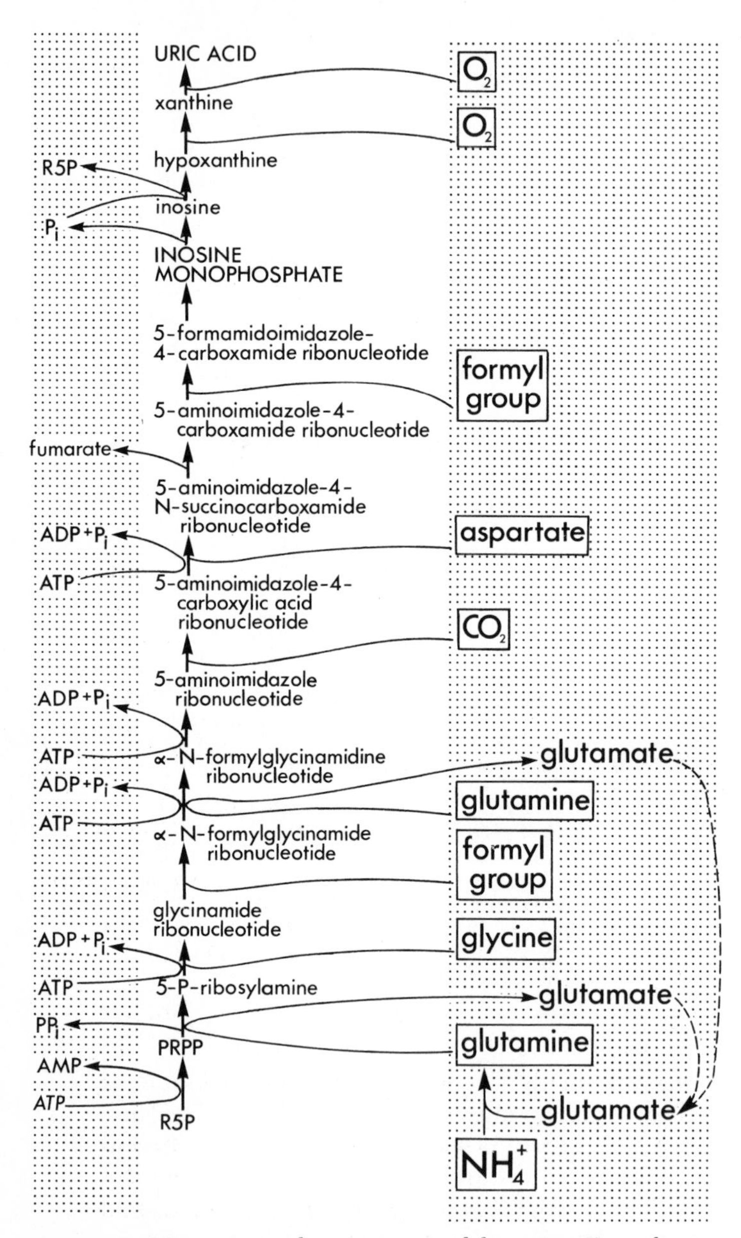

Figure 6–2 Major intermediates in uric acid formation. Key substrates contributing to the uric acid skeleton (Figure 6–1) are indicated on the right. Cosubstrates and products are shown on the left.

MULTIFUNCTIONAL ASPECTS OF NITROGEN EXCRETORY PATHWAYS IN THE MOLLUSCS

In Chapter 5, we colored our presentation of nitrogen excretory metabolism in the vertebrates with the impression that the chief function of each of the pathways of nitrogen metabolism is simply the production of the final, "waste" product for elimination to the outside. That is, the pathways are largely unifunctional. This simplest of possible situations also appears to occur in the cephalopod molluscs (octopus, squid, etc.). These are all marine forms, and in all of these ammonia constitutes ⅔ to ¾ of the excreted, waste nitrogen. In the octopus, the glutaminase hydrolysis of glutamine to yield glutamate and NH_4^+ is established as an important mechanism for release of this product at the gills. In all other aquatic forms of molluscs, both in fresh and in salt water, ammonia remains a significant form in which waste nitrogen is eliminated, although the percentage contribution varies in different species. But all these organisms can and do produce measurable quantities of uric acid; some can and do produce urea. Why do organisms living in a habitat favoring an ammonotelic mode of nitrogenous waste disposal continue to synthesize these more typically terrestrial waste products? Is the urea and/or uric acid produced by these molluscs fulfilling physiological functions other than waste disposal?

A clue to the functional significance of these different pathways of nitrogen metabolism arises from studies of terrestrial and semiterrestrial molluscs. In these organisms, there appear to be at least four distinct functions for the waste products of nitrogen metabolism:

1. Excretion. In terrestrial molluscs the major excretory product is uric acid. Under both normal and dehydrating conditions, uric acid accounts for the bulk of the nitrogen eliminated to the environment.

2. Shell Formation. Under normal physiological circumstances, i.e., in the absence of dehydration, urea is hydrolyzed and serves as a source of ammonia which, in turn, may play an important role in the control of shell formation.

3. Control of Blood Volume. Under the stress of dehydration, urea is no longer broken down to ammonia and CO_2, but instead accumulates in the blood and, by reducing the vapor pressure of the blood, aids in reducing evaporative water loss.

4. Control of Cell Volume. Again, under dehydrating conditions, uric acid is retained *within* the cells to balance the rising osmotic pressure of the blood.

Let us examine the biochemical mechanisms by which these functions of nitrogen metabolism are effected.

UREA CYCLE FUNCTION IN SHELL FORMATION OF LAND MOLLUSCS

Because of the variety of metabolic fates available to NH_4^+, the primary product of amino acid metabolism, we would expect and indeed find that the particular route through which nitrogen is channelled will depend upon the physiological state of the organism.

In the normally feeding and active gastropod, the percentage participation of the urea cycle is low. Under these conditions the urea cycle reactions function to produce urea, which in turn is quickly hydrolyzed by urease.

$$NH_4^+ + CO_2 \xrightarrow[\text{cycle}]{\text{urea}} \text{urea} \xrightarrow{\text{urease}} NH_3 + CO_2$$

Hence, urea concentrations in many species are also low. What is the function of this paradoxical arrangement which first synthesizes and then degrades urea almost simultaneously? The answer seems to be to supply the mantle with NH_3 rather than NH_4^+. Both the site of formation of NH_3 and the urine of these animals are alkaline, as is the pH optimum of the urease reaction. Hence, urea hydrolysis leads to the release of NH_3, not NH_4^+. Since urea (and for that matter uric acid) can accumulate in tissues other than the gastropod liver (concentrations in the mantle, for example, are substantially higher than they are in the hepatopancreas of the amphibious snail, *Pilo* sp.), Campbell has postulated that the role of the urea cycle is to deliver the relatively non-toxic urea to the sites of shell formation (Figure 6–3). Here, the urease enzyme in a controlled catalysis hydrolyzes urea to CO_2 and NH_3 and in this way delivers NH_3 to the mantle. The NH_3 acts as a proton acceptor in the dissociation of bicarbonate to yield carbonate for $CaCO_3$ precipitation in the shell:

$$NH_3 + HCO_3^- \rightarrow NH_4^+ + CO_3^=$$

These reactions may occur in several tissues, as the urease enzyme is widely distributed in the tissues of these organisms, but they would be of physiological significance only at sites of $CaCO_3$ deposition. An important point, which has not been given sufficient emphasis, *is that when the urea cycle plus the urease step* operate in this concerted manner, their overall function in nitrogen metabolism is *a cyclic and catalytic one*: NH_4^+ initiates the reaction pathway, and NH_4^+ is regenerated by it (Figure 6–3). There occurs no net drain of nitrogen into this pathway except under conditions when a significant amount of ammonia gas is lost to the outside.

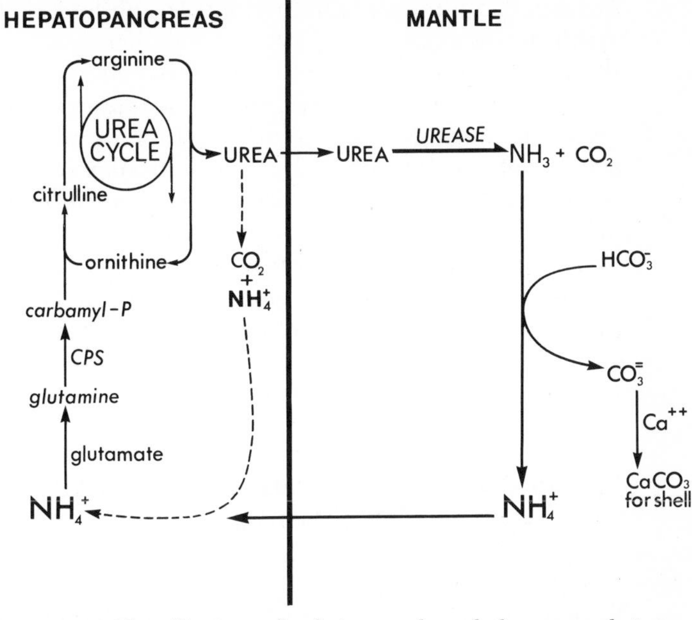

Figure 6–3 The effective path of nitrogen through the urea cycle in terrestrial gastropods under normal physiological conditions. The probable function of this arrangement is the controlled delivery of NH_3 to sites of shell formation.

FUNCTION OF UREA AND URIC ACID IN WATER BALANCE OF TERRESTRIAL MOLLUSCS

The above situation is drastically altered during estivation in terrestrial gastropods. In those species (e.g., *Strophocheilus*) that lose water under these conditions, urea synthesis appears to increase and urea *accumulates,* presumably to reduce the vapor pressure of the blood and thus to reduce evaporative water loss. Mechanisms accounting for the urea accumulation are not known, but clearly must involve an inhibition of the ubiquitous ureases in the organism. A simple mechanism might be increased acidity of fluids during estivation, a situation which might be brought about by the accumulation of acidic end products of anaerobic metabolism. Since urease has a distinctly alkaline pH requirement, such a mechanism could lead to the automatic inhibition of the enzyme.

In estivation, then, the urea cycle does not function in a catalytic manner. Its function now *constitutes a definite drain of nitrogen, which normally is channelled into the uric acid pathway.* This may appear to be somewhat paradoxical, for uric acid at this time accounts for *all* of the *excreted* nitrogen. However, this represents only an apparent paradox. During estivation, although all of the *excreted* nitrogen is in the form of uric acid, the rate of uric acid synthesis (as estimated by C^{14}-glycine incorporation into purines) is reduced to about 1/4 the rates found in active animals. This reduction in uric acid synthesis reflects a reduction in the estivating metabolic rate. Much of the uric acid synthesized during estivation accumulates within the cells of most tissues, and here presumably plays a role in maintenance of osmotic balance. Thus, the *amount* of uric acid actually excreted is greatly diminished. This is an important change since, under these conditions, blood osmolality rises owing to extensive water loss and to accumulation of urea in the blood for purposes of reduction of the vapor pressure. Uric acid is similarly used in osmoregulation during exposure of marine molluscs to hypersaline environments.

SUMMARY OF FUNCTIONS OF NH_4^+, UREA, AND URIC ACID IN LAND MOLLUSCS

To recapitulate (Figure 6–4), the situation in terrestrial gastropods appears to be this: during normal activity periods, the urea

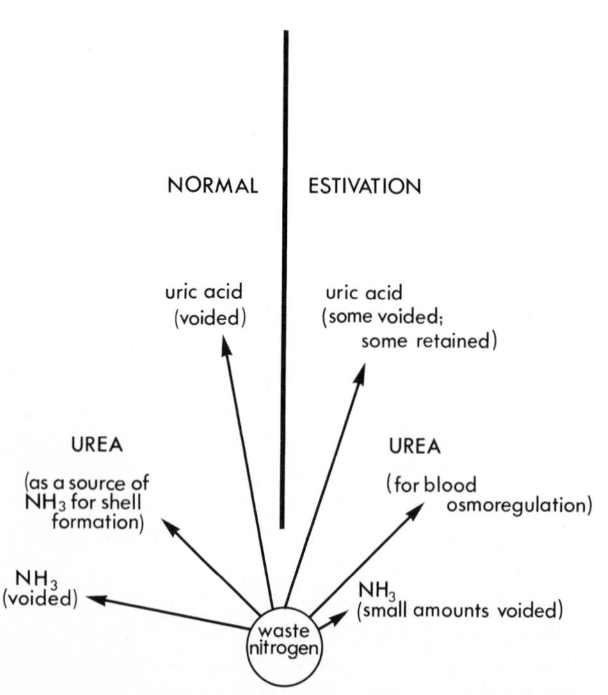

Figure 6–4 Summary of various fates of waste nitrogen in normal and in estivating gastropods.

cycle in concert with urease functions in a cyclic, catalytic manner. NH_4^+ initiates the sequence and NH_4^+ is regenerated by the sequence. At this time, the urea cycle represents little if any drain upon nitrogen metabolism. The bulk of waste nitrogen is channelled into the uric acid pathway; a small percentage may be released as NH_4^+ from glutamine by the action of glutaminase. In contrast, during estivation, *all* of the nitrogen excreted is in the form of uric acid, a situation of selective advantage, since the amount of water lost with the uric acid is minimal. But at the same time, because of an overall reduction in metabolic rate, the amount of uric acid synthesized is reduced to about 1/4 of normal rates. A part of the nitrogen that would normally end up in uric acid now appears in urea, for in estivation, urea degradation by urease is inhibited. The urea pathway is now not a cyclic, *catalytic* mechanism, but rather a *synthetic* one which produces significant quantities of urea and leads to the accumulation of urea in the blood. The mechanism presumably was selected as a device to lead to a reduction in vapor pressure and hence to a reduction in evaporative water loss from the blood. (Parenthetically, it might be added that by utilizing uric acid in maintenance of intracellular osmotic balance, the organism also saves water which would otherwise be lost in the excretion of that uric acid!) The question now arises as to what enzymic mechanisms are involved in these adjustments in nitrogen metabolism.

CHANNELLING OF THE FLOW OF NITROGEN TOWARDS URIC ACID OR TOWARDS UREA IN THE MOLLUSCS

As we indicated in Chapter 5, carbamyl phosphate synthetase, CPS-I, plays a pivotal role in the regulation of nitrogen flow into the urea cycle in vertebrate systems. It appears to be no less pivotal in the molluscs (Figure 6–5). In these organisms CPS *cannot utilize* NH_4^+ *directly as a substrate*; rather, *glutamine supplies the* NH_4^+ *and is an absolute substrate requirement* for the enzyme. The glutamate utilized in forming the glutamine is regenerated by the CPS reaction; hence, this scheme requires only catalytic quantities of glutamate (Figure 6–5). Unlike the situation in the vertebrates, CPS here is in direct competition for glutamine. In the purine biosynthetic pathway, the rate limiting reaction is the amidotransferase catalyzed production of 5-P-ribosylamine; one of the two substrates for this enzyme is glutamine. Hence, in these gastropods, *CPS and amidotransferase compete in an immediate manner for*

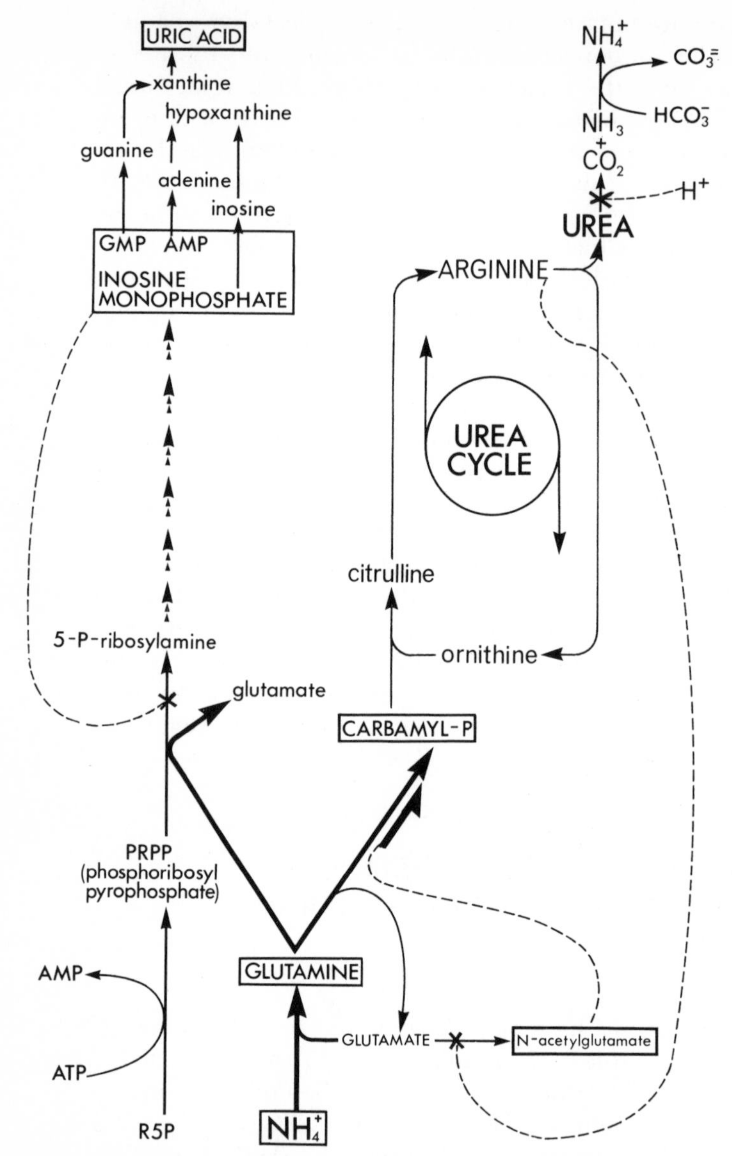

Figure 6–5 Control circuitry involved in the channelling of nitrogen towards urea or towards uric acid in terrestrial gastropods.

glutamine (Figure 6–5), the common amino carrier-molecule here and in most other animal species. As in the case of the vertebrate CPS-I, gastropod CPS displays an absolute requirement for N-acetylglutamate. Under the conditions of high glutamate concentrations found in gastropod tissues, N-acetylglutamate probably would also be produced in relatively high concentrations; *these would lead to a specific, potent activation of carbamyl phosphate synthetase and to a transient accumulation of carbamyl phosphate.* Because of the role of glutamine as an immediate and direct substrate for CPS, this regulatory interaction would *in effect channel glutamine nitrogen towards urea and away from uric acid.*

The carbamyl-P inhibition of glutamine synthetase, which appears to be an integral part of the control of NH_4^+ metabolism in vertebrates (Chapter 5), would not be a selected feature of control here (unless different regulatory isozymes of glutamine synthetase occur), for a block in this position would also block the further synthesis of carbamyl phosphate. A more favorable regulatory interaction would involve negative *CP modulation of the "bottleneck" amidotransferase reaction in the uric acid pathway.* Should this interaction occur (and at this time it has not been tested), it would favor the acetylglutamate-activated flow of nitrogen towards the urea cycle.

A further reduction in the flow of nitrogen towards uric acid might be expected at this time because of the feedback inhibition of the amidotransferase step by AMP, GMP, and IMP (Figure 6–5). In estivation, large accumulation of uric acid occurs concomitantly with the accumulation of the monophosphate precursors, IMP, AMP, and GMP. These accumulations occur in nearly all tissues of amphibious and terrestrial gastropods. The purine monophosphates are known to be specific feedback inhibitors of the first committed reaction in the uric acid pathway—the amidotransferase mediated conversion of PRPP to 5-P-ribosylamine. This regulatory effect can account for the established reduction in purine biosynthesis rates during estivation; *such feedback control of the amidotransferase, moreover, would spare glutamine for the urea pathway* (Figure 6–5). Upon termination of estivation, the reversal of these regulatory interactions would lead again to conditions favoring channelling of glutamine nitrogen towards uric acid. Arginine feedback inhibition of N-acetylglutamate formation may come into play under these conditions, as in the vertebrates.

One possible final level of control of nitrogen metabolism should be mentioned at this point. As we have seen in Chapter 5, control of NH_4^+ metabolism in the vertebrates during adaptation to differing degrees of water deprivation is to some extent *dependent upon control of the level of carbamyl phosphate synthetase.* The time-course of the adaptation in the vertebrates is in the order of days to weeks, ample time for modulation of synthesis rates of enzymes. The same situation may hold in the case of terrestrial gastropods. When these animals go into estivation, time periods involved are weeks to months; again ample time is available for control at the level of *enzyme concentration.* However, careful studies of this problem in the molluscs currently are not available.

THE WATER-LAND TRANSITION IN THE CRUSTACEA

Most of the animals now living successfully on land, including the terrestrial Protozoa, Turbellaria, Annelida, and vertebrates, are thought to have evolved from ancestors which lived in fresh water. Many of the land gastropods, too, may have evolved from freshwater or brackish water forms. Terrestrial Crustacea, in contrast, appear to have invaded the land directly across the marine littoral strip. In some cases, the ancestors may have come from landlocked saline lagoons. All the terrestrial Crustacea are members of one of three orders—the Decapoda, Amphipoda, and Isopoda. Of these, the Decapoda are the least emancipated from the sea. Most, if not all, land crabs have aquatic larval stages and, therefore, at least the females have to return to the sea to release their offspring. For this reason, the terrestrial range of crabs is limited to a few hundred yards from the shore.

The terrestrial amphipods and isopods retain their eggs in brood pouches and compress the development of young so that they are released as immature tiny adults. These organisms no longer have an aquatic stage in the life cycle and therefore they are no longer limited in their distribution by the necessity of remaining close to a large body of water. Thus, along the spectrum from fully aquatic to full terrestrial life styles within the class Crustacea, even within the order Decapoda, many degrees of independence from the sea can be observed. Terrestrial crustaceans consequently have aroused a great deal of interest in recent years.

MOST CRUSTACEANS EXCRETE AMMONIA

From our own point of view at the moment, a fundamental observation is that ammonia constitutes the major excretory end product of nitrogen metabolism *in all Crustacea, whether they be fully aquatic, semiterrestrial, or fully terrestrial.* In many crustacean species, urea and/or uric acid can be produced. In addition, amino acids and other amines also may be released. But, quantitatively, these are of little significance in the overall excretion of waste nitrogen.

POTENTIAL PATHWAYS OF AMMONIA PRODUCTION

Pathways for the release of ammonia are not known with certainty. Since glutamine occurs in quite high concentrations in the Crustacea, we are assuming that it is an important amino-carrier molecule in this group as it is elsewhere (Chapter 5), and that the glutaminase-catalyzed hydrolysis of glutamine is an important pathway for the release of ammonia. Since urease is present in all crustaceans thus far examined, its catalysis of urea to ammonia and CO_2 probably represents a second important pathway for the release of waste nitrogen.

The classical urea cycle is apparently absent in this group. None of the enzymes normally associated with this cycle have been detected except for arginase. The latter occurs in fairly high activities and is assumed to function primarily in the regulation of arginine levels. However, as we shall see, it may also have an important function in the generation of urea as substrate for urease. As do the gastropods, many crustaceans lay down significant quantities of $CaCO_3$ in the exoskeleton; this is a process that occurs actively for about 75% of the total molt cycle and hence may require specific adaptations for maintenance of conditions favoring carbonate formation and precipitation.

The uric acid pathway is assumed to be present in the crustaceans, but the basis for this assumption is argument by analogy with the insects; direct evidence of significant participation of this pathway in excretory nitrogen metabolism is absent. We tentatively assume, therefore, that its major function in crustaceans is the synthesis of purines for nucleic acids, but not the synthesis of uric acid for nitrogen excretion. Moreover, it should be stressed that the pathway of uricolysis (breakdown of uric acid to urea + glyoxylate, with the subsequent breakdown of urea via the urease reaction to ammonia and CO_2) apparently occurs in most crustaceans. It is probable that uric acid in these organisms represents a waste product of nucleic acid metabolism rather than of amino acid metabolism. If so, its participation in overall excretory nitrogen metabolism would be low, and this prediction indeed appears to be consistent with those observations which are available. In some species, uric acid is accumulated during intermolt and may be voided during molting.

As far as we can assess from the available literature, then, there are two reaction pathways available to the Crustacea for the production of urea. These are:

(1) the uricolysis pathway: uric acid $\rightarrow$ glyoxylate + urea
(2) the arginase pathway: arginine $\rightarrow$ ornithine + urea

Similarly, ammonia can be produced by two different mechanisms:

(1) the urease pathway: urea $\rightarrow CO_2 + NH_3$
(2) the glutaminase route: glutamine $\rightarrow$ glutamate + NH_3

The relative importance of these four reaction pathways will probably vary between species; in all cases it is easy to see how the organism can produce ammonia as the major end product of nitrogen metabolism, but these reactions do not render an insight into why the Crustacea release ammonia in preference to other excretory products.

FUNCTIONAL SIGNIFICANCE OF NH_3 PRODUCTION IN CRUSTACEA

As we mentioned in Chapter 5, in the invasion of land, other organisms (such as the vertebrates and the insects) developed enzyme pathways for urea or uric acid excretion *as mechanisms for conservation of water.* (Recall that on a molar basis, the least amount of water is lost during uric acid excretion; the most is lost during ammonia excretion.) Upon invasion of land, the Crustacea encounter the same problems of water conservation as do all other terrestrial forms; yet, in contrast, they do not excrete urea or uric acid in any significant quantities. How, then, do they deal with ammonia as the primary excretory product of nitrogen metabolism? The answer appears to be the *expulsion of NH_3 as a gas, which requires none of the water that would be lost during excretion of NH_4^+ in solution.*

Terrestrial crustaceans do not possess the epicuticular wax layer which so effectively decreases diffusion of substances across the body surface of insects. They consequently *can release waste nitrogen as ammonia gas directly across any or all parts of the exoskeletal surface.* In the case of isopods, the release of NH_3 gas can account essentially quantitatively for *all* nitrogen being excreted! In terrestrial decapods, the relative importance of this mechanism has not been quantified, but would appear to be potentially large. In addition to the obvious selective advantage of releasing NH_3 *without a necessary concomitant loss of large amounts of water,* there may be other functions for this mechanism.

Firstly, in order to volatilize NH_3, alkaline conditions are required at the periphery of the body to bring about the dissociation, $NH_4^+ \rightarrow NH_3 + H^+$. Under alkaline conditions, the two products of the urease reaction are ammonia gas and CO_2. The NH_3 formed under these conditions could either (1) serve as a proton acceptor to cause the dissociation of HCO_3^-, thereby providing $CO_3^=$ for $CaCO_3$ precipitation in the exoskeleton during periods of hardening of the exoskeleton or (2) simply be released as NH_3 gas to the outside.

In this connection, it is significant that the only other proven example of NH_3 gas evolution in the animal kingdom is that reported by Speeg and Campbell for terrestrial gastropods. Here, as in the Crustacea, $CaCO_3$ formation and precipitation is a vital component of mineral metabolism; also here, as in the Crustacea, the utilization of NH_3 as a proton acceptor from bicarbonate (forming carbonate in the process) is a vital component of $CaCO_3$ formation. And here, as in the Crustacea, when NH_3 gas is delivered at rates greater than required for $CaCO_3$ formation, it can be released directly to the outside without the necessary concomitant loss of water which occurs in aquatic organisms. In the land gastropods, urea is a major source of NH_3, for injected urea increases the rate of NH_3 gas liberation, and N^{15} labelled ammonia is formed from injected N^{15}-urea as is $C^{14}O_2$ from C^{14}-urea. Campbell and his coworkers further showed that the surface involved in the land molluscs is not that of the lung which communicates with the exterior. Rather, the NH_3 appears to exit through the shell, and hence it is the outer surface of the mantle underlying the shell which presumably is involved in the release of ammonia gas. This is not surprising, for it is the mantle which is also involved in the deposition of $CaCO_3$ during new shell formation. It is probable that a similar situation occurs in crustaceans, such as the terrestrial decapods, where hardening ($CaCO_3$ precipitation) occurs in all parts of the exoskeleton. However, in the land crustaceans, this latter contention remains to be tested.

THE AMMONOTELIC-URICOTELIC TRANSITION IN INSECTS

As in the case of vertebrates, water-to-land transitions represent important events in the evolutionary history and, often, the lifespan of invertebrate organisms. The earliest invertebrates were, of course, aquatic forms. And, for many contemporary species, there exists an obligatory aquatic larval stage. We would thus predict that important restructurings of the pathways of nitrogenous end-product metabolism would characterize these temporally diverse transitions.

The initial invasion of terrestrial habitats by insects must have occurred concomitantly with the acquisition of an active uric acid pathway of nitrogenous end-product metabolism. In these forms, uric acid is produced in the fat-body by the usual metabolic reactions (Figure 6–2). From this site the uric acid is actively secreted into the Malpighian tubules. The Malpighian tubule-hindgut-rectum system of insects is the effector organ for handling water, ion, and uric acid excretion. In land insects, uric acid accounts for well over 80 per cent of excreted nitrogen.

Aquatic insects, as we would expect, display an ammonotelic pattern of nitrogenous waste disposal. This ancestral pattern is also characteristic of larval stages of insects having terrestrial forms as adults. Surprisingly, no information is available on the mechanisms effecting the switch-over from ammonotelism to uricotelism in these organisms. This open question offers a promising research domain for future comparative biochemists.

SUGGESTED READING

Books and Proceedings

Comparative Biochemistry of Nitrogen Metabolism. I. The Invertebrates (1970). (Ed. J. W. Campbell) Academic Press, New York, 1–493.

Terrestrial Adaptations in Crustacea (1968). Amer. Zoologist 8, 307–685.

Reviews and Articles

Campbell, J. W., and S. H. Bishop (1970). Nitrogen metabolism in molluscs. In *Comparative Biochemistry of Nitrogen Metabolism* (Ed. J. W. Campbell), Academic Press, N.Y. Vol. 1, 103–206.

CHAPTER 7

TEMPERATURE

I. THE BASIC EFFECTS OF TEMPERATURE AT THE BIOCHEMICAL LEVEL

The atoms and molecules which constitute organisms are constantly in motion; that is, they possess kinetic energy. The parameter termed "temperature" can be regarded, for our intents and purposes, as the average kinetic energy of the atoms and molecules of a system. Temperature thus measures not the total heat content of the system, but rather the intensity of this heat.

Each organism and, in fact, each biological structure and process is tolerant of only a finite range of heat intensities. When too much or too little kinetic energy is possessed by the atoms and molecules of the organism, the rates of vital processes and the cellular structures upon which life depends may be adversely or even lethally disturbed. In this first section of the chapter dealing with temperature, we will discuss these basic thermal perturbations of biochemical reactions and structures. Once we have outlined the basic problems which arise when extremes of temperature and/or large and sudden changes in temperature impinge on the biochemistry of organisms, we will begin to examine some of the fundamental strategies utilized by different groups of organisms to overcome the effects of thermal stress.

We hope that this introduction to temperature phenomena will give the reader a coherent frame of reference for appreciating the wealth of information available on temperature effects – information which offers especially clear illustrations of the basic strategies of biochemical adaptation which form the theme of this book. More specifically, we trust that our introduction will accomplish two particular goals. Firstly, we want to give the reader a firm basis for appreciating why the biochemical machinery of the cell is so susceptible to the effects of temperature. And, secondly, we hope that the astute reader will glean, perhaps between our lines, some hints of the biochemical strategies which organisms might employ to circumvent the harmful effects of changes in temperature or extremes of temperature.

THE TWO BASIC CLASSES OF TEMPERATURE EFFECTS: "RATE EFFECTS" AND "WEAK-BOND-STRUCTURAL EFFECTS"

Changes in the temperature of an organism have but two basic effects on its biochemistry. Firstly, changes in the average kinetic energy of the atoms and molecules of the organism will be translated into changes in the rates at which the chemical reactions comprising metabolism occur. We term this temperature effect the "rate effect," and we will soon discuss the physical chemical basis for the rather large dependency of reaction rates on temperature.

The second basic effect of temperature involves changes in biochemical structures and, thereby, processes which are dependent for their *integrity* and *fidelity* on a class of chemical bonds termed "weak bonds" or "weak interactions" (Table 7–1). We will devote much of the latter part of this section to a discussion of weak bonds and the roles they play in biological systems. For the moment we will only stress two points concerning these bonds. Firstly, as indicated in Table 7–2, virtually all of the higher orders of biochemical structure (e.g., 3° and 4° structure of proteins, membrane structure, and nucleic acid structure), and most of the biochemical interactions which demand a high degree of stereochemical specificity (e.g., the binding of substrates to enzymes), are highly, if not entirely, dependent on weak bonds. Secondly, all classes of weak bonds are characterized by bond energies which are no more than an order of magnitude greater than the thermal energies present in organisms. In other words, weak bonds in biological systems are, *individually,* just what their name denotes. Thus, at any given instant, many of the weak bonds upon which life depends will be broken at physiological temperatures. And, as the temperature rises to the upper limits the organism can tolerate, a point may be reached at which one or more of the weak-bond-dependent struc-

TABLE 7–1 The classes of "weak" chemical bonds which are important in biological systems. Bond energies in aqueous solutions are also given. Bond energies of covalent bonds are at least one order of magnitude greater than the energies of weak bonds. The average thermal energy of molecules at room temperature is approximately 0.6 kcal/mole. (Data are from Watson, 1970, who gives an excellent treatment of the roles of weak bonds in biological systems.)

Class of Bond	Approximate Range of Bond Energies (Enthalpies) (kcal/mole)
van der Waals Forces	1
Hydrogen Bonds	3–7
Ionic Bonds	5
Hydrophobic Interactions	−(1–3)

TABLE 7–2 BIOCHEMICAL STRUCTURES AND PROCESSES WHICH ARE DEPENDENT ON "WEAK" CHEMICAL BONDS

1. Higher Orders of Protein Structure.
2. Membrane Structure.
3. Enzyme-Ligand Complexes.
4. Water Structure.
5. Lipid-Lipid Interactions.
6. Nucleic Acid: Nucleic Acid Interactions.
7. Nucleic Acid: Protein Interactions.
8. Hormone-Receptor Protein Binding.

tures and processes listed in Table 7–2 become, for the organism, lethally deranged.

In short, changes in environmental temperature, to the extent that they lead to alterations in body temperature, pose two distinct perturbations to the biochemical structures and functions of living systems. As just discussed, many of the most important biochemical structures and processes are absolutely dependent on chemical bonds which are easily broken by slight inputs of kinetic energy. And, secondly, changes in kinetic energy exert a large effect on the rates of metabolic reactions which involve the breakage and formation of strong, covalent bonds.

"RATE EFFECTS" OF TEMPERATURE: PROBLEMS OF TEMPERATURE CHANGES AND TEMPERATURE EXTREMES

The problems which organisms encounter due to the effects of temperature on the rates of their metabolic reactions fall into two categories. On the one hand, changes in the kinetic energies of metabolites lead to changes in the rates at which they are formed or degraded. Secondly, a very basic problem exists due to the low absolute temperatures characteristic of the biosphere: at the temperatures which permit life, the rates of covalent bond breakage and formation occur only very slowly in the absence of biological material. On an absolute temperature scale the earth is extremely cold, and the chemical transformations which comprise metabolism simply cannot occur at life-supporting rates in the absence of biological catalysts, the nature of which we will discuss shortly.

THE LOW TEMPERATURE ORIGIN OF METABOLISM AND THE CONCEPT OF ACTIVATION ENERGY

The primordial temperature adaptation "problem" which faced the chemical precursors of "life" can be stated as follows: If metabolism involves complex series of chemical reactions in which covalent bonds are made and broken, and if these bond formations/ruptures occur at very low (read: non-life-supporting) rates at earthly temperatures, how can these transformations be accelerated to the extent that living processes, at least as we know them, are possible? In short, how can metabolism and, therefore, "life" exist at low absolute temperatures?

We all have been taught one answer to the above questions: "Enzymes are essential for life because they speed up chemical reactions to the extent that life is possible." However, this answer really begs the question, for it fails to come to grips with the actual mechanisms by which enzymes conduct their catalytic function. To present an answer which will be satisfying to the more mechanistically minded reader, we must discuss in some detail the concept of activation energy, specifically the *free energy of activation* of a reaction, symbolized as $\Delta G^{\ddagger}$ (delta G "double dagger").

Covalent bonds are usually very stable at physiological temperatures. Even though a chemical reaction involving the splitting or formation of a covalent bond may be energetically favorable in a thermodynamic sense, i.e., the overall reaction may involve a decrease in the free energy of the system (Figure 7–1), the reaction may occur at negligible rates at physiological temperatures. *The kinetic feasibility of a chemical reaction is independent of its thermodynamic feasibility.* In other words, the overall free energy change (ΔG^0) involved in the conversion of reactants to products

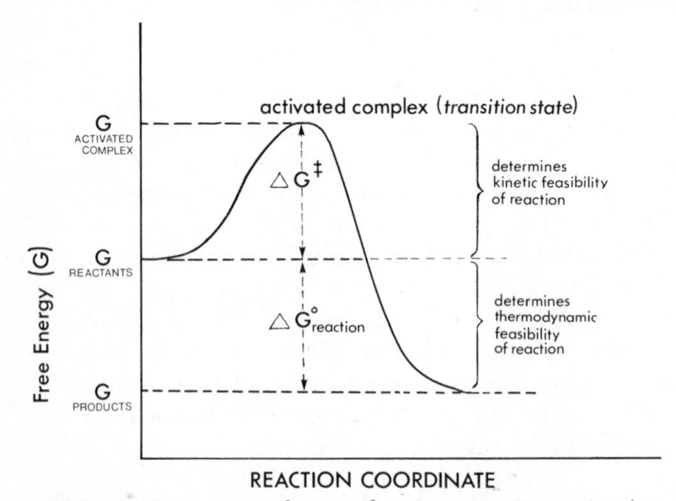

Figure 7–1 Free energy diagram for an exergonic reaction (reaction in which the free energy (G) of the product(s) is less than that of the reactant(s)). The overall free energy change of the reaction, $\Delta G^0{}_{reaction}$, determines the position of equilibrium (extent of the reaction). The rate of approach to equilibrium is determined by the energy "barrier" to the reaction, the free energy of activation ($\Delta G^{\ddagger}$).

tells us absolutely nothing about how rapidly this conversion will occur.

The index to the rate at which the reaction will occur is supplied by another free energy change, the aforementioned $\Delta G^{\ddagger}$, which occurs during the *activation* step of the chemical reaction. To best appreciate this second free energy function, we might approach the topic historically.

During the early decades of the past century, enough data had accumulated in studies of temperature effects to support the generalization that the rates of most chemical reactions, specifically those involving the breakage/formation of covalent bonds, were doubled or tripled with each 10°C increase in temperature. Thus, the temperature coefficient of these reactions, defined as the Q_{10} value, where

$$Q_{10} = \text{Velocity } (T + 10)^{\circ}\text{C}/\text{Velocity } (T)^{\circ}\text{C}$$

normally falls in the range from 2 to 3.

This observation was difficult to reconcile with the fact that a 10°C rise in temperature leads to only a 3 per cent increase in the average kinetic energy of the molecules of a system.[1] How then do we account for the fact that a 3 per cent increase in average kinetic energy leads to a 200 to 300 per cent increase in reaction velocity?

In the late 1880s, Svante Arrhenius proposed an answer to this question. Arrhenius argued that not all molecules in a given population are equally able to react. Temperature is an average property of all the molecules, whereas chemical reactivity may be a property only of those molecules which have energies much higher than average. Arrhenius proposed that, for a molecule to be reactive, it must have at least a certain threshold energy, which he termed the energy of activation (Ea). The relationship between the Arrhenius energy of activation, the absolute temperature, and reaction velocity is given by the Arrhenius Equation:

$$\frac{d \ln k}{dT} = \frac{Ea}{RT^2} \text{ or } \ln \frac{k_2}{k_1} = \frac{Ea}{R}\left(\frac{1}{T_1} - \frac{1}{T_2}\right)$$

where the k values represent velocity constants at each temperature (T_1, T_2), and R is the gas constant.

[1]On an absolute or Kelvin scale, absolute zero is taken as the true 0° point. At absolute zero there is no kinetic energy. On the Centigrade scale, 0°C is the freezing point of pure water under standard conditions. 0°C equals + 273°K (Kelvin), and a one degree increment on both scales is the same, i.e., 1°C=1°K. Thus it is seen that, since absolute temperature measures the average kinetic energy of a population of molecules, a 10°C change in temperature around room temperature (approximately 25°C or 298°K) represents about a 3% increase in the absolute temperature, and hence the kinetic energy of the system.

Arrhenius computed the effects of temperature on the fraction of molecules of a population which have the necessary energies to react (Figure 7–2). He found that a relatively small percentage change in the average kinetic energy of a population was accompanied by a relatively large change in the fraction of molecules having kinetic energies equal to or greater than the critical activation energy. In this way Arrhenius accounted for the apparent discrepancy between the effects of heat changes on temperature, on the one hand, and reaction velocity on the other.

Building on the conceptual foundation laid by Arrhenius, later physical chemists have refined the concept of activation energy to a considerable degree. Whereas the original development of the topic by Arrhenius dealt with an enthalpy based parameter (Ea equals the enthalpy of activation plus the gas constant R times the absolute temperature, i.e., $Ea = \Delta H^{\ddagger} + RT$), later treatments are formulated on the basis of $\Delta G^{\ddagger}$, an energy value which incorporates both the enthalpy *and* the entropy changes which occur during the activation process.

The key tenet of modern treatments of activation theory can be summarized as follows: for a chemical bond to be made or broken, the reacting molecules must be strained and contorted in such a manner that their internal bonds are weakened, and thus made more susceptible to reaction. Only those molecules which have suf-

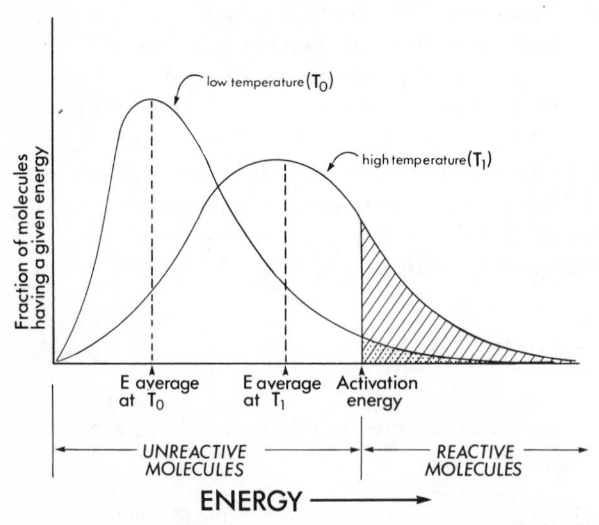

Figure 7–2 Energy-distribution curves computed from the Maxwell-Boltzmann equations, for a population of molecules at two temperatures (T_0 and T_1). The average energy of the molecules equals the temperature in °K. Only those molecules having energy equal to or greater than the activation energy are reactive (indicated by stippling at T_0 and shading at T_1).

ficient energy to undergo such bond straining can react. The minimal energy which must be possessed by a molecule or molecules to react, and form what is termed an *activated complex*, is $\Delta G^{\ddagger}$ (Figure 7–1).

ACTIVATION ENERGY EFFECTS ARE THE BASIS OF TWO THERMAL "PROBLEMS"

From the above discussion of activation energy theory, it should be obvious that two distinct rate "problems" arise owing to the dependence of reaction velocities on the concentration of molecules having a certain threshold energy, the free energy of activation.

Firstly, and most basically, we can appreciate the tremendous energy barrier which is posed to reactions which involve the formation and breakage of covalent bonds. For most of the chemical reactions of intermediary metabolism, the energy which is needed to first stretch and strain molecular bonds to form the activated intermediate is not available in sufficient quantities to permit life-supporting rates of chemical activity, at least not in the absence of enzymes. Enzymes, protein catalysts, lower the free energy of activation values of their reactions to the extent that the thermal energy which is present in the organism is sufficient to activate the reactants (Figure 7–3). At typical physiological temperatures, the rates of enzymic reactions exceed those of non-catalyzed reaction by 8 to 12 orders of magnitude. Thus, the primordial "problem" of temperature adaptation was "solved" by the evolution of enzyme catalysts.

The second basic implication of activation energy relationships to biological systems involves the effects of changes in temperature rather than the absolute temperature *per se*. Since the rates of metabolic reactions are dependent on the concentration of activated molecules, and since these concentrations rise 2- to 3-fold for each 10° C increase in temperature, serious problems may arise if the body temperature of an organism varies by more than a few degrees C. Temperature decreases may slow down metabolism to the extent that processes required for life cannot be maintained at life-supporting rates. Conversely, increases in temperature may accelerate metabolism to the extent that the supply of needed fuels, e.g., food and oxygen, cannot be maintained at necessary levels. Furthermore, not all reactions have the same Q_{10} values. Thus, changes in temperature will have differential effects on metabolism, i.e., metabolism may be thrown out of balance by temperature changes. Because these "problems" have existed for living systems from the beginning of evolutionary history, we might expect that diverse

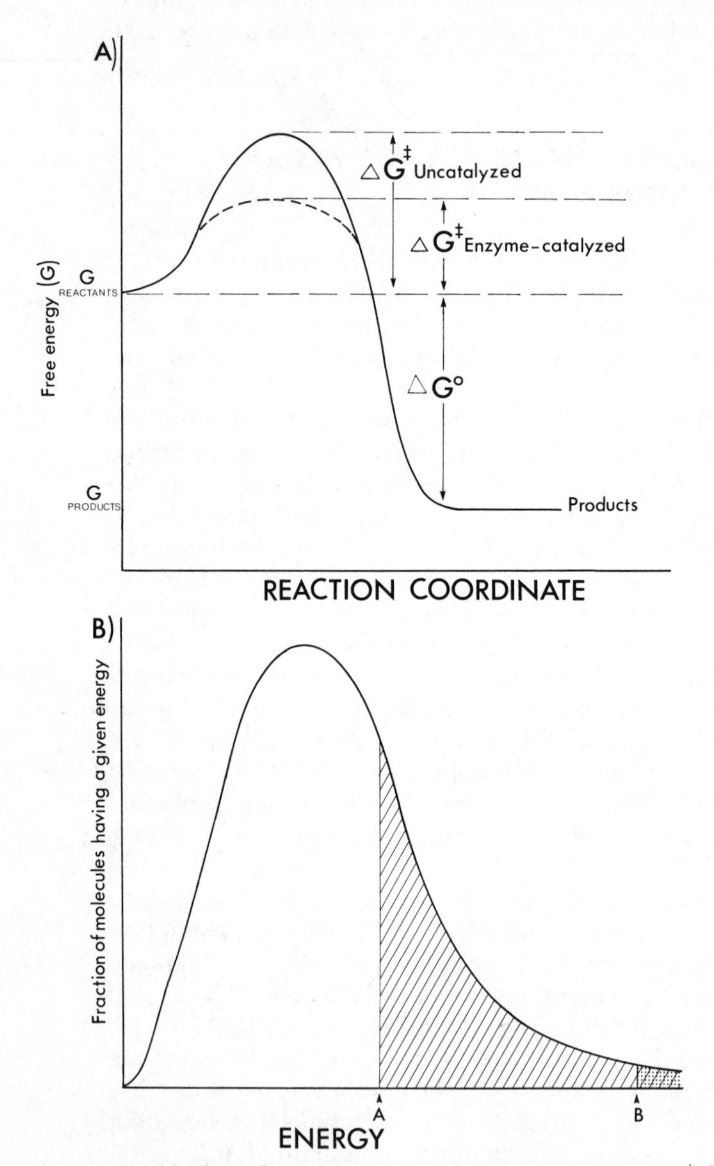

Figure 7–3 (a) The reduction in the free energy of activation ($\Delta G^{\ddagger}$) by an enzyme. The "energy barrier" in the absence of enzymic catalysis is the large value, $\Delta G^{\ddagger}_{\text{uncatalyzed}}$. In the presence of an enzyme, this "barrier" is greatly reduced. (b) Energy distribution curve at a given temperature, illustrating the increase in the fraction of reactive molecules under conditions of enzymic catalysis. In the absence of catalysis, only those molecules having energies equal to or greater than B are reactive (indicated by stippling). In the presence of an enzyme, all molecules having energies greater than or equal to A are reactive (indicated by shading).

mechanisms have evolved which permit organisms to gain and maintain self-control of the rates of their vital processes. In fact, the gaining of a capability to keep the rates of metabolic processes independent of the environmental temperature has been an evolutionary accomplishment which characterizes species in most phylogenetic lines. While the strategy employed will vary—some organisms maintain a constant body temperature and thus shield themselves from the direct impingement of environmental temperature on their cellular chemistry, whereas other organisms adjust their biochemistry following its subjection to thermal change—the end result is the same. In the words of the eminent physiologist Joseph Barcroft, "... nature has learned so to exploit the biochemical situation as to escape from the tyranny of a single application of the Arrhenius equation. She can manipulate living processes in such a way as to rule, and not be ruled by, the obvious chemical situation."[2] Sections II and III of this chapter will discuss the strategies whereby different animal species have circumvented the "obvious chemical situations" discussed on the preceding pages.

WEAK BONDS AND STRONG BONDS: SOME BIOLOGICALLY IMPORTANT DIFFERENCES

In the preceding discussion of "rate effects," we were primarily concerned with the influence of temperature on reactions which involve the formation or breakage of covalent bonds. Relative to the amount of kinetic energy present at physiological temperatures, these bonds are strong, and for them to be made or broken at rapid, life-supporting rates, the assistance of catalysts is required. Thus, our discussion of rate effects was in terms of enzyme-catalyzed reactions which, when summed together, add up to "metabolism."

The second class of chemical bonds in our scheme, the so-called "weak" bonds or "weak" interactions, have low bond energies and, most importantly, the free energies of activation associated with the formation or breakage of these interactions are also small. Consequently, weak bonds are readily broken by the kinetic energy present in living systems. No enzymic assistance is needed to facilitate the formation or the rupture of weak bonds, in contrast to the situation which exists for covalent bonds. (There is some "irony" in the interdependence of strong (covalent) and weak-bonded systems: for covalent bonds to be ruptured or formed at physiological temperatures, a catalytic system is required which owes much of its functional capacity to an enormous number of weak bonds.)

---

[2]Barcroft, J., 1934. *Features in the Architecture of Physiological Function.* Cambridge University Press, London, p. 40.

A second important difference between biochemical systems which involve covalent bonds and those which depend on weak chemical bonds lies in the number of bonds which are necessary for the process or structure in question. For example, the peptide bonds which hold amino acid residues together in a polypeptide chain are covalent bonds. One of these strong bonds is sufficient to link an amino acid to a growing polypeptide chain. In contrast, the bonds involved in stabilizing the tertiary structure of a polypeptide chain are weak bonds, and these bonds are effective only when they act together in large numbers. Dozens of weak bonds stabilize the tertiary structure of a polypeptide chain.

WHY DEPEND ON WEAK BONDS?

In light of the thermal lability of weak chemical bonds, one justifiably might ask why living systems have come to rely so heavily on something so fragile. Consider the case of enzymes, for example. If these catalysts are of such vital importance in making metabolism possible at low absolute temperatures, would it not be "sensible" to stabilize the higher orders of protein structure using covalent bonds which are thermally stable? Or is there some underlying "sense" in the fact that thermal lability is a concomitant of efficient enzyme function?

Thanks to very recent studies of the stereochemical changes which occur during enzymic catalysis and regulation, we can give fairly conclusive answers to the above questions. There is indeed a great amount of "sense" in proteins' reliance on weak chemical bonds. Most, if not all, interactions between enzymes and substrates, and enzymes and their modulators, involve changes in the 3° and 4° structures of the enzymes. Stereochemically, these changes may be large or small; biologically, they are absolutely vital. The rate at which an enzyme catalyzes a chemical reaction probably is dependent on how rapidly the enzyme can undergo a reversible conformational change which results from enzyme-substrate interactions. The proper response by an enzyme to the binding of a regulatory metabolite depends on the enzyme's ability to alter one or more of its higher orders of structure. In some cases the structural change involves the enzyme's tertiary conformation. In other cases—for example, glycogen phosphorylase—the regulatory effect involves a change in 4° structure.

Thus, in catalysis *per se* and in the modulation of the rates and directions of catalysis, the structural flexibility of proteins is essential. And it is this inherent, necessary structural flexibility which renders enzymes so sensitive to changes in temperature. Much as weak bonds can be broken by the small energies released during the binding of substrates and modulators, so can they be broken by additions of kinetic energy. We can, of course, imagine proteins with highly inflexible structures, for example, 3° and 4° structures which are stabilized by vast numbers of disulfide linkages. These proteins would be extremely stable in the face of temperature changes. However, as enzymes they would almost certainly be non-functional since they could not respond to substrates and regulatory metabolites in the necessary ways. In the case of enzymes and other proteins involved in regulatory processes—for example, repressor proteins—the need for structural flexibility is translated into a dependence on weak bonds. Enzymes have had to "accept" a relatively high degree of thermal lability in order to gain their phenomenal capacities to catalyze and control the reactions which comprise metabolism.

TEMPERATURE EFFECTS ON PROTEIN STRUCTURE: GENERAL CONSIDERATIONS

Among the temperature effects which we have grouped under the heading "weak bond effects," the temperature-dependent changes which occur in the higher orders of protein structure are uniquely interesting. They at once reveal how thermal disruption of weak bonds can lead to gross changes in the functional properties of macromolecules and, of equal importance, they offer an intriguing suggestion of how organisms might circumvent some of the problems raised by changes in, or extremes of, temperature. Thus, the changes in 3° and 4° protein structure which we will discuss must be viewed from a dual perspective. On the one hand, these changes in higher order structure can be viewed as a threat to the function of the proteins. On the other hand, we must not exclude the possibility that the different 3° or 4° structure formed at the new temperature might have new functional attributes which particularly suit the protein for function under the new thermal regime.

HOW WEAK BONDS STABILIZE PROTEIN STRUCTURE

While there are certain enzymes which rely in part on covalent bonds, especially disulfide linkages (-S—S-), for stabilizing their higher orders of structure, in all cases examined to date, the major determinants of 3° and 4° structure have been found to be four classes of weak bonds/interactions: hydrophobic interactions, hydrogen bonds, van der Waals forces, and ionic bonds. Whereas hydrogen bonds are critically important in stabilizing 2° structure, e.g., α-helical regions, the most important determinant of 3° and 4° structure appears to be hydrophobic interactions.

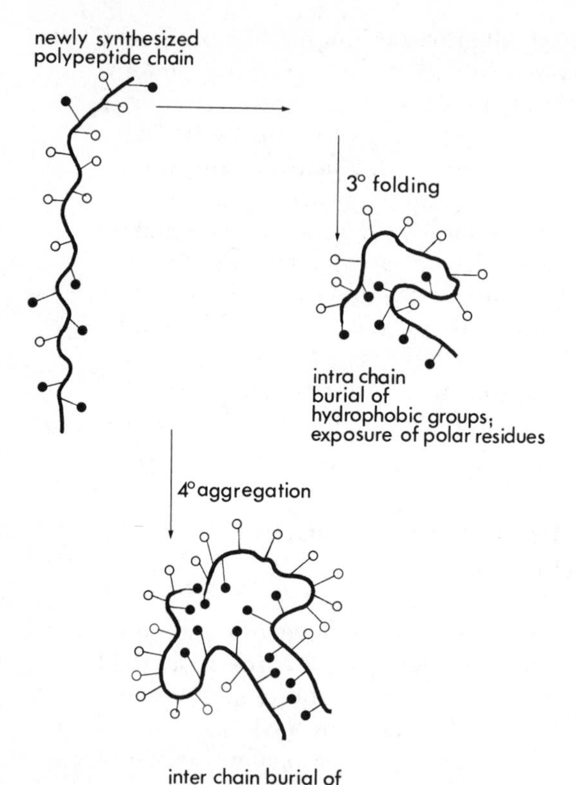

Figure 7–4 A simplified representation of the factors involved in the formation and stabilization of protein 3° and 4° structure. The different classes of amino acid residues are represented as follows: ○, polar and ionic residues (e.g., threonine, serine, lysine); ●, hydrophobic residues (e.g., leucine, isoleucine, valine).

Hydrophobic interactions can be best defined in terms of what they are not. Non-polar amino acid residues, e.g., those of leucine, isoleucine, valine, methionine, and phenylalanine, tend to avoid interacting with water, and, instead, bury themselves in the non-aqueous interior of the polypeptide chain or, in the case of 4° structure, between the faces of adjoining subunits. This type of "bond" is illustrated in Figure 7–4. What this figure illustrates, in layman's terms, is that oil and water do not mix. The reason polar and non-polar substances remain in separate phases is that an unfavorable (negative) entropy change occurs when non-polar molecules are hydrated. A significant amount of ordering of water occurs around the non-polar molecules or residues. This unfavorable entropy change, i.e., the increase in order of the system, makes a two-phase system energetically preferable to a one-phase system.

The contribution which hydrophobic interactions make to the stabilization of protein 3° and 4° structures is sizeable: probably more than half of the stabilizing force for 3° structure is due to hydrophobic interactions. For at least some multimeric enzymes, hydrophobic reactions appear to play the dominant role in maintaining subunit aggregation.

From the standpoint of temperature effects, two properties of hydrophobic interactions should be remembered. Firstly, like other weak bonds/interactions, hydrophobic interactions can easily be disrupted by the amount of thermal energy present in the living system. Secondly, unlike other classes of weak bonds, hydrophobic interactions are more effective at temperatures in the neighborhood of 25°C than near 0°C. (If the reader is curious about the basis of this latter effect, he is urged to read the paper by J. Brandts listed in the bibliography.)

TEMPERATURE EFFECTS ON PROTEIN QUATERNARY STRUCTURE

Loss of 4° structure at low temperatures appears to be a relatively common phenomenon, especially in the case of enzymes from mammals and their microbial intestinal flora. Table 7–3 lists examples of multi-subunit enzymes which disaggregate as temperature is lowered. Interestingly, in at least some cases, the homologous form of an enzyme from a "cold-blooded" organism does not exhibit loss of 4° structure at low temperature. When loss of quaternary structure occurs, there is usually a concomitant loss of catalytic and/or regulatory function of the enzyme.

The fact that many enzymes lose their 4° structure at low temperatures has been taken as an indication that hydrophobic interactions are of particular importance in stabilizing the aggregated state of polypeptide subunits. As mentioned, of all the weak bonds/interactions which stabilize the higher orders of protein structure, only hydrophobic interactions weaken as temperature is lowered. It has even been suggested that the existence of multisubunit enzymes is due to the presence of more hydrophobic residues in the polypeptide chains than can be buried within the single tertiary-folded chain (Figure 7–4). Some estimates of the proportion of hydro-

TABLE 7–3 SOME MULTIMERIC ENZYMES OF MAMMALS WHICH LOSE THEIR QUATERNARY STRUCTURES AT LOW TEMPERATURES (FROM BEYER, 1972).

Mitochondrial ATPase
Glyceraldehyde-3-phosphate Dehydrogenase
Pyruvate Carboxylase
Arginosuccinase
AcetylCoA Carboxylase
Glycogen Phosphorylase

phobic residues which is necessary to lead to 4° structure imply that approximately 30 per cent hydrophobic residues will almost always necessitate the aggregation of polypeptide chains to form a 4° structure. Amino acid analysis of single chain and multisubunit proteins reveals that the former class of enzymes has 13 to 31% hydrophobic residues, while the multisubunit enzymes have 29 to 38% hydrophobic residues.

It should also be pointed out that loss of quaternary structure is likely to be preceded in many cases by a change in 3° conformation. Thus an initial change in 3° structure weakens the 4° structure enough to permit subunit disaggregation.

In light of the fact that interspecific differences exist in the cold lability of 4° structure of different forms of a particular enzyme, it would be interesting to determine whether cold- and warm-adapted organisms rely on different types of bonding for the maintenance of protein 4° structure. At present there appear to be too few data of comparative amino acid contents (especially in the regions where subunits interact) to allow an answer to be given to this question.

TEMPERATURE EFFECTS ON PROTEIN TERTIARY STRUCTURE

Although the study of temperature-induced changes in enzyme 3° structure is in its infancy, there are enough data available to support the contention that temperature changes over the biological thermal range may cause significant changes in the tertiary conformation of enzymes. Unlike the saltatory changes which occur during loss of 4° structure, the structural (and functional) changes which mark temperature-induced alterations in tertiary conformation appear to be somewhat gradual and moderate. Change in 3° structure appears to occur over a rather broad range of temperatures, and the alteration in protein conformation often leads only to a change in, but not a loss of, catalytic capacity.

From the standpoint of temperature adaptation, it should not be too difficult to appreciate the potential importance of such temperature-induced changes in 3° structure. If gradual changes can be induced in 3° conformation as temperature varies over the biological temperature range of the organism, and if these structural changes lead to alterations in the enzyme's kinetic properties rather than loss of catalytic or regulatory function, then the possibility arises that these structural/functional changes may lead to an enzyme which is uniquely suited for function at the new temperature at which it must operate. Thus, in a functional sense, at each new temperature the organism may have "new" enzymes in its cells. We will soon consider the extent to which these functionally "new" enzymes confer advantages and disadvantages to the organism.

TEMPERATURE AND PROTEIN QUINTINARY STRUCTURE

While there are virtually no data describing the effects of temperature on protein quintinary structure, there are several strong *a priori* arguments favoring the supposition that temperature changes may have important consequences at this level of macromolecular organization. A particularly strong case can be made for membrane-associated enzymes. The attachment of enzymes to membranes, whether the attachment involves a surface-plating or the actual incorporation of the enzyme protein into the interior of the membrane, almost certainly involves weak bonds/interactions. Consequently, the data presented above concerning temperature effects on the weak bondings which stabilize 3° and 4° structure may turn out to describe quite closely the effects of temperature on 5° structure. A similar case can be raised for enzymes which exist in multienzyme complexes, e.g., the pyruvate dehydrogenase complex. Again, if weak bonds are involved in stabilizing the complex, then temperature extremes may be expected to disrupt structure and, perhaps, function.

Lastly, whereas we have little information concerning the role of temperature in favoring or disrupting protein 5° structure, it is known that, for at least some enzymes, the enzyme is more thermally stable when associated with a membrane than when removed from the membrane. Thus, if temperature can lead to the loss of 5° structure, the subsequent loss of 4° and/or 3° structure may be rapid.

TEMPERATURE AND THE CONFORMATION OF NASCENT POLYPEPTIDES

If the higher orders of protein structure can be altered by temperature changes, then we might predict that the conformation of a growing (nascent) polypeptide chain, bound to the polysome complex, might also be thermally labile. In fact, a nascent polypeptide which has not folded into its final conformation may be relatively more sensitive to thermal effects than a fully folded polypeptide.

In a study of a mutant enzyme of a strain of *E. coli*, Roodman and Greenberg observed that the enzyme thymidylate synthetase was synthesized normally at 25°C, whereas at 37°C the translation process did not go to completion. The authors suggested that the partially completed polypeptide chain underwent a temperature-dependent conformational change at the higher temperature, and that the new conformation somehow interfered with the completion of translation. They also observed that in cells which had first been incubated at 25°C before growth at the higher temperatures, enzyme synthesis at 37°C could occur satisfactorily. The explanation given for this latter phenomenon is as follows: the enzyme subunits synthesized at 25°C remain in the cells at 37°C and can act as "conformational templates." The 25°C-synthesized subunits, which are folded properly, interact with the nascent polypeptide chains and induce in them the correct conformation to permit completion of translation.

While the above explanation of these data is hypothetical, we feel it is worth presenting since it raises an interesting question regarding mechanisms of biochemical adaptation: Can a growing polypeptide chain assume different conformations at different temperatures and, as a consequence, have distinct enzymic properties at different temperatures?

TEMPERATURE EFFECTS ON ENZYME-LIGAND INTERACTIONS

Before we say any more about the effects of temperature on the weak bonds which support the higher orders of macromolecular structure, we will briefly consider some of the important, weak-bond-based interactions which occur between macromolecules and relatively low molecular weight ligands. As indicated in Table 7–2, enzyme-substrate and enzyme-modulator complex formations likely involve one or more weak chemical bonds. Thus, we would expect that changes in temperature would influence these interactions, in much the same way as temperature affects the higher orders of macromolecular structure.

We should immediately see that temperature-dependent interactions between enzymes and their substrates and modulators bode both "good" and "evil" for organisms. If the strengthening of a weak-bond interaction enhances the ease with which a process can occur, then a temperature change which stabilizes the weak bond(s) involved may be of benefit to the organism. For example, if an enzyme's ability to bind a substrate molecule enhances its activity—which we know is the case at low substrate concentrations—and if the weak bonds holding the enzyme-substrate complex intact are strengthened at low temperature, then there exists the possibility that a partial reversal of the rate-decelerating effects of reduced temperature can be obtained. Conversely, if either high or low extremes of temperature should render a weak-bonding interaction between enzyme and substrate (or modulator) very unstable, then the reaction and the organism may be adversely affected.

TEMPERATURE EFFECTS ON NUCLEIC ACID SECONDARY STRUCTURE

Thermal disruption of the secondary structure of double-stranded nucleic acids is a classic example of a loss of higher order structure due to the rupture of large numbers of weak chemical bonds. The two nucleic acid strands of a DNA molecule are held together by a combination of hydrogen bonds, between guanine-cytosine (G-C) and adenine-thymine (A-T) base pairs, and hydrophobic interactions, which tend to force the hydrophobic ring structures of the purines and pyrimidines into the water-free interior of the molecule. This type of secondary structure also places the charged phosphate groups in contact with water, an energetically favorable situation. These structural features are illustrated in Figure 7–5.

As in the case of proteins, the addition of heat to the system containing the macromolecule will tend to disrupt hydrogen bonding. At moderate temperatures, the most likely event to follow bond disruption is bond reformation. However, beyond a certain temperature there will be a net loss of weak bonds, especially hydrogen bonds, and the DNA molecule will begin to lose its secondary structure.

The loss of double-strandedness, often termed the "melting" of DNA, is marked by a sharp increase in the UV absorbance of the molecules (due to the exposure of the ring structures of the purine and pyrimidine bases) which is strongly sigmoidal, indicating that the rupture of hydrogen bonds during melting is a cooperative phenomenon; i.e., the rupture of each bond makes it progressively easier for the next bond to be broken.

The melting temperature of double-stranded DNA is directly proportional to the guanine-cytosine content, on a percentage basis, of the DNA molecule. Each G-C base pair is stabilized by three hydrogen bonds, whereas an A-T pair is stabilized by only two hydrogen bonds. The potential for evolving more thermally stable double-stranded nucleic acids is thus present, assuming that organisms can accumulate high percentages of G-C base pairs in their genomes.

In studies of the base composition of nucleic acids from different organisms, it has been found that, while considerable variation is noted in G-C content, there is no strong correlation between G-C percentage and adaptation temperature in the case of DNA. However, a different situation is observed for ribosomal RNA (rRNA). At least in the case of thermophilic bacteria, there appears to have been selection for relatively high levels of guanine and cytosine. This difference between DNA and rRNA may stem from the fact that, while DNA is a fully double-stranded molecule, the secondary

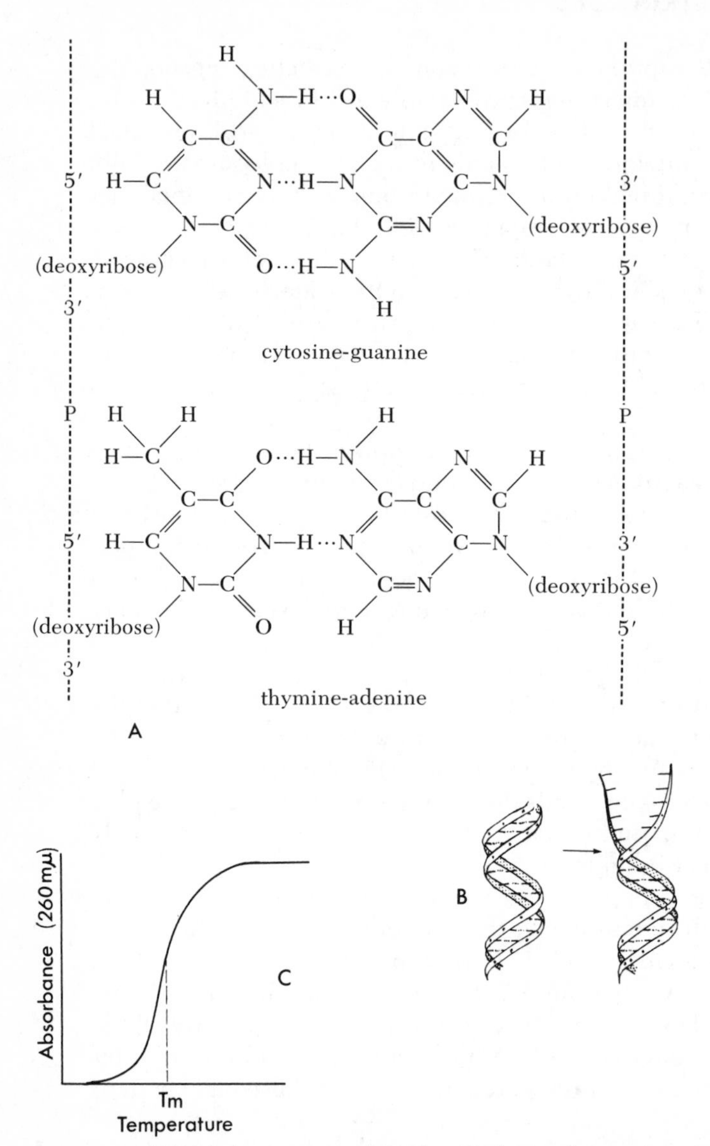

Figure 7–5 DNA base-pairing and "melting" of secondary structure. (a) Base-pairing between guanine-cystosine and adenine-thymine ring structures. These "weak" bonds, when present in large numbers, are a major stabilizing force for the double-stranded, secondary structure of the DNA molecule. (b) Melting of the double-helix. At a sufficiently high temperature, the rate of hydrogen bond rupture exceeds that of hydrogen bond reformation, and the two DNA strands separate ("melt"). (c) Optical density (absorbance) change during DNA melting. The exposure of the previously buried ring structures, which absorb strongly in the ultraviolet region, leads to an increase in the absorbance of a DNA solution. The melting temperature (Tm) is proportional to the G:C content of the DNA molecule.

structure of rRNA is more limited, and only short regions are involved in base-pairing. Since weak-bonded systems are dependent for their integrities on populations of weak bonds, an individual weak bond involved in stabilizing the secondary structure of rRNA may be relatively more important than a DNA weak bond.

TEMPERATURE EFFECTS ON THE 3° STRUCTURE OF TRANSFER RNA (tRNA)

The temperatures at which double-stranded nucleic acids "melt" are usually much higher than the temperatures at which the organism itself can survive. Thus, at least in the case of DNA, thermal disruption of the 2° structure is probably of limited biological significance. One clear example of biologically important alterations in higher orders of nucleic acid structures is known, however. This is the change in tRNA tertiary structure with temperature.

tRNA molecules (Figure 7–6) are known to possess a 3° structure which is essential for their activity in protein synthesis. Alteration of this 3° structure is manifested in a failure of the tRNA molecules to function normally in the enzymic reaction wherein amino acids are joined to their proper tRNA acceptors. This reaction is catalyzed by a class of enzymes termed *aminoacyl tRNA synthetases.* The overall reaction may be written as follows:

$$\text{amino acid} + \text{ATP} + \text{tRNA} \rightarrow \text{aminoacyl-tRNA} + \text{AMP} + \text{PP}_i.$$

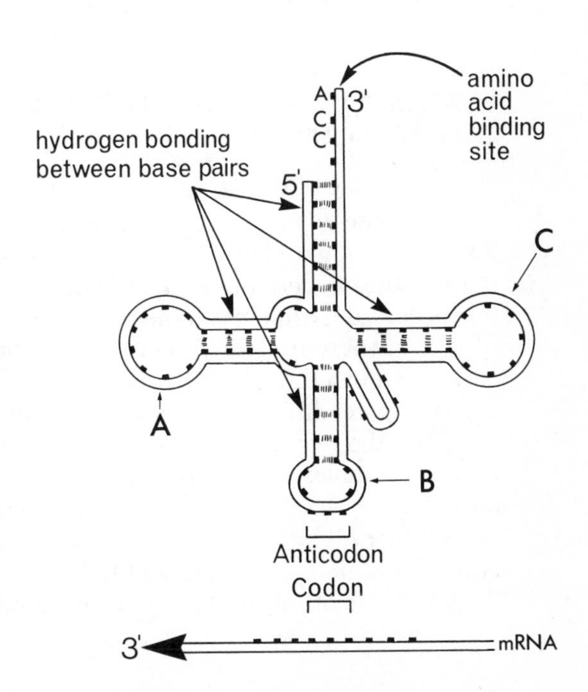

Figure 7–6 The structure of a transfer RNA (tRNA) molecule. The diagram illustrates the conventional "cloverleaf" structure of the molecule which occurs when the 80 or so nucleotides are maximally hydrogen bonded. The 3′ hydroxyl end of the molecule always ends with the triplet ACC. Loop A is thought to be involved in binding to the synthetase enzyme. Loop C is thought to be involved in binding to the ribosome. Loop B contains the anticodon base triplet which binds to the messenger RNA (mRNA) molcule. Hydrogen bonds are responsible for the stabilization of the cloverleaf structure and the tRNA-mRNA complex.

Each amino acid has a unique aminoacyl tRNA synthetase enzyme and, of course, a tRNA acceptor (albeit there may be more than one tRNA for each amino acid). The synthetase enzymes have a complex set of substrates, and three recognition sites (for a single amino acid, ATP, and a specific tRNA) must be present on the enzyme surface. Loss of the ability to recognize and bind any of these three substrates could prove devastating for protein synthesis and, thereby, lethal for the organism.

In a study of a cryophilic microbe (*Micrococcus cryophilus*), Malcolm observed that temperature-dependent changes in tRNA 3° structure were a partial basis for the inability of the microbe to grow at high temperatures. When the tRNA molecules specific for glutamic acid and proline were incubated at high temperatures, they lost their abilities to bind to their aminoacyl tRNA synthetase. Thus, these two tRNA species appear to exist in high and low temperature conformations, and only the latter conformation can be recognized by the synthetases of the cryophilic microbe. However, the high temperature conformations of the two tRNA species do exhibit full biological activity under certain conditions. When the synthetase enzymes from a thermophilic bacterium were added to the *in vitro* protein synthetic system containing the heated tRNAs from the cryophile, full synthetic activity was observed. It therefore seems that the enzymes from the heat-adapted microbe can recognize the high temperature conformation of the tRNAs, while the enzymes from the cold-adapted microbe can recognize only the low temperature conformation of the tRNAs.

TEMPERATURE EFFECTS ON THE ACCURACY OF TRANSLATION

The effects of temperature on the translation process may be less drastic than those just discussed. For example, extremes of temperature appear to be capable of making the translation process less accurate, even though the process itself continues at a high rate.

In vitro studies of protein synthesis, using the synthetic machinery from a thermophilic bacterium, showed that at low temperatures the reading of a synthetic mRNA molecule was much less accurate than at the temperatures approximating the microbe's normal habitat temperatures. In a comparable study, using the protein synthesis apparatus from a mesophilic bacterium, it was observed that higher than normal temperatures led to a loss in translational accuracy: an isoleucyl-tRNA synthetase began catalyzing the formation of valine-tRNA and serine-tRNA complexes, using the tRNA "specific" for isoleucine. Thus, a certain fraction of isoleucyl residues were replaced by the wrong amino acids.

The molecular basis of translational error at extremes of temperatures is not known. However, in light of our foregoing discussions of other weak-bond-dependent systems, it appears likely that the protein synthetic machinery—the tRNA molecules, the mRNA strand, and the associated enzymes and the ribosomes—would all be susceptible to temperature-dependent conformational changes which, in turn, could lead to decreased fidelity of function.

TEMPERATURE EFFECTS ON GENE REGULATION

As in the case of mRNA translation, the regulation of gene expression involves a number of weak-bond-dependent interactions between nucleic acid and protein molecules. Thus, there is a strong probability that temperature can exert important influences on gene expression. And, as has often been true when we have considered thermal effects on weak-bonded systems, there arises potential for "good" and "evil." These we will consider after a brief discussion of some examples, taken entirely from bacterial systems, in which temperature effects at the gene regulation level are known to occur.

Example 1. In a strain of *E. coli*, the enzyme glutamate decarboxylase is inducible at 37°C, but is partly constitutive at 30°C. That is, at the higher temperature the enzyme needs to be induced via gene activation, whereas at the lower temperature the gene coding for the enzyme is at least partially turned-on all of the time.

While the precise mechanism underlying this observation is not known, the data are consistent with the following hypothesis: the repressor molecule which blocks transcription of the gene coding for glutamate decarboxylase binds more tightly to the operator region of the gene at high temperatures than at low temperatures. This difference in binding may be due to temperature-dependent conformational changes in the repressor protein.

Example 2. In another *E. coli* strain, β galactosidase is synthesized constitutively at 43.5°C, but needs to be induced at 14°C.

To account for these data we need only slightly modify the above hypothesis. In this mutant the repressor protein binds more tightly to the operator region of the gene at low temperatures than at high temperatures. This difference may be due to differences in repressor protein conformation or a strengthening of weak bonding at low temperatures.

Example 3. Another mutant of *E. coli* displays the following temperature-dependent regulatory phenomenon: at 30°C, tryptophan can induce the synthesis of tryptophanase, while at 15°C induction is not possible.

Here, perhaps, the binding between the amino acid and the repressor protein occurs only at the higher temperature. Since the repressor protein will not leave the operator region unless it is bound to tryptophan, regulation is blocked at the lower temperature. Alternatively, perhaps the binding of tryptophan to the repressor does occur, but the needed change in repressor conformation to permit the operator-repressor interaction(s) to be broken cannot occur.

We raise these examples to illustrate the problems and potentials which might arise owing to temperature effects at the level of gene regulation. On the "problem" side of the ledger, we can easily appreciate how temperature changes might inhibit the proper regulatory responses involved in the control of gene expression. Thus, a change in temperature might prevent the turning-on or turning-off of a given gene. On the positive side, the possibility is raised that temperature might directly activate genes which code for proteins or RNA species needed at the new temperature. For example, if new classes of enzymes are required at the new habitat temperature, perhaps direct thermal activation of the genes coding for these enzymes might occur.

WEAK BONDS AND THE PHYSICAL STATE OF BIOCHEMICAL SYSTEMS

In biological systems, transitions from a liquid to a solid state invariably are due to a critical stabilization of large numbers of weak chemical bonds. As the kinetic energy of water is lowered, the hydrogen bond interactions among water molecules become more stable and, below a certain temperature, ice forms. Less familiar liquid-to-solid state transitions occur in the case of lipids. Again, a lowering of temperature leads to an increase in the stability of weak bonds, and an eventual solidification occurs.

Neither of these liquid to solid phase transitions is beneficial to organisms under most circumstances. Thus, we will see that organisms have "invented" a wide array of biochemical mechanisms to reduce the dangers of freezing—whether the solidification involved is aqueous or organic.

HORMONE-RECEPTOR PROTEIN INTERACTIONS AT DIFFERENT TEMPERATURES

Last, least understood, and especially fascinating is the topic of hormone-receptor protein interactions. The relatively new field of hormone receptor research has given us several examples of systems where temperature affects the binding of a hormone to its receptor protein, e.g., the thyroxin-thyroxin receptor system.

As in the case of temperature effects on the weak-bonded systems which control gene expression, the possible regulatory significance of temperature-induced changes in hormone-receptor protein interactions is vast. For example, if the biochemical changes which occur during thermal acclimation, as discussed in Section III of this chapter, are hormonally controlled, might there not be regulatory switches which involve temperature-dependent hormone binding and, as a result, gene or metabolic activations? If, for example, thyroid hormones trigger certain of the metabolic changes known to occur during temperature acclimation in fishes, might this hormone action be mediated through an increased binding affinity of a hormone receptor protein for thyroxin? With the possible exception of temperature effects at the level of transcription, no weak-bonded system is so poorly understood and so fascinating a topic for study.

REFERENCES

Books, Symposia, and Review Articles

Beyer, R. F. (1972). Effects of low temperature on cold sensitive enzymes from mammalian tissues. In *Hibernation-Hypothermia: Perspectives and Challenges,* pp. 17–54 (F. E. South, J. P. Hannon, J. R. Willis, E. T. Pengelley, and N. R. Alpert, eds.). Elsevier, Amsterdam.

Brandts, J. F. (1967). Heat effects on proteins and enzymes. In *Thermobiology,* pp. 25–72 (A. Rose, ed.). Academic Press, New York.

Farrell, J., and Rose, A. H. (1967). Temperature effects on micro-organisms. In *Thermobiology,* pp. 147–218 (A. H. Rose, ed.). Academic Press, New York.

Friedman, S. M. (1968). Protein-synthesizing machinery of thermophilic bacteria. *Bacteriological Reviews 32*:27–38.

Miller, G. T., Jr. (1971). *Energetics, Kinetics, and Life: An Ecological Approach.* Wadsworth, Belmont, California, 360 pp. (An excellent treatment of thermodynamics and kinetics for the student with a minimal background in physical chemistry.)

Watson, J. D. (1970). The importance of weak chemical interactions. In *The Molecular Biology of the Gene,* pp. 102–141. Benjamin, New York.

Journal Articles

Gavin, J. R. III, Roth, J., Jen, P., and Freychet, P. (1972). Insulin receptors in human circulating cells and fibroblasts. *Proceedings of the National Academy of Sciences* (U.S.) *69*:747–751.

Malcolm, N. L. (1969). Molecular determinants of obligate psychrophily. *Nature 221*:1031–1033.

Massey, V., Curti, B., and Ganther, H. (1966). A temperature-dependent conformational change in D-amino acid oxidase and its effect on catalysis. *Journal of Biological Chemistry 241*:2347–2357.

Roodman, S. T., and Greenberg, G. R. (1971). Conditions allowing synthesis of thymidylate synthetase at the nonpermissive temperature in a temperature-sensitive *thy* mutant normally blocked in its translation. *Journal of Biological Chemistry 246*:4853–4858.

II. THE REGULATION OF BODY TEMPERATURE

When an organism encounters an unfavorable environmental temperature, it can do one of two things to avoid the harmful effects discussed in the preceding section. Firstly, the organism may utilize behavioral, anatomical, or physiological means for keeping its body temperature essentially constant. Or, alternatively, if the organism cannot prevent the external temperature from acting directly on its internal biochemistry, a different adaptive strategy may be employed, one which involves compensatory changes in the chemistry of the cell. This second, biochemical, approach to temperature adaptation will be considered in the third section of this chapter.

BEHAVIORAL REGULATION OF BODY TEMPERATURE

While we are primarily concerned with biochemical adaptations in this volume, it would be misleading and unfair to restrict our attention solely to the biochemical level of biological organization. This approach might imply that most organisms are lacking in important non-biochemical defenses against the effects of thermal stress. In fact, we know that this is untrue. It is probably even fair to say that biochemical adaptations to temperature are frequently "last resort" responses which are made if and only if the organism lacks adequate behavioral, anatomical, and/or physiological avenues of escape from thermal stress.

The behavioral adaptations to temperature observed among widely different groups of organisms may well be the most varied and successful of all types of temperature adaptations. There can be little doubt that the vast majority of motile organisms seek to avoid temperature extremes. This behavioral regulation takes many forms and occurs over a number of time courses. Many organisms migrate diurnally and seasonally to minimize the range of temperatures they must cope with. For example, intertidal fishes may seek deeper, cooler pools of water when the tide recedes and the temperature of the intertidal zone rises. Similarly, during the summer many pelagic fishes migrate to deeper water where temperatures are lower than those of shallow waters.

Other patterns of behavioral regulation of body temperature involve body orientation in the sunlight and the aggregation of individuals to effect group-regulation of body temperatures. Insects and reptiles furnish us with especially good examples of thermal regulation via body orientation. These animals may bask in the sun during cool periods and seek shade during especially hot periods. Reptiles may also alter the shapes of their bodies to change the area of the light absorptive surface.

Aggregation phenomena, like other patterns of behavioral thermal regulation, are phylogenetically diverse. Many species of birds huddle together during periods when the temperature is low and/or the wind is strong enough to create a significant chill factor. An analogous behavior pattern is observed with newborn mammals, which often huddle closely together as a means of retaining metabolically generated heat. Behavioral thermal regulation thus involves both the acquisition of heat from the environment and the control of heat dissipation to the environment. Given that many organisms have effective ways of regulating the flow of heat energy between their bodies and the environment, let us consider how the rates of heat production and flow within the organism can be controlled so as to isolate the organism's biochemistry from changes in ambient temperature.

PHYSIOLOGICAL REGULATION OF BODY TEMPERATURE: ENDOTHERMY AND ECTOTHERMY

Before we examine some of the mechanisms of physiological regulation of body temperature, it is imperative that we carefully define several terms which deal with thermal regulation. The terms we are about to define are in common use in biological literature. Unfortunately, this does not mean that their usage has been especially consistent, or even particularly logical.

Some workers choose to categorize animals into two major groups on the basis of the species' ability to maintain a relatively constant body temperature. One group encompasses the "homeotherms," organisms which, as their name implies, tend to maintain a constant or near-constant body temperature under most circumstances. Birds and mammals are the only organisms normally grouped under this heading. We will use the term "homeotherm" only in its most basic, *etymological* sense; i.e., the term "homeotherm" will be used in reference to any organism which can hold its body temperature relatively constant—regardless of the phylogenetic status of that organism.

The second class of animals in this scheme are the "poikilotherms" ("poikilo-" is a Greek prefix meaning "varied"). These organisms have varying body temperatures and, in addition, usually have only minimal capacities to regulate their body temperatures physiologically.

Although the authors of this book, like the majority of physiologists, have used these two terms in the past, we wish to depart from the traditions of the majority and utilize different and, we feel, more meaningful, terminology in our discussion of thermal relationships. We prefer to use the terms "ectotherm" and "endotherm," terms which denote the characteristic *major source of body heat.* This terminology has several advantages. Firstly, the two terms indicate clearly where the organism obtains the greater share of its heat energy. Endotherms, as the name suggests, obtain their body heat largely from their own metabolic activities. Ectotherms obtain their heat from the environment; i.e., they cannot maintain a body temperature much different from the ambient temperature, owing merely to their own metabolic activity.

Another worthwhile advantage of categorizing organisms in terms of their major source of body heat is that we need not fall into semantic difficulties when dealing with ectotherms having constant body temperatures. Thus, an abyssal fish probably has a more nearly constant body temperature (i.e., is more "homeothermic" in the strict sense of this term) than a human. Similarly, most so-called homeotherms exhibit considerable variations in body temperatures diurnally or seasonally. Thus, hummingbirds may maintain a constant and high body temperature during the day, while at night they may largely cease to thermoregulate.

Finally, for our purposes of discussing basic adaptive strategies, we will see that it is advantageous to employ terms which do not have precise phylogenetic connotations. Thus, while the term "homeotherm" has normally been equated with mammals and birds, we will find that within the category of endothermic organisms we must group not only birds and mammals, but also such diverse organisms as tuna fishes, sharks, and moths. The terms "ectothermy" and "endothermy" therefore help us appreciate the fact that phylogenetically diverse organisms can rely on identical adaptive mechanisms for avoiding thermal stress.

ENDOTHERMY IN LARGE FISHES: REGULATION OF BODY TEMPERATURE IN TUNAS

Because all metabolizing systems generate heat, all organisms are at least *potentially* endothermic. In reality, however, most organisms have neither a sufficient rate of metabolism nor the means for retaining enough metabolically generated heat to allow them to attain the endothermic state. Particularly in the case of aquatic organisms, it is highly difficult to maintain the body's temperature different from the ambient temperature. This difficulty is especially acute in gill breathing forms such as fishes. The rate of thermal equilibration is approximately ten times as rapid as gaseous (oxygen and CO_2) equilibration. Consequently, if efficiency of gas exchange is given precedence over the retention of body heat—as must be the case for all gill breathers—then there would seem to be little way of avoiding an equilibration of body temperature with the ambient water temperature.

TABLE 7–4 MEASURED BODY TEMPERATURES OF SEVERAL FISH SPECIES (FROM CAREY ET AL., 1971)

	Muscle Temperature °C	Water Temperature °C
Grouper (*Epinephelus* sp.)	28.3	28.0
Mackerel (*Scomber scombrus*)	23.8	22.5
Bigeye tuna (*Thunnus obesus*)	28.9	21.0
Bluefin tuna (*Thunnus thynnus*)	29.5	19.0
Mako shark (*Isurus oxyrhynchus*)	24.9	20.4
Blue shark (*Prionace glauca*)	21.9	21.9

While most gill breathers do in fact have body temperatures essentially identical to the temperature of the surrounding water, there are some fascinating examples of partially endothermic gill breathing organisms. The rapidly swimming tuna fishes are perhaps the most impressive of these "unexpected" endotherms. As the data of Table 7–4 illustrate, several species of tuna, and some large and active sharks, succeed in maintaining their body temperatures several degrees C above ambient temperature. And, at least in the case of the large bluefin tuna (*Thunnus thynnus*), this endothermic capacity leads to a large degree of homeothermy; i.e., this species not only can maintain its body temperature above ambient, but also can hold the temperatures of several key body regions essentially constant.

The mechanisms whereby tuna succeed in retaining a significant fraction of their body heat, in spite of a gill breathing habit, are illustrated in Figure 7–7. The basic principle involved in the "design" of the tuna heat retention (=exchanger) system is not unique to this one physiological "application." Countercurrent exchange systems are also utilized in the exchange of gases (e.g., in fish gills), and comparable heat exchangers are found in other endothermic organisms (e.g., in the legs of sea birds, where the heat exchangers prevent loss of deep body heat to the environment through the poorly insulated [and often submerged] legs and feet).

In heat exchangers such as those found in tuna, the warmed venous blood passes into close contact with cooled, aerated arterial blood arriving from the gills. The intimate contact between the arterial and venous circulations enables a significant amount of heat to be passed into the arterial blood, i.e., to be retained in the tissue.

There are several advantages of endothermy for the tuna. By maintaining its muscle temperatures at a relatively high level, the power of the swimming musculature is greatly increased. The rate of the contraction-relaxation cycle (at least in frogs) exhibits a Q_{10} of approximately 3. Thus, by warming its musculature by about 10°C,

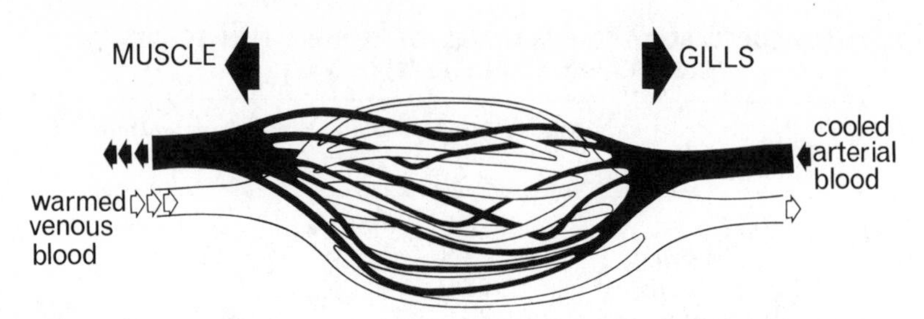

Figure 7–7 A diagrammatic representation of the anatomy of the tuna heat exchanger. Blood warmed in the rapidly metabolizing musculature passes into close contact with cooled, aerated blood arrived from the gills. Heat is passed from the venous circulation to the arterial circulation in the small blood vessels of the heat exchanger. See Carey *et al.* (1971) for further details.

the tuna likely gains a 3-fold increase in the power it can extract from a given mass of muscle fibers. This gain in power, and thus in swimming speed, enables the tuna to "break into" a whole new class of food organisms: the rapidly swimming pelagic organisms such as squid and mackerel.

Certain tuna species also display abilities to hold the temperatures of their viscera above ambient. This no doubt speeds digestion and absorption. The ability to process ingested food rapidly and efficiently may allow the tuna to get along with a relatively reduced stomach capacity which, in turn, may be a further aid to swimming efficiency.

It has recently been shown that the retinal and brain temperatures of some tuna are also maintained above ambient temperature. Whether this aspect of tuna endothermy leads to increased perceptual acuity and a better "thinking" prowess remains an open question.

Finally, large tuna species like the bluefin, which are homeothermic as well as endothermic, gain another key advantage over their ectothermic relatives: the bluefin species *can select its environment on other than thermal criteria.* Thus, the particular habitat which is chosen may be rich in food or especially suited for breeding; its thermal properties may or may not be particularly suitable for a less homeothermic ectotherm.

ENDOTHERMY IN THE INVERTEBRATES: THE SPHINX MOTH

We have already stated that the key factor in determining whether or not a potential for endothermy exists is *not* the phylogenetic status of the organism. Rather, it is the abilities of the organism to *generate* sufficient metabolic heat and to *retain* enough of this heat to keep its body temperature above ambient. To the reader accustomed to regarding only mammals and birds as true endotherms, it may come as a mild shock to learn that even some small invertebrates can display impressive capacities for endothermic control of their body temperatures. The sphinx moth is a case in point.

Like other rapidly flying insects, the sphinx moth (*Manduca sexta*) generates large amounts of heat in the musculature which drives the wings. These muscles are located in the thorax of the organism and, in the sphinx moth, much metabolic heat is retained within this region. The thoracic region of this moth exhibits a high degree of endothermy and homeothermy.

Although we are unaccustomed to thinking of insects as capable of the degree of endothermy found among "higher" organisms, in reality it should be clear that insects like the sphinx moth have relatively few problems to solve on their way to at least a partially endothermic state. Unlike fishes, insects do not respire via water-bathed gills. Instead, they obtain oxygen and release CO_2 by means of a tracheal system which does not involve aqueous gas transport. As a consequence of this gaseous transport system, insects do not suffer the rapid heat loss encountered by most gill breathers.

Non-aquatic insects also can insulate their bodies with relative ease and efficiency. The thoracic region of the sphinx moth is covered by a thick layer of "hair" which creates a dead-air space and permits a great reduction in the rate of heat flow to the environment. In contrast, the abdomen of the moth is poorly insulated.

Considering the raw material available to the sphinx moth, its mechanism of endothermy should be readily deducible. Heinrich has found that the key factor which couples the aforementioned raw material into an efficient endothermic apparatus is a control system which regulates the rate of blood flow from the thorax to the abdomen. The rate of pulsation of the abdominal blood vessel was found to be directly proportional to the thoracic temperature. Thus, the flow of warmed hemolymph from the thickly insulated and metabolically active thorax to the poorly insulated abdomen was closely regulated to insure that enough heat was retained in the thorax to maintain its temperature at approximately 41°C, while excess heat was efficiently exhausted to the environment. The homeothermic efficiency of this system is most impressive: a thoracic temperature of approximately 41°C could be maintained over an ambient temperature range from 15° to 35°C.

ENDOTHERMY IN THE INVERTEBRATES: THE BEES

The bumblebee (*Bombus vagans*) also displays a considerable ability for endothermic control of temperatures. Like many flying insects, bumblebees warm up their thoracic (flight) musculature by shivering (muscular contraction which is not coupled to work). Bees also have well-insulated thoracic regions, permitting a considerable reduction in heat flow from the body to the environment. When feeding at temperatures below about 26°C, bees resume shivering when they are stationary on a flower. The continuation of shivering allows the thorax to maintain a temperature of 32 to 33°C over an ambient temperature range of 9° to 24°C. At higher temperatures the thorax temperature increases as the ambient temperature rises. This would seem to indicate that the bumblebee does not have a mechanism for regulating heat flow from thorax to abdomen, as in the case of the sphinx moth. In fact, the stationary feeding bees cannot use their abdomens efficiently as heat exchangers, as can the steadily flying sphinx moth. When stationary, the loss of heat via convection is substantially reduced and, therefore, so is the bee's capacity for homeothermy.

Thermogenesis in the stationary bumblebee may involve, in addition to shivering, a heat-generating reaction scheme which is a metabolic "short circuit." In several species of the genus *Bombus*, Newsholme and co-workers have demonstrated extraordinarily high levels of fructose diphosphatase in the flight musculature. FDPase concentrations were found to equal the concentrations of phosphofructokinase, a most unusual situation for a non-gluconeogenic tissue (see pp. 40–42). Total FDPase activity was up to 40 times that observed in vertebrate tissues.

As we have already stressed in the earlier discussion of carbohydrate metabolism, the simultaneous activity of PFK and FDPase leads to a "short-circuit" of energy metabolism and the wholesale splitting of ATP. Thus, as quickly as FDP is generated via the reaction F6P + ATP = FDP + ADP, the molecule is split to regenerate F6P and inorganic phosphate. The net reaction is thus ATP = ADP + P_i + heat. Clearly, under almost all physiological circumstances, FDPase and PFK must not be operative simultaneously, and, in fact, the two enzymes are normally activated or inhibited in diametrically opposed manners. FDPase is usually inhibited by AMP, whereas PFK is sharply activated by a low energy charge of the adenylate system. The bumblebee FDPase, in contrast to all other FDPase enzymes which have been examined, is refractory to AMP.

The flight musculature of the bumblebee therefore appears to represent one of the relatively rare instances in which the "wastage" of high energy phosphate bonds is to the organism's advantage. Newsholme and colleagues propose that during stationary phases of food gathering, the bumblebee musculature is kept warm largely by the heat produced in the FDPase-PFK "short circuit." This thermogenic ability enables the bumblebee to forage on cool and/or damp days which are inimical to food gathering activities in the case of

other bees (e.g., members of the genus *Apis*, which lack high le
of AMP-insensitive FDPase). One major question remainin
be answered concerns the mechanisms which turn off the bum
bee's FDPase during flight, a regulatory effect which, of course, is vital under conditions where large levels of glycolytically formed ATP are required for mechanical work.

Much as an individual bee can partially regulate its thoracic temperature, so can bee colonies control to a large extent the temperature of their hives. In hot weather, honeybees (*Apis mellifera*) are widely spaced within the hives, and the temperature of the hive interior is lowered via evaporative cooling resulting from fanning. At low ambient temperatures, bees cluster together and conserve body heat by reducing conduction. It has been found by Southwick and Mugaas that a honeybee hive maintained a temperature of 18 to 32°C over an ambient temperature range of −17°C to +11°C.

HEAT GENERATION IN MAMMALS: THYROXIN ACTIVATION OF THE Na^+/K^+ ATPase SYSTEM

The sources of the heat utilized for endothermic control of body temperature are varied. For active organisms such as tuna, flying insects, running mammals, and flying birds, enough heat is generated by the working muscles to supply the energy required for endothermic control of the body temperature. When the organism is stationary, heat may often be generated by shivering responses, in which ATP is split (and heat is released) by the usual muscle actin-myosin ATPase system, but no work is accomplished. However, shivering thermogenesis is seldom a chronic mechanism for heat generation, and in mammals shivering thermogenesis is usually superseded by non-shivering thermogenesis if the cold stress is prolonged.

The mechanisms of non-shivering thermogenesis have long eluded physiologists and biochemists. While for some time it has been clear that the thyroid hormones, thyroxin and triiodothyronine (T_3), stimulate this process, the mechanistic basis of their activities has until recently been unknown.

Ismail-Beigi and Edelman now have found that the source of heat in thyroid-stimulated thermogenesis is the terminal "high energy" bond of ATP, and that the splitting of this bond is effected by the Na^+/K^+ ATPase system of the membranes of calorigenic tissues, including liver, skeletal muscle, and kidney (brain, gonad, and smooth muscle are notably refractory to the calorigenic effects of thyroid hormones). Administration of thyroid hormone is accompanied by an increase in the activity of this ATPase system, with a concomitant elevation in the rate of body heat production.

While this discovery is important because it appears to answer a long-standing question in the field of physiology, it also is significant in terms of illustrating an important evolutionary strategy: in the "design" of this vital heat-generating system, an enzyme system which must have appeared at the dawn of biological (or at least cellular) evolution has, some billion years later, come to serve a second and highly important function.

BROWN ADIPOSE TISSUE: AN EXAMPLE OF A TISSUE ADAPTED FOR THERMOGENESIS

In most of the examples of endothermy we have discussed, the heat utilized to regulate body temperature is a by-product of work activities such as muscle contraction. The work activities of tissues such as muscle are obviously not directed primarily towards the production of heat for thermoregulation. Instead, endothermy (or the potential for it) is a "spinoff" of the metabolic reactions involved in activities like contraction and ion transport.

In mammals we find one of the few examples known* of a tissue which functions chiefly as a biological furnace. Brown adipose tissue is not known to perform any useful work in the sense of coupling chemical energy to drive mechanical processes or biosyntheses. The metabolic heat generated in brown adipose tissue is known to be important in mammals during (i) early post-natal life, (ii) cold acclimation, and (iii) arousal from hibernation.

The anatomical localization and ultrastructural properties of brown adipose tissue reveal how well suited this tissue is for thermogenic function. Brown adipose tissue either surrounds vital organs, such as the heart, or is located "upstream" of these organs, in terms of circulatory patterns. Thus, the heat generated by brown adipose tissue is efficiently channelled towards two key organs (heart and brain) whose function must be maintained for survival.

Ultrastructurally, brown adipose tissue is characterized by large numbers of circular mitochondria which have elaborately folded cristae. During cold acclimation these mitochondria are surrounded by fat droplets. Thus, the "fuel" and the "machinery" for a vigorous oxidative metabolism are present in this tissue.

The actual pattern of oxidative metabolism in brown adipose tissue differs significantly from that of other highly oxidative tissues such as heart muscle. Whereas the oxidation of fatty acids usually leads to the generation of large quantities of ATP, in brown adipose tissue it appears that oxidative phosphorylation is uncoupled (Figure 7–8). Thus, the energy released during acetylCoA oxidation is not trapped in "high energy" bonds of ATP, but rather is dissipated as heat.

---

*See Bendell and Bonner (1971) *Plant Physiol.* *46*:236–245, and Schonbaum *et al.* (1971) *Plant Physiol.* *47*:124–128.

Regulation of brown adipose tissue metabolism is still unclear. The tissue is definitely under hormonal control by catecholamines such as epinephrine. However, the uncoupling mechanism is in doubt, although free fatty acids may be the most significant uncouplers, as suggested in Figure 7–8.

From an evolutionary standpoint, perhaps the most interesting aspect of brown adipose tissue function is that, as in the case of thyroid-activated non-shivering thermogenesis, a pre-existing biochemical system is slightly modified and put to a quite different physiological use. The oxidative phosphorylation machinery is regulated to permit the energies of electrons removed from acetylCoA residues in the Krebs citric acid cycle to be "wasted." However, when this "waste" is viewed in the context of the whole organism, it is seen that the activity of brown adipose tissue provides an effec-

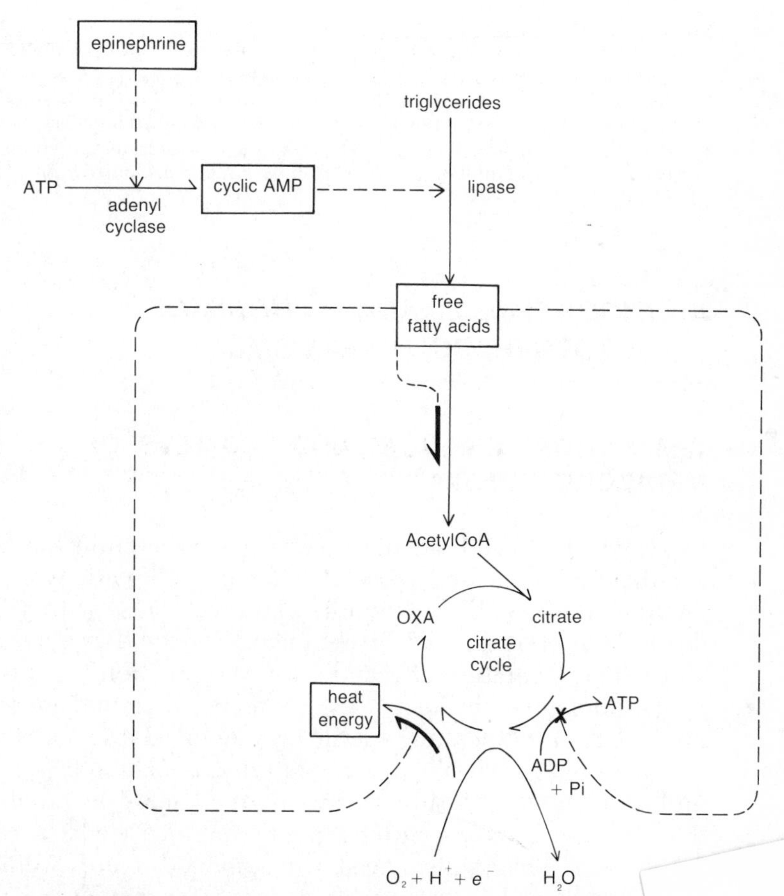

Figure 7–8 Metabolism of brown adipose tissue. Illustrated ar pathway of heat generation and the likely control circuitry. Free are thought to serve as the "signal" for activating respiration (anc production) and for uncoupling respiration from phosphorylation.

tive and rapid—even if energetically expensive—means for promoting endothermy. The fact that mammals "squander" large amounts of energy for thermoregulation, by the several means discussed, indicates how great are the advantages to be gained from shielding the cellular chemistry from changes in temperature.

REFERENCES

Journal Articles

Carey, F. G., Teal, J. M., Kanwisher, K. W., and Lawson, K. D. (1971). Warm-bodied fishes. *American Zoologist 11*:137–145.

Heinrich, B. (1971). Temperature regulation of the sphinx moth, *Manduca sexta.* II. Regulation of heat loss by control of blood circulation. *Journal of Experimental Biology 54*:153–166.

Heinrich, B. (1972). Temperature regulation in the bumblebee *Bombus vagans*: a field study. *Science 175*:185–187.

Ismail-Beigi, F., and Edelman, I. S. (1970). Mechanism of thyroid calorigenesis: role of active sodium transport. *Proceedings of the National Academy of Sciences (U.S.) 67*:1071–1078.

Newsholme, E. A., Crabtree, B., Higgins, S. J., Thornton, S. D., and Start, C. (1972). The activities of fructose diphosphatase in flight muscles from the bumble-bee and the role of this enzyme in heat generation. *Biochemical Journal 128*:89–97.

Southwick, E. E., and Mugaas, J. N. (1971). A hypothetical homeotherm: the honey-bee hive. *Comparative Biochemistry and Physiology 40*:935–944.

Stevens, E. D., and Fry, F. E. J. (1971). Brain and muscle temperatures in ocean caught and captive skipjack tuna. *Comparative Biochemistry and Physiology 38*:203–211.

III. BIOCHEMICAL ADAPTATIONS OF ECTOTHERMIC ORGANISMS

TEMPERATURE CHANGES AND PROBLEMS OF METABOLIC CONTROL

Environmental temperature changes pose unique problems for ectothermic organisms, since changes in the kinetic energy of the environment usually are rapidly and quantitatively transferred to the cellular chemistry of the organism. For ectothermic organisms, all of the temperature effects discussed in Section I of this chapter apply. Among these effects, perhaps none are as fundamental as the influences of temperature on the rates of metabolic reactions.

The "rate effects" of temperature on metabolic systems fall under two general headings. Firstly, the overall metabolic rate of the organism will vary with temperature. Since the basic metabolic functions of the organism must be maintained at rates falling within certain more or less narrow limits, it is clear that changes in body temperature can pose severe threats to the organism's survival.

Thus, it is not surprising to find that many ectothermic organisms have impressive abilities to hold their rates of metabolism relatively stable in the face of often quite large changes in environmental (body) temperature. We term this pattern of metabolic homeostasis "*metabolic rate compensation.*" The biochemical bases of this compensatory response will provide much of the subject matter of this section.

The second major effect of temperature on metabolism is of a differential nature. Not all enzymic reactions are accelerated or decelerated by a temperature change to the same extent. Under conditions of saturating substrate concentration, the temperature coefficients (Q_{10} values) of different enzymic reactions range between approximately 2 and 4; the range of Q_{10} values, as we shall see, is much greater when the enzymic reactions occur under conditions of physiological (non-saturating) substrate concentration. As a consequence of these differing temperature sensitivities of various enzymic reactions, changes in body temperature can lead to serious imbalances in the relative activities of different metabolic reactions. Thus, we would predict that the regulatory properties, like the total catalytic capacity, of ectothermic enzyme systems will be found to display unique adaptations which permit function in a varying thermal environment. To reiterate a major theme of this volume, successful metabolic function demands (i) the proper amounts of catalytic potential and (ii) the close regulation of the rates at which, and the directions in which, this potential is put to use.

THE RELATION OF TIME AND ADAPTIVE STRATEGY

The challenges posed by an environmental change—and the potential means available to the organism for circumventing the potential ill effects of the change—are closely related to the rate at which the environmental alteration occurs. In general, the more rapid the environmental change, the more serious is its impact on organisms. Changes which occur within matters of seconds or minutes may not give organisms time to behaviorally adapt to (e.g., escape from) the altered environment, much less allow the organism to erect whatever physiological and biochemical defenses are necessary to insure continued success in the altered habitat. On the other hand, environmental changes which occur gradually, say over time periods of weeks or months or, especially, during the time spans of many generations, may allow the organism ample time to develop compensatory changes in its cellular chemistry. In other words, to restate again a fundamental theme of this volume, the greater the time available to the organism to adapt, the more fundamentally might it restructure its basic chemistry.

The mechanisms utilized by ectotherms to control their rates of metabolic function under circumstances of varying temperature illustrate clearly this key interplay between time and adaptive strategy.

METABOLIC COMPENSATION: THREE TIME-COURSES OF RESPONSE

For more than a century, biologists have been aware that ectothermic organisms have prodigious abilities to hold their rates of metabolism relatively constant (i.e., more constant than the laws of physics and chemistry appear to predict), despite large temperature changes. Early naturalists who sailed into high latitude waters observed that the rates of activity of polar aquatic ectotherms often were not appreciably lower than the rates characteristic of similar species from temperate and tropical seas. Later workers found that oxygen consumption and growth rates were also quite similar among ectotherms from widely different latitudes. Especially among aquatic species which, unlike many terrestrial forms, normally must remain active throughout the year, some degree of *evolutionary rate compensation* to temperature appears to be the rule, rather than the exception (Figure 7–9).

These observations, made on different species widely separated latitudinally, were repeated when biologists began examining populations of a single species which were acclimated to different temperatures. Thus, at their respective acclimation temperatures,

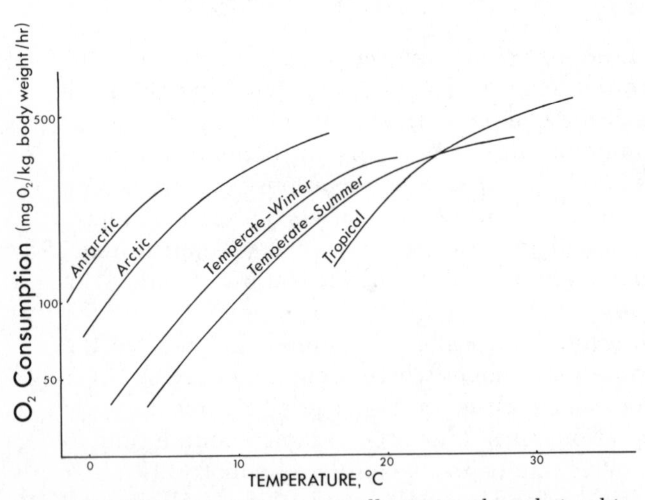

Figure 7–9 Representative curves illustrating the relationship between temperature and standard metabolic rates of fishes adapted or acclimated to different temperatures (modified after Brett, 1971).

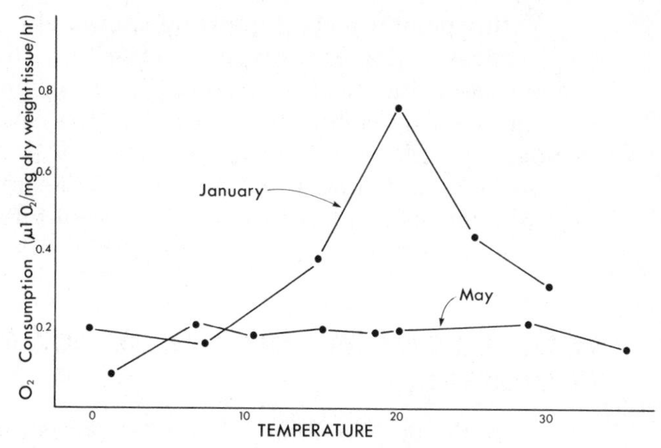

Figure 7–10 Seasonal variation in the effect of acute temperature change on the standard rate of oxygen consumption of the winkle *Littorina littorea.* Ambient air temperature in January was approximately 4°C; the May air temperature was approximately 15°C. Note the temperature-independent metabolic rates over the range of temperatures the organism is likely to encounter in its habitat. (Data from Newell and Pye, 1970.)

warm- and cold-acclimated ectotherms were found to have markedly similar metabolic rates—the cold-acclimated specimens metabolized much faster than one would have predicted by extrapolating the metabolic rates of the warm-acclimated specimens down to low temperatures (Figure 7–9). *Seasonal rate compensation,* like its evolutionary counterpart, also was found to characterize the metabolism of isolated tissues. This finding suggested that the basis of the rate-compensatory responses lay in fundamental changes in the cellular chemistry.

Quite recently, physiologists have come to recognize a third time-course of metabolic compensation to temperature: *immediate rate compensation.* At least some organisms, notably intertidal invertebrates which encounter large and fairly rapid changes in body temperature on a diurnal basis, exhibit essentially temperature-independent rates of metabolism (Figure 7–10). Furthermore, the temperature range over which the metabolic Q_{10} approximates unity can be modified seasonally (Figure 7–10).

All three time-courses of temperature compensation, but especially immediate compensation, raise a key question: How do ectothermic organisms succeed in maintaining the rates of their enzymic processes at least somewhat independent of changes in the kinetic energies of the reacting molecules? In other words, how do ectothermic organisms apparently "disobey" the fundamental laws describing temperature effects, such as the Arrhenius Equation, which we treated in Section I of this chapter?

At this point a second question should also be raised: Are the biochemical strategies utilized in these three time-courses of rate compensation identical, or must different compensatory mechanisms be used in each of these temporally disparate, yet phenomenologically similar, processes?

To frame potential answers to these questions, let us first consider the means available to ectotherms for effecting temperature compensatory changes in their rates of enzymic activity.

POTENTIAL ENZYMIC MECHANISMS FOR RATE COMPENSATION

Since the basic problem of rate compensation to temperature is one of regulating the rates of enzymic activity, most of the metabolic control mechanisms known to be important under conditions of constant temperature should be of potential use to the thermally compensating individual. In particular, three major ways of adjusting rates of enzymic activity seem especially important:

(i) Changes in the concentrations of pre-existing enzymes (the "Quantitative Strategy").

(ii) Changes in the types of enzymes present in the system (the "Qualitative Strategy").

(iii) Modulation of the activities of pre-existing enzymes (the "Modulation Strategy").

These three strategies, of course, are not mutually exclusive, and successful temperature compensation may often involve all three types of adaptive response.

RATE COMPENSATION VIA CHANGES IN ENZYME CONCENTRATION: BASIC POTENTIALS AND LIMITATIONS OF THE "QUANTITATIVE STRATEGY"

The strongest *a priori* argument for an important role of enzyme concentration changes in metabolic compensation to temperature is based on our knowledge of enzyme induction phenomena in bacteria and eucaryotic organisms. In many well-studied procaryotic and eucaryotic systems, changes in the chemical environment of the cell, e.g., in the types of nutrient molecules present, are followed by large scale alterations in the levels of intracellular enzymes. For example, the addition of a new species of sugar to the culture medium in which bacteria are growing may be followed by the appearance of large concentrations of the enzyme species which enable the bacteria to take up and metabolize this new energy and carbon source.

In mammalian cells the mechanisms for changing enzyme concentrations are somewhat more complicated and much less clearly understood. While dietary and hormonal changes, for example, may stimulate enzyme synthesis, mammals have an additional "handle" on the situation: rates of enzyme degradation are also subject to control, although this control is likely much less precise than the regulation of enzyme synthesis. Nonetheless, mammals and, undoubtedly, other eucaryotes can regulate their enzyme concentrations in two ways. Bacteria, in contrast, do not appear to degrade "unneeded" enzymes but, instead, reduce the enzyme concentrations via dilution during cell division. This, at least, appears to be the pattern in rapidly multiplying bacterial cultures.

While eucaryotic ectotherms have two major ways of increasing enzyme concentrations during cold adaptation—or of reducing enzyme levels during warm adaptation—it is important to realize that these two mechanisms differ in their usefulness. Degradation can affect only the types of enzymes already present in the cells, whereas synthesis (involving transcription and translation of new genetic information) can lead to the appearance of new enzyme variants, a point to which we will return below.

Whatever the means used for altering the concentrations of enzymes present in the cells, the "quantitative strategy" of rate compensation can be appreciated as having considerable potential. Especially in the case of rate-limiting enzymes, increases in enzyme concentration in the cold and decreases in concentration at warm temperatures could bring about major stabilization of metabolic rates.

There are, however, some obvious limitations to the "quantitative strategy" of metabolic compensation. Firstly, it is hardly conceivable that the concentrations of *all* enzymes, perhaps even all rate-limiting enzymes, could be increased during cold adaptation. The solvent capacity of the cell is limited (p. 97), and the number of binding sites for membrane-associated enzymes is likewise limited. In addition, the time required for changing enzyme concentrations via synthesis of new proteins is almost certainly too great to render the "quantitative strategy" of much usefulness in immediate temperature compensation.

There is even a more critical limitation to this mechanism of metabolic compensation. Because the structures of enzymes are thermally labile, as discussed earlier, it is likely that some of the types of enzymes present in the cell may not function well, if at all, under the new thermal regime. When this is the case, there seems little point in synthesizing more of the same types of enzymes if they function only poorly under the new environmental conditions. When a change in temperature adversely affects the structures and functions of enzymes, selective forces are likely to be especially acute for the generation of new enzyme variants.

CHANGES IN ENZYMIC ACTIVITIES DURING THERMAL ACCLIMATION

Given the advantages and limitations of the "quantitative strategy," what do available data allow us to conclude about the importance of this mechanism in thermally compensating ectotherms? Do ectotherms living at low temperatures contain more enzyme molecules than their warm-adapted or warm-acclimated counterparts?

Unfortunately, no studies to date have attacked these questions directly. The data presented in Table 7–5 are *activity* estimates, not enzyme concentration values. Thus, even though it is abundantly clear that certain enzymes exhibit higher activities in cold-acclimated organisms than in warm-acclimated ones, we cannot rigorously conclude that concentration differences are responsible for the observed changes. A further ambiguity stems from the fact that most of the data summarized in Table 7–5 are maximal velocity (V_{max}) values obtained in studies with crude tissue homogenates. Consequently, we have no information about the relative enzymic activities which would be observed under physiological substrate concentrations. Furthermore, except for a small number of cases, it is not known whether the same enzyme variants (isozymes) were present in the warm- and cold-acclimated tissues. Therefore, we cannot exclude the possibility that some of the activity differences are due to differently efficient enzymes in the warm- and cold-acclimated tissues. Lastly, the available data tell us nothing about possible changes in the local environment in which the enzymes operate. The possibility therefore exists that some of the activity differences may be the result of alterations in the intracellular milieu, e.g., in ionic concentrations or, in the case of lipoprotein enzymes, in the structure of enzyme-associated fatty acids.

In spite of the ambiguities associated with the data of Table 7–5, two conclusions are clear. Firstly, the activities of many of the most important "mainline" catabolic reactions (e.g., those of oxidative phosphorylation) are elevated during cold acclimation. Secondly, not all enzymes exhibit the same pattern of activity change; i.e., acclimation to different temperatures involves differential effects on different enzymic pathways.

DIFFERENTIAL CHANGES IN ENZYMIC ACTIVITY DURING ACCLIMATION

The enzymic reactions which exhibit the highest degrees of temperature compensation are primarily those involved in generating the energy "currency" (ATP, NADH, etc.) needed by the

TABLE 7–5 ENZYMIC ACTIVITIES IN TISSUES OF DIFFERENTLY ACCLIMATED ECTOTHERMS (FROM HAZEL AND PROSSER, 1970)

A. Enzymes exhibiting temperature compensation.

Enzyme	Organism
Glycolytic Enzymes:	
Phosphofructokinase	Goldfish (*Carassius auratus*)
Aldolase	Crucian carp (*Carassius vulgaris*)
	Golden orfe (*Idus idus*)
Lactate dehydrogenase	Goldfish
	Brook trout (*Salmo fontanalis*)
	Lake trout (*Salmo namaycush*)
Hexose Monophosphate Shunt Enzymes:	
6-Phosphogluconate dehydrogenase	Crucian carp
Krebs Cycle and Electron Transport Enzymes:	
Succinic dehydrogenase	Eel (*Anguilla vulgaris*)
	Goldfish
	Golden orfe
	Earthworm (*Lampito mauritii*)
Malate dehydrogenase	Golden orfe
Cytochrome oxidase	Goldfish
	Golden orfe
	Crab (*Uca pugnax*)
Succinate-Cytochrome c reductase	Goldfish
NADH-Cytochrome c reductase	Goldfish
Protein Synthetic Enzymes:	
Amino acyl transferase	Toadfish (*Opsanus tau*)
Miscellaneous Enzymes:	
Na^+-K^+ ATPase	Goldfish
Mitochondrial ATPase	Fly (*Musca domestica*)
	Cockroach (*Periplaneta americana*)

B. Enzymes exhibiting inverse compensation or no compensation (* indicates no compensation was observed).

Enzyme	Organism
Peroxisomal and Lysosomal Enzymes:	
Catalase	Crayfish (*Cambarus affinis*)
	Eel (*Anguilla vulgaris*)
	Crucian carp
	Goldfish
Peroxidase	Goldfish
Acid phosphatase*	Goldfish
D-amino acid oxidase*	Goldfish
Miscellaneous Enzymes:	
Mg^{2+}-ATPase	Goldfish
Choline acetyl transferase	Goldfish
Acetylcholine esterase	Goldfish
	Killifish (*Fundulus heteroclitus*)*
	Rainbow trout (*Salmo gairdneri*)*
Glucose-6-phosphate dehydrogenase	Crucian carp

cells at all times. Thus, the activities of the enzymes associated with the Krebs citric acid cycle and electron transport are found to rise significantly during cold acclimation. In light of the fact that the oxygen consumption rate of the organism also rises during cold acclimation, these enzymic findings should not be too surprising.

Other metabolic reactions display an "inverse" pattern of temperature compensation, i.e., their activities are higher in tissues of warm-acclimated specimens. As indicated in Table 7–5, many of these inversely compensating reactions are involved in degradative processes, e.g., the reactions associated with the peroxisomes and lysosomes. Hazel and Prosser speculate that at high temperatures, when the metabolic rate of the organism is elevated, there will be a sharp increase in the need for ridding the cells of metabolic breakdown products. Thus, enzymes which are involved in degradation reactions might be expected to exhibit this inverse pattern of temperature compensation.

Studies of differently acclimated fish have also revealed that the activities of competing metabolic pathways change during thermal acclimation. For example, the pentose phosphate pathway (hexose monophosphate shunt) appears to assume a greater role in glucose catabolism, relative to glycolysis, in cold-acclimated fishes. This change in the relative activities of pathways involved in glucose catabolism was the first discovered example of what might be termed "*seasonal metabolic reorganization*," the differential rechannelling of metabolic flow at high and low temperatures. Further manifestations of metabolic reorganization include a shift toward a higher level of fatty acid synthesis in cold-acclimated fish and, possibly, an increased reliance on aerobic metabolism in the cold-acclimated state.

The physiological functions of these seasonal metabolic reorganizations can be at least partially understood. Because there is a reduction in the basal or maintenance metabolism at low temperatures, organisms likely do not need to maintain as high a metabolic rate at low temperatures as they do at higher temperatures. This reduction in metabolic demands at low temperatures is illustrated in Figure 7–9; low temperature compensation is not "perfect," but rather the warm-acclimated or warm-adapted organisms display at least slightly higher levels of metabolism at their normal temperatures compared to their cold-acclimated (cold-adapted) counterparts.

If the maintenance metabolism of ectotherms is reduced at low temperatures, then the organism may be able to channel a larger share of its food into the synthesis of energy storage compounds, such as fats, and into protein synthesis ("growth"). We can thus appreciate one reason why an increased pentose phosphate pathway activity would be advantageous at low temperatures: the NADPH generated by this pathway supplies the reducing power for fatty acid synthesis, and the pentose moieties are important precursors for nucleic acid synthesis.

These seasonal reorganizations of metabolism can thus be viewed as exploitative adaptations to an annually changing thermal regime. If temperature decreases permit a reduction in maintenance metabolism, then selection might strongly favor organisms with the ability to redirect their metabolic flow in ways consistent with optimal usage of their ingested food. Reorganizing metabolism is not a rate-compensatory process in the strictest sense of the term. Rather, it is a process which takes into account the potential benefits accruing from an environmental change and attempts to exploit these potentials with a high degree of biochemical efficiency.

THE "QUALITATIVE STRATEGY" AND RATE COMPENSATION: GENERAL CONSIDERATIONS

We mentioned earlier that the "quantitative strategy" has some important limitations. Firstly, in our discussion of this strategy, we implicitly assumed that the types of enzymes present in the cell will continue to function satisfactorily under the new thermal regime. In light of what we have said in Section I of this chapter concerning the thermal lability of the higher orders of protein structure and of protein-ligand interactions, this assumption probably is not valid for all enzymes. Consequently, one *a priori* argument in support of the "qualitative strategy," i.e., the utilization of different variants of a given enzyme at different temperatures, can be based on the likelihood that a single enzyme variant will not be adequate, functionally, over the entire range of temperatures at which catalysis and regulation must be performed. *This argument can be seen to apply to all three time-courses of temperature compensation.*

A second *a priori* argument in support of the "qualitative strategy" of enzyme adaptation is based on questions of cellular "architecture." Even if the types of enzymes already in the cells do continue to function satisfactorily when the temperature is changed, and, therefore, temperature compensation can be effected by varying enzyme concentrations, there may be a limit to the number of enzyme molecules which can be packaged into the cellular water and membranes.

Thirdly, we can fault the "quantitative strategy" on grounds of cellular energy "economy." Because protein synthesis is an energetically costly process, the synthesis of large quantities of enzymes at low temperatures may be highly demanding on the cell's energy resources. Thus, it would appear to be more efficient to produce a "better" catalyst at low temperatures than to synthesize larger numbers of relatively poor catalysts. Let us now consider how this type of "qualitative" adaptation might be effected by examining possible loci for "improving" the functional characteristics of enzymes for operation at low temperatures.

ENZYME VARIANTS IN TEMPERATURE COMPENSATION: SITES OF SELECTION FOR IMPROVED FUNCTION

On the basis of earlier discussions of enzyme function, the ways in which an enzyme might be improved for low temperature function should be obvious. On the one hand, an enzyme will work at a faster rate if its ability to bind substrates or cofactors is enhanced. This, in fact, is what usually happens when positive modulators stimulate enzymic activity. Secondly, the catalytic rate of an enzyme will be raised if the enzyme succeeds in further reducing the free energy of activation "barrier" to its reaction. And, lastly, the splitting of the enzyme-product complex to generate free enzyme and product(s) can be facilitated as a means of increasing catalytic efficiency. In short, each general step in the catalytic process is *potentially* sensitive to selection for increased efficiency. We will now examine the "use" to which this potential raw material has been put by thermally adapting ectotherms.

ACTIVATION ENERGY DIFFERENCES AMONG ENZYME VARIANTS

Historically, comparative biochemists began their search for a "better" enzyme at the activation step of the catalytic process. Since the primary catalytic function of enzymes is to reduce the energy "barriers" to chemical reactions, an enzyme's catalytic efficiency is inversely proportional to the free energy of activation ($\Delta G^{\ddagger}$) of its reaction. Thus, for a given metabolic reaction, the best catalyst will be the one which reduces the $\Delta G^{\ddagger}$ value the most. Since it is known that different isozymes of a given enzyme which are present in a single individual can differ markedly in their catalytic efficiencies by this criterion (Table 7–6), it seems plausible to hypothesize that selection will favor the evolution of enzymes with greater abilities to lower $\Delta G^{\ddagger}$ values in low-temperature-adapted ectotherms. In other words, this hypothesis predicts a positive correlation between the free energy of activation of an enzymic reaction and the temperature to which the organism (enzyme) is adapted. This hypothetical relationship and its effect on catalysis is illustrated in Figure 7–11.

TABLE 7–6 "Activation Energy" Characteristics and Substrate Turnover Numbers of Lactate Dehydrogenase (LDH) Isozymes of the Rabbit. H_4-LDH is the characteristic "heart-type" isozyme; M_4-LDH is the "muscle-type" LDH. (Data given by Wroblewski, F., and Gregory, K. F. (1961). *Annals of the New York Academy of Sciences. 94*:912–932.)

LDH	"Activation Energy" (Ea) = $\Delta H^{\ddagger} + RT$ (calories/mole)	Predicted Relative Rates on Basis of Ea Differences[1]	Observed Substrate Turnover Numbers[2]
H_4	8,300	3,300	43,000
M_4	13,200	1	12,000

[1]Assuming the entropy of activation values of the two reactions are identical; i.e., the free energy of activation in each case is assumed to be primarily determined by the enthalpy of activation (Ea-RT).

[2]Measured at 30°C.

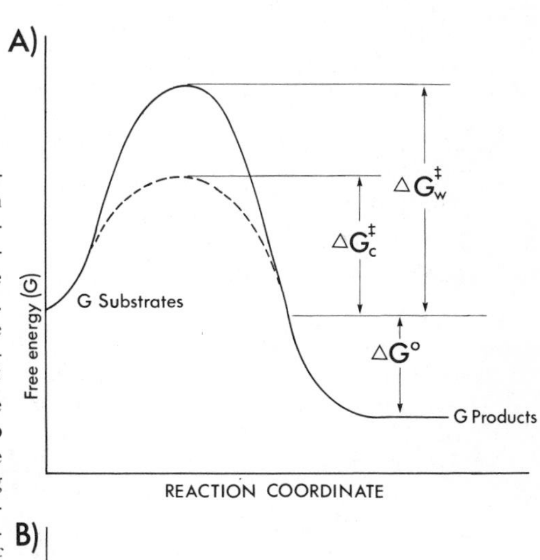

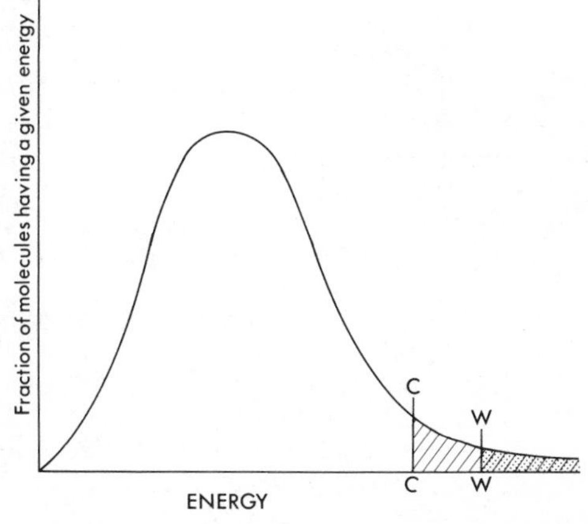

Figure 7–11 (A) A schematic representation of the free energy changes which occur when a particular enzymic reaction is catalyzed by two enzyme variants having differing abilities to reduce the free energy of activation ($\Delta G^{\ddagger}$). The hypothetical enzyme from the warm-adapted organism is characterized by the solid curve and the $\Delta G^{\ddagger}$ value $\Delta G^{\ddagger}_{W}$. The cold-adapted enzyme is able to reduce the energy barrier to $\Delta G^{\ddagger}_{C}$. Note that both reactions have the same $\Delta G°$ value; i.e., the differing abilities to speed-up the transformation are not accompanied by differences in the equilibrium position of the reaction. (B) Theoretical energy distribution curve, illustrating how a lowering of the free energy of activation by the cold-adapted enzyme increases the fraction of the population of molecules which is reactive at any given temperature. The cold-adapted enzyme, characterized by the $\Delta G^{\ddagger}_{C}$ value, reduces the energy "barrier" to the reaction to "C". All molecules having energy equal to or greater than "C" are potentially reactive. The warm-adapted enzyme reduces the energy barrier only to "W"; thus, only those molecules having energies equal to or greater than "W" (indicated by dots) can react.

The appeal of this hypothesis has led to numerous comparative studies of activation energies of enzymic reactions. Unfortunately, aside from a very small number of instances, these "tests" of the hypothesis have involved the measurement of the enthalpy of activation ($\Delta H^{\ddagger}$), not the free energy of activation. The former parameter can be readily obtained from the slope of an Arrhenius plot, which graphs log V_{max} of the reaction versus the reciprocal of the absolute temperature. The free energy of activation can only be computed if the entropy of activation, as well as the enthalpy of activation, is known, since $\Delta G = \Delta H - T\Delta S$. Because the entropy of

TABLE 7–7 FREE ENERGY OF ACTIVATION ($\Delta G^{\ddagger}$) VALUES AND SUBSTRATE TURNOVER NUMBERS OF ENZYMES FROM DIFFERENTLY ADAPTED ORGANISMS

Enzyme/Organism	Assay Temperature (°C)	$\Delta G^{\ddagger a}$	Turnover Number	Reference
Glyceraldehyde-3-phosphate dehydrogenase				
Rabbit	5	15.3	14.7[b]	1
	35	14.9	436	
Lobster	5	14.6	55.0	
	35	14.8	532	
Cod	5	14.7	44.8	
	35	14.8	544	
Phosphorylase-b				
Rabbit	0	15.9	2.4[b]	2
	30	15.2	180	
Lobster	0	15.0	13.5	
	30	15.1	212	
Lactate dehydrogenase-M_4				
Duck	25	–	82[c]	3
Pheasant	25		88	
Bullfrog	25	–	86	
Dogfish	25	–	109	
Halibut	25	–	159	
Tuna	25	–	160	
Bovine	25	–	80	

[a] kcal/mole
[b] μmoles substrate converted to product/μmole enzyme/second
[c] (moles of DPNH oxidized per mole of enzyme per minute) × 1000

References:

1. Cowey, C. B. (1967). *Comparative Biochemistry and Physiology 23*: 969–976.
2. Assaf, S. A., and Graves, D. J. (1969). *Journal of Biological Chemistry 224*:5544–5555.
3. Pesce, A., Fondy, T. P., Stolzenbach, F., Castillo, F., and Kaplan, N. O. (1967). *Journal of Biological Chemistry 242*:2151–2167.

activation is difficult to determine, there are few cases in which all three activation parameters are known for different variants of a given enzyme. Some of the better experimental data are summarized in Table 7–7.

The data of Table 7–7 show that enzymes from ectothermic organisms can have much higher efficiencies than the homologous enzymes from mammals. The relative catalytic capacities of mammalian and ectothermic enzymes are best reflected in the characteristic *substrate turnover numbers* of the enzymes. A *seemingly* small difference in $\Delta G^{\ddagger}$ between homologous forms of an enzyme leads to a very large difference in the actual rate of catalysis.

WHY ARE MAMMALIAN ENZYMES LESS EFFICIENT THAN ECTOTHERMIC ENZYMES?

If we accept, even tentatively, the conclusion that ectotherms have enzymes with higher catalytic efficiencies than the homologous enzymes from mammals, we face a most intriguing question: Assuming that mammals evolved from an ectothermic ancestor which already possessed highly efficient catalysts, how do we account for an apparent loss of enzymic efficiency during the course of mammalian evolution? How, in other words, can we account for the loss of something which appears to be to an organism's advantage?

Like many—perhaps most—questions about evolutionary changes, we can only speculate as to the basis for this particular outcome of the selective process. One explanation of this seeming loss of an adaptive property is based on fundamental considerations of protein structure. If an enzyme is well adapted to function at temperatures near, let us say, 5°C, it will have the abilities to undergo reversible conformational and, perhaps, aggregational changes at quite low temperatures on the biological scale. To be highly flexible at low temperatures, the protein may have an inherently nonrigid tertiary or quaternary structure, since certain of the weak bonds which stabilize the higher orders of protein structure will be relatively strong at lower temperatures where, by definition, less kinetic energy is present to disrupt these bonds. In contrast to this cold-adapted enzyme, the enzymes present in mammals and birds must undergo reversible changes in 3° and 4° structure in the presence of large amounts of kinetic energy. Mammalian and avian enzymes therefore may have more rigid higher order structures than ectothermic enzymes. This is not to imply that ectothermic enzymes will be totally denatured at temperatures near 37°C; we know that this is not the case. However, these enzymes might be so "loose" in their higher orders of structure that their functions which depend on a specific protein geometry would be seriously impaired. Support for this conjecture comes from studies of the temperature-dependence of substrate binding, a topic we will discuss shortly.

The final assumption needed for this argument is that the rate of catalysis of which an enzyme is capable, at least under *saturating* concentrations of substrate, may be directly proportional to the rate at which reversible conformational changes in and around the catalytic site can occur. An enzyme with a flexible 3° structure may therefore be especially efficient, and we would predict that an ectothermic enzyme would exhibit a higher V_{max} than a mammalian or avian enzyme at all temperatures. However, the flexibility which makes catalysis rapid may make substrate binding more difficult. Consequently, selection may have favored the binding of substrate, rather than the activation of substrate, during the evolution of mammalian enzymes. The maintenance of a specific geometry at the site where substrate is bound may require a relatively rigid protein structure which, as a result, leads to a lesser degree of catalytic efficiency with respect to $\Delta G^{\ddagger}$ reduction.

"CLIMBING" THE $\Delta G^{\ddagger}$ BARRIER: EVOLUTIONARY DIFFERENCES IN THE TYPE OF ENERGY USED

Since $\Delta G^{\ddagger} = \Delta H^{\ddagger} - T\Delta S^{\ddagger}$, the free energy of activation barrier to an enzymic reaction may be "climbed" either by increasing the enthalpy (heat content) of the enzyme-substrate system or by increasing the degree of structure or order of the system (decreasing its entropy). Both types of effect will obviously increase the free energy of the enzyme-substrate complex.

When we examine the contributions of activation enthalpy ($\Delta H^{\ddagger}$) and activation entropy ($\Delta S^{\ddagger}$) to $\Delta G^{\ddagger}$ for homologous reactions, we find that ectothermic and avian-mammalian systems differ significantly (Table 7–8). For all reactions which have been studied, the $\Delta H^{\ddagger}$ and $\Delta S^{\ddagger}$ values of the ectothermic reactions are lower than the corresponding parameters of the homologous avian-mammalian reactions.

The explanation for this difference may lie in the different energy characteristics of the local, intracellular environments in which ectothermic and avian-mammalian enzymes conduct their functions. In ectotherms, the heat content of the cell and the entropy of the cell are low, relative to the situation present in "warm-blooded" birds and mammals. Thus, for ectothermic enzymes it may be relatively feasible to increase the structure or order of the enzyme-substrate system and thereby partially "climb" the $\Delta G^{\ddagger}$ barrier with $T\Delta S^{\ddagger}$ energy.

In birds and mammals, with body temperatures near 40°C, the increased thermal energy present in the system, relative to ectotherms, may render the ordering or structuring of the enzyme-sub-

TABLE 7–8 THERMODYNAMIC ACTIVATION PARAMETERS FOR ENZYMES FROM DIFFERENTLY ADAPTED ORGANISMS

Enzyme/Organism	Assay Temperature (°C)	$\Delta G^{\ddagger}$ [a]	$\Delta H^{\ddagger}$ [b]	$\Delta S^{\ddagger}$ [c]	Reference
M_4 LDH					
Rabbit	5	13.20	12.55	−2.3	1
	35	13.25	12.50	−2.5	
Chick	5	12.85	10.55	−8.4	
	35	13.15	10.45	−8.7	
Halibut	5	12.5	8.75	−13.5	
	35	12.9	8.65	−13.7	
Glyceraldehyde-3-phosphate dehydrogenase					
Rabbit	5	15.3	18.45	11.4	1,2
	35	14.9	18.40	11.3	
Lobster	5	14.5	13.95	−2.2	
	35	14.8	13.90	−2.9	
Cod	5	14.7	13.95	−2.6	
	35	14.8	13.90	−2.9	
Phosphorylase-b					
Rabbit	0	15.95	20.65	17.2	1,3
	30	15.20	20.60	17.8	
Lobster	0	15.05	15.35	1.1	
	30	15.10	15.30	0.8	

[a] Kcal/mole
[b] Kcal/mole
[c] entropy units/mole

References:

1. Low, P., Bada, J., and Somero, G. (1973). *Proceedings of the National Academy of Sciences* (U.S.), in press.
2. Cowey, C. B. (1967). *Comparative Biochemistry and Physiology 23*: 969–976. Data recalculated by Low, Bada and Somero.
3. Assaf, S. A., and Graves, D. J. (1969). *Journal of Biological Chemistry 224*:5544–5555. Data recalculated by Low, Bada and Somero.

strate system more difficult. However, owing to the higher heat content of avian-mammalian cells, the activation of the enzyme-substrate complex by the addition of enthalpy may be less difficult, compared to ectothermic systems.

In brief, it appears that each of these groups of organisms utilizes the energetic characteristics of its intracellular environment to optimal advantage. In ectothermic systems it may be relatively simple to keep entropy increases to a minimum (e.g., glycogen phosphorylase-b) or, in fact, to significantly increase the degree of order of the enzyme-substrate system during activation (lactate dehydrogenase and glyceraldehyde-3-phosphate dehydrogenase). In avian and mammalian systems, the higher entropy of the cell appears to be "accepted," and the $\Delta G^{\ddagger}$ barrier is "climbed" almost entirely, if not completely, by the addition of enthalpy to the enzyme-substrate complex.

EVOLUTIONARY CHANGES IN $\Delta G^{\ddagger}$ VALUES: SOME FINAL COMMENTS

In light of the foregoing discussion and the data of Table 7–7, what can we conclude about the adaptive significance of changes in the free energy of activation of a reaction? Considering the likely evolutionary *direction* of the changes listed in Table 7–7, we must immediately realize that a low free energy of activation was almost certainly the "primitive state" of enzymic reactions. Thus, it is illogical to portray the high efficiencies of ectothermic enzymes, relative to mammalian enzymes, as something gained by ectothermic species specifically to enable them to exist at low temperatures, except in the general sense that this is why enzymes are here at all.

The ability of an enzyme to reduce the $\Delta G^{\ddagger}$ value of a reaction is probably a property which is not independent of other enzymic parameters. Thus, as we have speculated earlier, heightened catalytic efficiency, in terms of $\Delta G^{\ddagger}$ reduction, may be accomplished only at the expense of reduced efficiency at other steps in the reaction process, such as in E-S interactions. Selection may therefore balance changes in $\Delta G^{\ddagger}$ values with changes in enzyme-ligand interactions. The extent to which $\Delta G^{\ddagger}$ can be reduced may be determined only after the enzyme has acquired the correct affinity for substrate.

Lastly, we must raise several unanswered questions concerning the importance of changes in $\Delta G^{\ddagger}$ values in temperature compensation. Are changes of this nature important in evolutionary adaptation of different ectothermic species? Do the enzymes from cold-acclimated ectotherms reduce $\Delta G^{\ddagger}$ values more than the enzymes from warm-acclimated specimens? Can immediate temperature compensation of metabolic rates be effected by instantaneous, temperature-dependent changes in the catalytic efficiency of an enzyme? This third question, which, like the other two, cannot be answered in the context of available information, leads us to consider a second important locus of enzyme evolution: temperature-dependent changes in enzyme-substrate affinities.

BASIC TEMPERATURE EFFECTS ON ENZYME-SUBSTRATE AFFINITIES

The affinity of the substrate-binding site for its substrate(s) largely depends on the geometry of the site (can the substrate "fit"?) and the charge properties of the site (will the substrate be attracted or repulsed?). These two criteria are illustrated for acetylcholinesterase (AChE) in Figure 7–12.

A The Acetylcholinesterase Reaction

$$H_3C-\overset{\oplus}{N}(CH_3)(CH_3)-CH_2-CH_2-O-\underset{CH_3}{C}{=}O \longrightarrow H_3C-\overset{\oplus}{N}(CH_3)(CH_3)-CH_2-CH_2-OH + {}^{\ominus}OOC-CH_3 + H^+$$

Acetylcholine — Choline — Acetate

B The Substrate-Binding Site of AChE

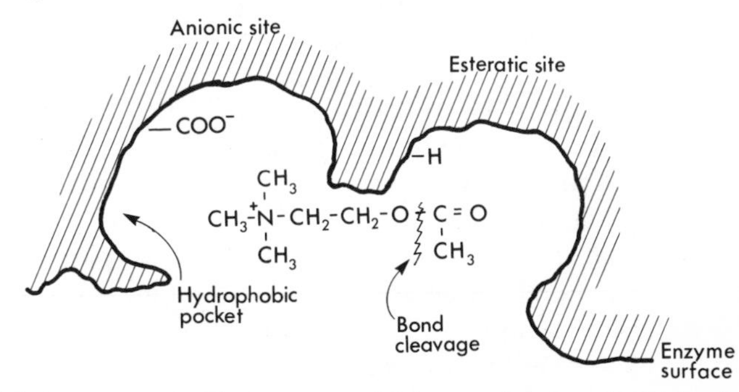

Figure 7–12 The acetylcholinesterase (AChE) reaction and the acetylcholine (ACh) binding site. The major portion of the stabilizing energy holding the AChE-ACh complex intact comes from the hydrophobic interactions between the methyl (CH_3) groups of ACh and the hydrophobic amino acid residues found in the hydrophobic "pocket" of the "anionic site." Splitting of the bond between the choline and acetate moieties occurs at the "esteratic site."

Note that the acetylcholine (ACh) binding site has two distinct regions. An "anionic site," possessing a negative charge, attracts the positively charged quaternary ammonium nitrogen. In the immediate vicinity of the "anionic site" are hydrophobic amino acid residues which provide a suitably water-free environment for the three methyl residues bound to the nitrogen atom. Indeed, the major source of stability of the AChE-ACh complex are the hydrophobic interactions between the enzyme and the three methyl groups. Replacement of the methyl groups by hydrogen leads to a sharp reduction in E-S affinity. A seven-fold decrease in binding affinity occurs when only a single hydrogen-for-methyl substitution is made. In contrast to the effect on E-S affinity, this substitution does not affect the V_{max} of the reaction. Thus, once the substrate is correctly positioned, catalysis proceeds with the same efficiency in both cases because the actual cleavage of the ACh molecule takes place at a separate site, the "esteratic site" (Figure 7–12). This separation of the sites of substrate attraction and substrate cleavage again illustrates the separate roles of substrate binding, on the one hand, and of catalytic action of enzymes, on the other.

On the basis of the chemistry of the AChE substrate binding site, we can appreciate how temperature changes can alter E-S affinities. Firstly, since the weak bonds which stabilize the E-S complex are temperature-sensitive, changes in temperature could either hinder or facilitate E-S complex stabilization, even in the absence of any temperature-caused changes in enzyme structure.

Secondly, because the same classes of weak bonds used to stabilize the E-S complex also stabilize the higher orders of protein structure, changes in temperature might be expected to alter the structure of the enzyme and, thereby, cause changes in the geometry of the substrate-binding site. We can easily visualize the likelihood that the active sites of enzymes may have the proper geometrical configuration only over a small range of temperatures.

It is important to realize that increases *and* decreases in temperature could interfere with E-S complex formation. Whereas interactions which are stabilized by ionic bonds will be strengthened at low temperatures, hydrophobic interactions may be substantially weakened as temperature falls. In the case of AChE-ACh interactions, it is difficult to predict the precise effects of temperature on E-S complex stability, since the "anionic site" involves both ionic and hydrophobic bonds. However, since the hydrophobic interactions appear to provide the major share of complex stabilization, we might predict that low temperatures could severely hinder AChE-ACh complex formation (see Figure 7–14).

Lastly, there exists the possibility that temperature changes may promote alterations in the charge properties of the substrate-binding site. Since the dissociation of weakly acidic and weakly basic groups is somewhat temperature sensitive, with low temperatures generally tending to impede dissociation, an enzyme's active sites might lose their requisite charge properties if the temperature exceeds a certain range. Again, either high or low temperatures could adversely affect an enzyme.

Let us now examine some of the patterns of variation in E-S affinity with temperature and try to determine how important these changes are in enzymic adaptation and metabolic compensation to temperature.

POSITIVE THERMAL MODULATION

Over an organism's normal range of habitat temperatures, the most common pattern of variation in E-S affinity with temperature is illustrated by the theoretical curves of Figure 7–13. Some examples of temperature dependence of apparent K_m values for different enzymes are illustrated in Figures 7–14, 7–15, and 7–16. The temper-

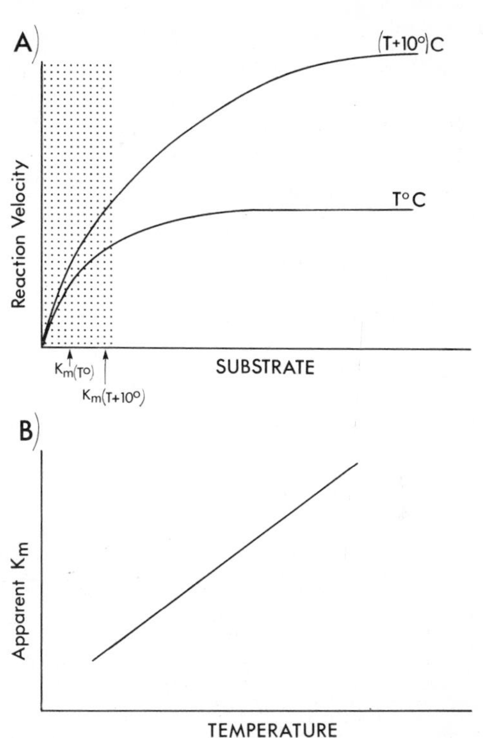

Figure 7–13 Positive thermal modulation of enzymic activity. (a) The relationship between reaction velocity and substrate concentration for a single enzyme at two different assay temperatures. The affinity for substrate (measured by the reciprocal of the apparent K_m value) approximately doubles as the temperature is decreased by 10°C. Physiological substrate concentrations are indicated by stippling. Note that the rate-compensating effects of positive thermal modulation occur only at non-saturating substrate concentrations. (b) A plot of apparent K_m versus assay temperature, illustrating positive thermal modulation.

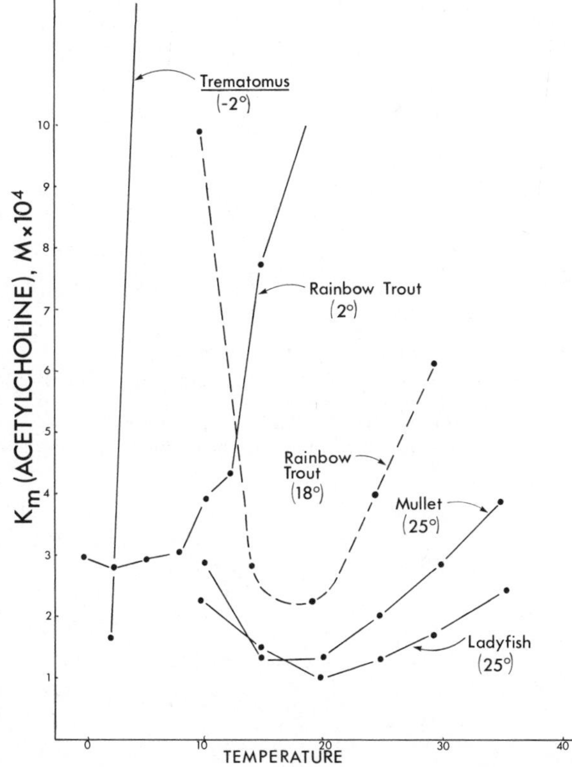

Figure 7–14 The effect of temperature on the K_m of acetylcholine (AChE) for acetylcholinesterase (AChE) enzymes of four species of fish: rainbow trout, acclimated to 2° and 18°C; *Trematomus borchgrevinki*, an Antarctic species; and the mullet (*Mugil cephalus*) and the ladyfish (*Elops hawaiensis*), two tropical species. The approximate temperature to which the fish is adapted or acclimated is given in parentheses below the fish's name. Data from Baldwin (1971) and Baldwin and Hochachka (1970).

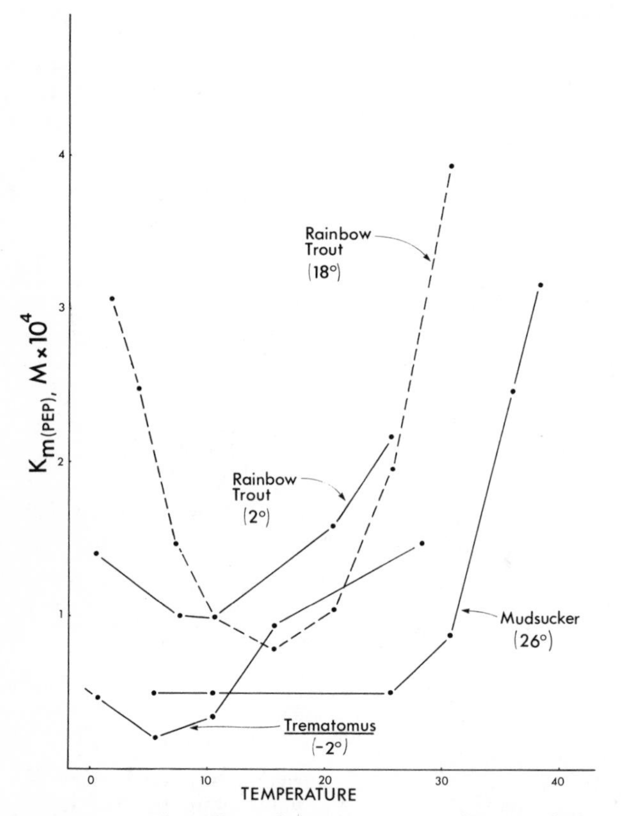

Figure 7–15 The effect of temperature on the K_m of phosphoenolpyruvate (PEP) for pyruvate kinase (PK) enzymes of three species of fish: rainbow trout (*Salmo gairdneri*), mudsucker (*Gillichthys mirabilis*), and *Trematomus borchgrevinki*. The approximate temperature to which the fish is adapted or acclimated is given in parentheses below the fish's name.

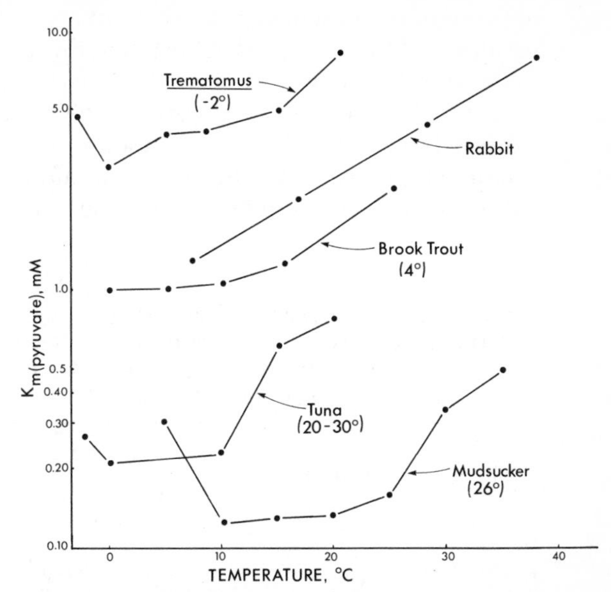

Figure 7–16 The effects of temperature on the K_m of pyruvate for lactate dehydrogenases from muscle (skeletal) of different organisms. Data from Hochachka and Lewis (1971), Somero (1969, 1973), and Somero and Johansen (1970).

ature-dependent K_m change illustrated in Figure 7–13—the direct relationship between temperature and apparent K_m—we term "*positive thermal modulation*" in order to stress the similar effects which positive modulators and temperature decreases have on the affinity parameters of enzymes.

Note again that affinity parameters are instrumental in regulating the rates of catalysis only at non-saturating substrate concentrations (Figure 7–13a). Thus, the rate-compensating effects of changes in E-S affinity values are noted only at non-saturating substrate concentrations; under V_{max} conditions the temperature-dependence of the enzymic reaction rate, of course, remains a function of the enthalpy of activation.

DETERMINANTS OF Q_{10} VALUES IN LIVING SYSTEMS

The exact temperature dependence (Q_{10} value) of an enzymic reaction *in vivo* will be determined by a number of factors in addition to the inherent temperature dependence set by the $\Delta H^{\ddagger}$ value of the enzymic reaction. Firstly, the temperature-dependence of the affinity parameter will in large measure determine how greatly the reaction velocity changes with temperature. The larger the temperature-dependent change in the apparent K_m or $S_{0.5}$ value, the greater will be the extent of "positive thermal modulation;" i.e., the lower will be the Q_{10} value of the reaction at any given concentration of substrate below saturating levels. Since different enzymes exhibit large differences in temperature-versus-affinity parameter relationships, we find wide variation among enzymes in their Q_{10} values at non-saturating substrate concentrations (Table 7–9).

TABLE 7–9 TEMPERATURE COEFFICIENT (Q_{10}) VALUES FOR SEVERAL ENZYMIC REACTIONS AT DIFFERENT CONCENTRATIONS OF SUBSTRATE. PHYSIOLOGICAL SUBSTRATE CONCENTRATIONS ARE APPROXIMATELY 0.50 mM OR LOWER IN ALL CASES

Substrate Concentration (mM)	Q_{10} Organism (Temperature Range of Measurements)			
Lactate Dehydrogenase	King Crab[1] (5–15°)	Trout[2] (5–15°)	Trout[2] (10–20°)	Zoarcid (Arctic fish)[1] (5–15°)
2.00	1.8	1.7	1.6	1.7
1.00	1.5	1.7	1.6	1.6
0.10	1.2	–	–	1.2
0.05	0.8	1.4	1.0	–
Pyruvate Kinase	King Crab[3] (5–15°)		Trout (summer-acclimated)[4] (10–20°)	
1.00	2.7		5.3	
0.10	2.5		4.2	
0.05	1.9		–	
0.02	1.6		–	
6-Phosphogluconate Dehydrogenase	King Crab[1] (5–15°)			
0.30	3.8			
0.20	3.4			
0.10	3.0			
0.01	1.9			
Isocitrate Dehydrogenase		Trout (summer)[5] (5–10°)		Trout (winter)[5] (5–10°)
1.00		5.14		4.0
0.10		4.0		4.0
0.075		4.0		3.4
0.05		7.1		3.4

References:
1. Somero, G. N. (1969). *The American Naturalist 103*:517–530.
2. Somero, G. N., and Hochachka, P. W. (1969). *Nature 223*:194–195.
3. Somero, G. N. (1969). *Biochemical Journal 114*:237–241.
4. Somero, G. N., and Hochachka, P. W. (1968). *Biochemical Journal 110*:395–400.
5. Moon, T. W., and Hochachka, P. W. (1971). *Biochemical Journal 123*:695–705.

Another major determinant of the Q_{10} value of an enzymic reaction is the concentration of substrate present near the substrate-binding site. When E-S affinity increases as temperature drops, for any enzyme the extent of "positive thermal modulation" is inversely proportional to substrate concentration. That is, the Q_{10} value of the reaction will fall as substrate concentrations are reduced. This relationship is documented by the data of Table 7–9. Note that at sufficiently low concentrations of substrate, the rate of the reaction may be higher at low temperatures (where E-S affinity is high) than at higher temperatures (where E-S affinity is relatively low). These data perhaps give us the most powerful illustration of the key importance of E-S affinity in governing the rates of enzymic catalysis under physiological substrate concentrations.

POSITIVE THERMAL MODULATION: A MECHANISTIC BASIS FOR IMMEDIATE COMPENSATION?

The importance of positive thermal modulation as a rate-compensatory mechanism is likely to be greatest in the case of immediate temperature compensation. Immediate compensation occurs too rapidly for changes in enzyme concentrations to be involved, and, therefore, a nearly instantaneous increase in enzymes' abilities to bind substrates as temperature is lowered represents a compensation mechanism of tremendous potential.

To determine whether positive thermal modulation does, in fact, underlie the immediate temperature compensation phenomena which have been observed, one would have to perform careful studies of the rate-limiting enzymes of the metabolic pathways under study, determining their affinity values at different temperatures and, in addition, measuring the intracellular concentrations of substrate available to the enzymes. To date, this rigorous an analysis of immediate temperature compensation has not been accomplished. Nonetheless, there are some lines of evidence which are completely consistent with the hypothesis that positive thermal modulation is important in effecting immediate metabolic compensation.

Newell and co-workers have conducted detailed studies of the thermal responses of intertidal organisms such as the winkle *Littorina littorea,* which exhibits a high degree of temperature-independence in its rate of oxygen consumption (Figure 7–10). To determine the molecular bases of this metabolic compensation, Newell and co-workers "dissected" the metabolic apparatus of this organism. They observed, firstly, that the Q_{10} values characteristic of whole organism respiration were also observed in studies of iso-

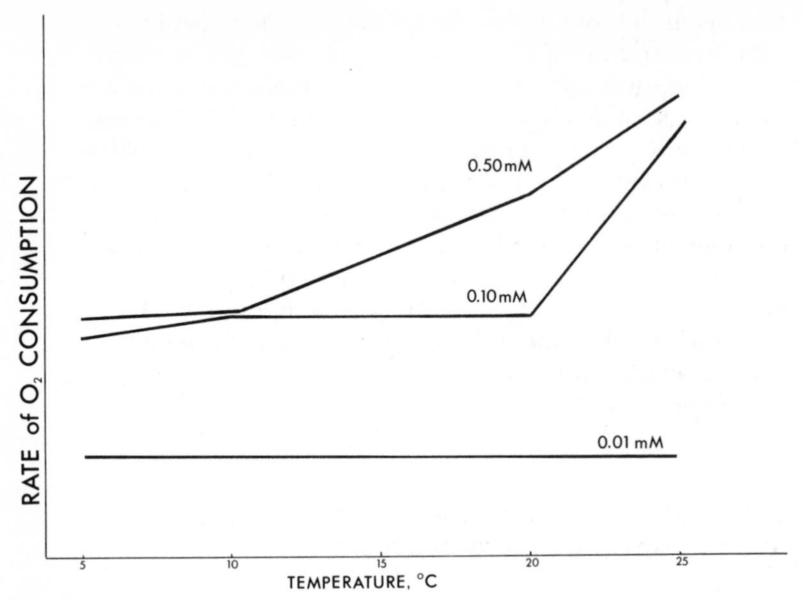

Figure 7–17 The effect of temperature on the rate of oxygen consumption of mitochondria from *L. littorea* in the presence of varying concentrations of pyruvate. Note the differing temperature dependencies of mitochondrial respiration at high and low substrate concentrations. (Data from Newell and Pye, 1971.)

lated tissues and tissue homogenates. Most importantly, they also found that the respiration of isolated mitochondria was temperature-independent when measurements were made using physiological (i.e., non-saturating) concentrations of substrate (pyruvate) (Figure 7–17). Under conditions of saturating pyruvate concentrations, the respiration of the mitochondria exhibited a sharp increase with temperature. These are precisely the results predicted on the basis of the positive thermal modulation data found for isolated enzymes.

Similar evidence supporting the role of positive thermal modulation in immediate temperature compensation has come from studies of *in vitro* tissue preparations in which physiological substrate concentrations were employed. The works of Dean, with rainbow trout tissues, and of Hochachka and colleagues, using a variety of ectothermic tissues, have shown that at least certain metabolic pathways can exhibit virtually temperature-independent rates when the relevant substrates are present at physiological concentrations. It is unfortunate that most studies of isolated tissues have been conducted with saturating substrate concentrations, for the true *in situ* thermal properties of the tissues' metabolism would probably not be measured under these unnatural circumstances.

BASAL METABOLISM AND ACTIVE METABOLISM: DIFFERING Q_{10} VALUES

The work of Newell and other physiologists working with ectotherms has revealed that the temperature-dependence of basal metabolism may be considerably lower than that of active metabolism. In other words, the Q_{10} of metabolism may be proportional to the metabolic rate *per se* (Figure 7–18).

The biochemical basis of this interesting pattern of temperature-dependence is not known. However, on the basis of the above discussions of positive thermal modulation and immediate temperature compensation, one hypothesis should quickly come to mind: the changing Q_{10} values observed as the metabolic rate varies may reflect changes in the intracellular concentrations of substrate. Under basal, non-active conditions, substrate levels may be relatively low. A sharp increase in the metabolic rate may be accompanied by rises in substrate concentration. Consequently, positive thermal modulation would be a more potent rate-stabilizing influence under basal conditions.

Unfortunately, like many other hypotheses concerning the biochemical bases of environmental adaptation, there are insufficient data available to allow a critical test of this hypothesis. Recent

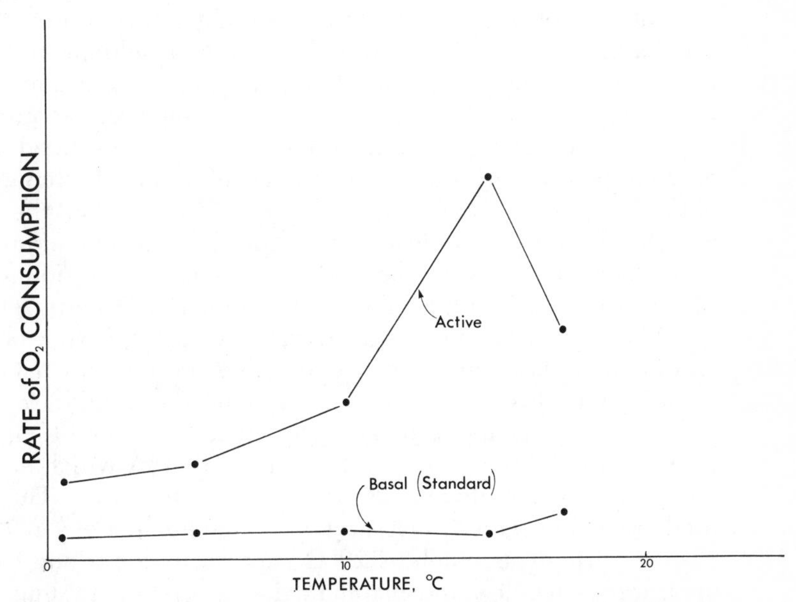

Figure 7–18 The effect of temperature on active and standard (basal) rates of metabolism of *L. littorea*, for organisms adapted to temperatures of approximately 12°C. (Data from Newell and Pye, 1971.)

studies of insect flight muscle metabolism have found relatively short-term fluctuations in the concentrations of glycolytic intermediates. On the other hand, studies of working vertebrate muscle have revealed that muscle lactate and pyruvate levels can increase and remain high durving vigorous activity.

Whatever the biochemical basis for the metabolic rate dependence of Q_{10}, it is clear that the stress imposed on an organism by a temperature change may be more or less severe, depending on the activity level of the organism.

IMMEDIATE TEMPERATURE COMPENSATION: GENERALITY OF THE RESPONSE, OR, WHY AREN'T ALL ORGANISMS "PERFECT" COMPENSATORS?

If all organisms were capable of the degree of immediate compensation which is characteristic of some intertidal organisms, we (or, more precisely, ectotherms) would not have to worry about the effects of seasonal acclimation and evolutionary adaptation to temperature. However, as the respiration data in Figure 7–9 illustrate, many ectotherms (e.g., almost all fishes which have been studied) exhibit fairly sharp Q_{10} values for their metabolic rates. Immediate temperature compensation as shown by many invertebrates of the intertidal region thus is probably an exception to a general pattern.

Another observation worth mentioning at this point is that seasonal acclimation and evolutionary adaptation seldom lead to complete or "perfect" metabolic compensation. For example, even though polar fishes do better at their low habitat temperatures than one would predict by extrapolating a tropical fish's metabolic rate down to temperatures near 0°C, the absolute metabolic rate of the polar fish at its habitat temperature is still lower than the rate exhibited by the tropical fish at its higher habitat temperature. The same lack of "perfect" metabolic compensation is found when differently acclimated ectotherms are compared (Figure 7–9).

These observations raise several questions. Firstly, how general is immediate temperature compensation; i.e., in what types of organisms is this time-course of compensation found? As already indicated, the available data suggest that immediate temperature compensation is most pronounced in ectotherms which encounter rapid and predictable changes in habitat temperature. Thus, intertidal organisms which must survive diurnal fluctuations in temperature of perhaps as much as 20°C may be under strong selective pressure to develop and maintain the means for holding at least their basal metabolic rates relatively constant, regardless of tide and temperature.

A second and related question might be phrased as follows: Since immediate temperature compensation has been shown to be possible, why don't all ectotherms display temperature-independent metabolic rates on this short time-course basis? Furthermore, why don't cold-compensating ectotherms "go all the way" and compensate "perfectly"? These two related questions are likely to have a common answer. As we stated earlier, reductions in temperature have been shown to lower maintenance energy requirements, at least in the case of certain fishes. Thus, an ectothermic organism may not need to maintain as high a level of metabolism at low temperatures, and, for this reason, it may not exhibit "perfect" compensation—even though there is no reason why, in theory, the organism could not "perfectly" compensate. In addition to being able to reduce the necessary metabolic level at low temperatures, winter conditions may also mean a reduction in the food supply and, therefore, there may be a second reason why a reduction in metabolic rate is desirable.

What these considerations suggest is that organisms "choose" an advantageous Q_{10} value for their metabolic processes. We should not view "imperfect" metabolic compensation as failure of natural selection to put polar ectotherms or winter-acclimated ectotherms on an equal metabolic footing with their more warm-adapted (warm-acclimated) counterparts. Furthermore, the Q_{10} of each metabolic pathway is undoubtedly "chosen" in such a way as to provide the proper metabolic balance (e.g., between lipid deposition and lipid mobilization) at a given temperature. Differential Q_{10} values among different metabolic pathways could provide a mechanism for seasonal (or even diurnal) redirection of metabolic flow—what we earlier termed "metabolic reorganization"—without the necessity of changes in the relative concentrations of the enzymes of the various pathways.

Lastly, the above points suggest that it may be meaningless to speak of a Q_{10} value of metabolism *per se*. If different pathways exhibit markedly different responses to temperature, then it becomes difficult to speak meaningfully about the temperature sensitivity of the entirety of metabolism. Even when taking the rate of oxygen consumption ("respiration") as an index of the overall metabolism of the organism we do not avoid ambiguities, for the products of respiratory metabolism (e.g., ATP) may be utilized by distinctly different metabolic reactions, or at least be put to different uses at different temperatures.

A good example of the need to examine the temperature dependencies of each metabolic process to gauge the influence of temperature on metabolism is provided by the elegant studies of van Handel. Using starved mosquitoes, van Handel followed the activities of glycogen and triglyceride syntheses following feeding of the specimens with C^{14} glucose. Each of the two processes exhibited specific temperature dependencies which varied sharply over the range of measurement; one process could be essentially independent of temperature over a thermal range through which the other process varied widely in rate. These whole organism experiments clearly show the complex effects of temperature on the many constituent pathways of metabolism and, further, reveal the ambiguities which can arise when one attempts to discuss the effects of temperature on "overall" metabolism.

NEGATIVE THERMAL MODULATION: A MAJOR PROBLEM FOR ECTOTHERMS

For at least some enzymes, such as AChE and PK (Figures 7–14 and 7–15), positive thermal modulation occurs only over a certain range of temperatures. Below this range, further decreases in temperature lead to losses of E-S affinity. We term this effect "*negative thermal modulation*" to indicate that temperature decreases are now affecting the enzyme like negative modulators (Figure 7–19).

The biological impact of negative thermal modulation is great. Simultaneous reductions of E-S affinity and kinetic energy act

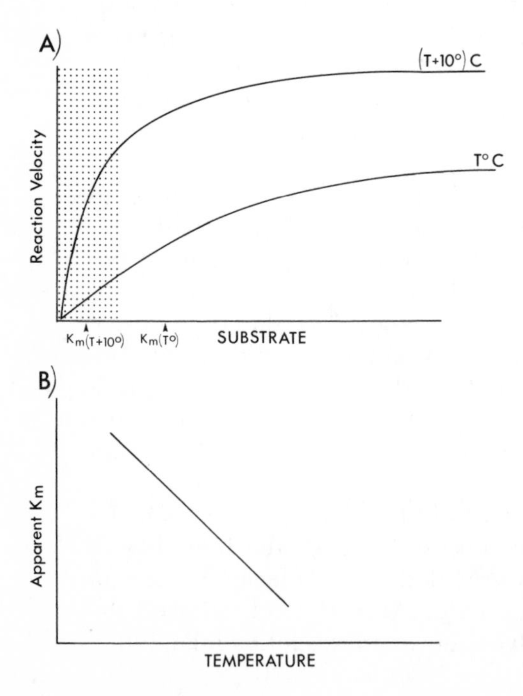

Figure 7–19 Negative thermal modulation of enzymic activity. (a) The relationship between reaction velocity and substrate concentration for a single enzyme at two different assay temperatures. The affinity of the enzyme for substrate (measured by the reciprocal of the apparent K_m value) decreases as temperature falls. Physiological substrate concentrations are indicated by stippling. Note that the rate-destabilizing effects of negative thermal modulation increase as the substrate concentration is reduced. (b) A plot of apparent K_m versus temperature, illustrating negative thermal modulation.

TABLE 7–10 The effect of temperature on the velocity of the pyruvate kinase reactions catalyzed by the "warm" and "cold" variants of the enzyme (see Figure 7–15), at a series of substrate (PEP) concentrations. Physiological PEP concentrations probably fall below 0.10 mM. Note the effect of negative thermal modulation in the case of the "warm" PK variant. (Data from Somero and Hochachka, 1971.)

	PEP Concentration	Q_{10} (3–7°C)
"warm" PK	1×10^{-3} M	9.8
	1×10^{-4} M	19.8
	5×10^{-5} M	24.6
"cold" PK	1×10^{-3} M	2.2
	1×10^{-4} M	3.5
	5×10^{-5} M	3.6

synergistically to reduce the rate of enzymic activity precipitously. Q_{10} values under conditions of negative thermal modulation may rise to enormous levels (Table 7–10) and, as we would expect, the value of the temperature coefficient rises as substrate concentrations fall. Thus, negative thermal modulation, like positive thermal modulation, exerts its strongest influences under physiological substrate concentrations.

The molecular basis of negative thermal modulation is not known for any given enzyme. However, in the case of AChE, a strong theoretical argument can be made on the basis of the role played by hydrophobic interactions in stabilizing the AChE-ACh complex (refer to Figure 7–12). Owing to the fact that hydrophobic interactions weaken as temperature is reduced from approximately room temperature to 0°C, it is not unreasonable to expect that for a given variant of AChE there may be a critical temperature at which the hydrophobic interactions between enzyme and substrate weaken sufficiently to cause a sharp rise in the apparent K_m of substrate.

THE ROLES OF AFFINITY-VARIANTS OF ENZYMES IN TEMPERATURE ADAPTATION

The establishment of satisfactory values of E-S affinity parameters is a major outcome of protein evolution. In the case of temperature adaptation, two major "goals" characterize the evolution of this important attribute of enzymes. Firstly, as the K_m-versus-temperature curves illustrated in Figures 7–14 and 7–15 reveal, *negative thermal modulation is largely avoided.* When temperature decreases do promote the loss of E-S affinity, this occurs only at temperatures near or below the lower limits of the species' habitat temperature. Over the long course of evolutionary history, selection has either "removed" or "pushed to one side" negative thermal modulation. This same adaptive strategy is noted for certain rainbow trout enzyme systems which change during thermal acclimation (Figures 7–14 and 7–15). PK and AChE exist in "warm" and "cold" variants which are preferentially formed during warm- and cold-acclimation, respectively. The major difference between the "warm" and "cold" isozymes is, much as we observed when comparing interspecific enzyme variants, a difference in the K_m-versus-temperature relationship. The "warm" isozymes display high affinity for substrates at temperatures near the organism's upper thermal range (approximately 15 to 20°C), and a rapid loss of E-S affinity occurs at winter temperatures (approximately 10°C and below). In contrast, the "cold" isozymes have their highest affinity for substrate at temperatures below 10°C and, at higher temperatures, bind substrate less well than the "warm" variants.

The second important consequence of enzyme temperature adaptation with regard to affinity parameters is *the maintenance of E-S affinity values within certain optimal ranges.* This "goal" characterizes evolutionary adaptation and seasonal acclimation. If E-S affinity varies too widely, owing to high degrees of thermal modulation, the all-important regulatory functions of the enzyme may be severely upset. Figure 7–20 illustrates the necessity of maintaining E-S affinity within a rather narrow range.

In terms of catalysis *per se*, an enzyme having a very low affinity for substrate will be able to utilize only a small fraction of its catalytic potential. In addition, large increases in substrate concentration are required to promote only small rises in reaction velocity.

If E-S affinity is extremely high, the danger exists of the enzyme's becoming quickly saturated with substrate as substrate levels rise. Under these conditions, cellular substrate concentrations could build up exponentially since, by definition, the saturated enzyme cannot further increase its rate of catalysis with further increases in substrate concentration.

The regulatory effects of positive and negative modulators would also be adversely affected under conditions where E-S affinities are either extremely high or very low. For an enzyme with an already high affinity for substrate, the addition of a positive modulator might so reduce the apparent K_m value as to lead to the immediate saturation of the enzyme, with the negative results just discussed. For an enzyme with an already low affinity for substrate, the action of a negative modulator might lead to a complete cessation of the enzyme's activity.

For these reasons, selection has "tailored" each enzyme to have an inherent substrate affinity which permits the enzyme to function in appropriate catalytic and regulatory manners at the

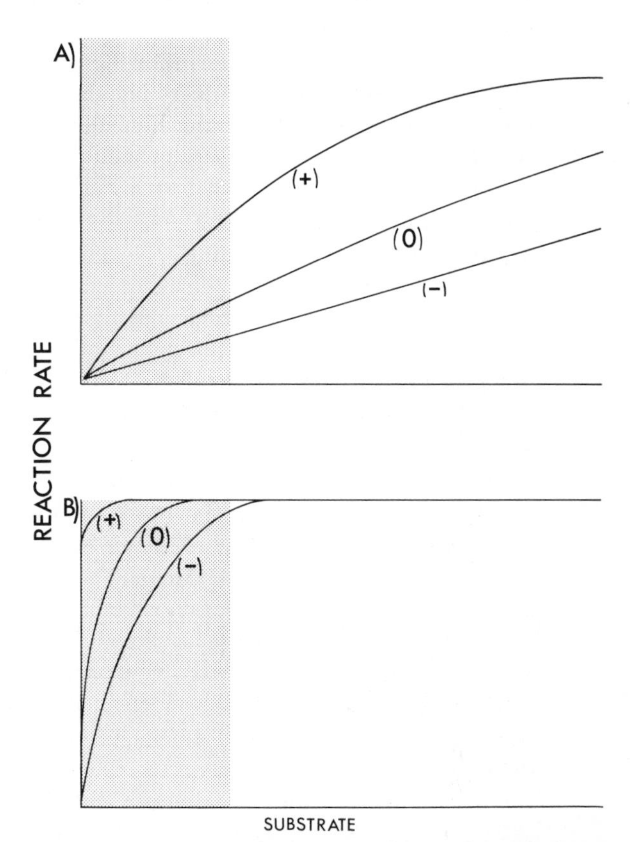

Figure 7–20 Substrate saturation curves for enzymes having E-S affinity values which are either too low or too high to permit optimal catalytic rates and/or regulatory responses. (a) An enzyme with an extremely low affinity for substrate. In the absence of modulators (curve marked "0"), the catalytic potential of the enzyme is barely tapped at physiological substrate concentrations (indicated by stippling). Even when a positive modulator is present (curve marked "+"), the enzyme does not function at a very high rate; i.e., the enzyme is relatively unresponsive to the activating signal and thus may not supply the necessary function required by the cell. When a negative modulator is present (curve marked "–"), the enzyme is virtually shut-off due to the further reduction of the already low E-S affinity. Under this condition a signal indicating the need for a deceleration, but not the cessation, of enzymic activity may lead to a complete blockage of the pathway's activity. (b) An enzyme with extremely high affinity for substrate. In the absence of modulators, the enzyme will be operating at a V_{max} rate over much of the physiological substrate range. Thus, increases in substrate supply will lead to rapid, exponential rises in substrate levels since the enzyme is already "flooded" with substrate. When a positive modulator is present, this situation is worsened. When a negative modulator is present, the activity of the enzyme is only slightly reduced; i.e., its affinity for substrate remains so high that a signal calling for a reduction or cessation of function may not be adequately responded to.

These curves demonstrate why wholesale changes in E-S affinity values are not a likely mechanism for effecting rate compensation over evolutionary and seasonal time courses, and why positive thermal modulation must be held within certain limits.

temperatures experienced by the organism. Thus, the potential rate-compensating benefits of positive thermal modulation must be balanced against the benefits of holding E-S affinity values within certain ranges. For eurythermal organisms such as the mudsucker (*Gillichthys mirabilis*), which may experience temperatures over the range from 4° to 30°C, the benefits which accrue from holding E-S affinity relatively stable may well outweigh the benefits of positive thermal modulation (Figures 7–15 and 7–16), at least over the lower range of habitat temperatures. *Eurythermal enzymes thus might have substrate-binding sites which have highly temperature-stable geometries.*

TEMPERATURE-DEPENDENT E-S AFFINITY PARAMETERS: MECHANISTIC BASIS

There are two fundamental questions we must deal with if we are to attempt to explain the mechanisms by which temperature-dependent changes in E-S affinity occur. Firstly, and most importantly, we must try to determine the types of weak-bonded systems which are either strengthened or weakened by temperature change. Secondly, we must consider a closely related question which asks what sorts of differences exist between, for example, the "warm" and "cold" variants of rainbow trout AChE and PK, such that the enzyme variants display different temperature-versus-K_m relationships.

Speculating about answers to the first question is relatively easy, for, as already discussed, we can single out a number of weak-bonded systems which might be thermally labile. For example, an E-S complex stabilized largely by ionic interactions is likely to be strengthened as temperature is reduced, since ionic interactions become stronger at low temperatures. Conversely, as in the case of AChE, an E-S complex which is stabilized largely by hydrophobic interactions may become very labile as temperatures approach 0°C.

Because the same classes of weak bonds which stabilize the E-S complex also stabilize the higher orders of protein structure, temperature changes may alter E-S affinity through changes in the geometry of the substrate-binding site.

There is thus good reason to believe that several mechanisms are involved in establishing the thermal dependence of E-S affinity. In fact, were this not the case, it would be difficult to understand the complex, U-shaped K_m-versus-temperature curves noted for some enzymes.

The types of amino acid changes which must occur to alter an enzyme's K_m-versus-temperature response are not known. However, we might predict that an enzyme which is adapted for function at low temperatures would benefit from an increased reliance on ionic, as opposed to hydrophobic, bonds for stabilizing the E-S complex and, perhaps, the 3° and 4° levels of protein structure. Although studies of the hydrophobic amino acid percentages of proteins from differently adapted organisms have revealed no correlation between adaptation temperature and the "hydrophobicity" of a given type of protein, until higher resolution studies of the amino acid residues which are actually responsible for setting E-S affinity or stabilizing the 3° and 4° levels of structure are conducted, it seems wise to regard this question as unanswered.

We can envision a mechanism for modifying the charge characteristic of the substrate-binding site with regard to the temperature-dependence of the charged group's ionization. For instance, if a carboxyl (-COOH) group must be dissociated for the substrate-binding site to be attractive to the substrate, and if this dissociation is impeded at low temperatures, then, during the course of evolution, the local environment of the carboxyl group might be modified to facilitate its dissociation. This might be accomplished by increasing the number of hydrophilic amino acid residues in the carboxyl group's environment, thereby supplying increased concentrations of water molecules to act as proton acceptors.

As high resolution studies of protein structure advance, for example through X-ray diffraction studies and other physical methods, we will no doubt begin to discern the classes of amino acid changes which are responsible for the important functional differences observed among proteins from differently thermally-adapted organisms.

MECHANISMS FOR GENERATING ENZYME VARIANTS

During the long time periods of evolutionary history, new enzyme primary structures may appear which provide the organism with functionally adaptive catalysts. In the case of AChE, for example, there are almost certainly a number of amino acid differences between the *Trematomus* enzyme and, say, the mullet enzyme, and these differences in primary structure must underlie the observed differences in substrate-binding characteristics.

During seasonal acclimation or immediate compensation, the organism must of course rely on phenotypic changes; i.e., it must make do with whatever gene products it already has in its cells or which it can generate by means of enzyme induction processes. Thus, the shorter the time-course of temperature compensation, the more limited is the enzymic repertoire available to the organism for fabricating its adaptive response.

On the basis of available data, it would appear that four distinct mechanisms are utilized by ectotherms to supply the cells with enzymes that have the proper E-S affinities during the processes of seasonal acclimation and immediate compensation. These are:

(1) The clean, on-off synthesis of season-specific isozymes, such that winter- and summer-acclimated organisms have distinctly different isozymes in their cells.

(2) The maintenance of a complex set of isozymes within the cell, with some forms being specifically suited for function in specific thermal ranges. Quantitative variation in the isozyme pattern may occur seasonally.

(3) In the case of lipoprotein enzymes, the protein component may remain unchanged, whereas the lipid component varies and promotes changes in the functional characteristics of the enzyme.

(4) Direct temperature-mediated changes in the conformational state of the enzyme, such that "instantaneous isozymes" are produced which are kinetically distinct variants of a single primary structure.

THE "ON-OFF" PRODUCTION OF SEASON-SPECIFIC ISOZYMES

Rainbow trout brain AChEs are the best currently available example of a system in which a dichotomous, "on-off" production of different isozymes occurs seasonally. The kinetic differences between the AChE isozymes of warm- and cold-acclimated trout (Figure 7–14) apparently are the result of distinct protein species (Figure 7–21) which are formed on a seasonal basis, for at 18°C a single AChE is synthesized which differs in its rate of electrophoretic migration from the AChE produced in 4°-acclimated specimens. Thus, at these two temperature extremes, there appears to be an *either-or* production of AChE isozymes. Notice, however, that at

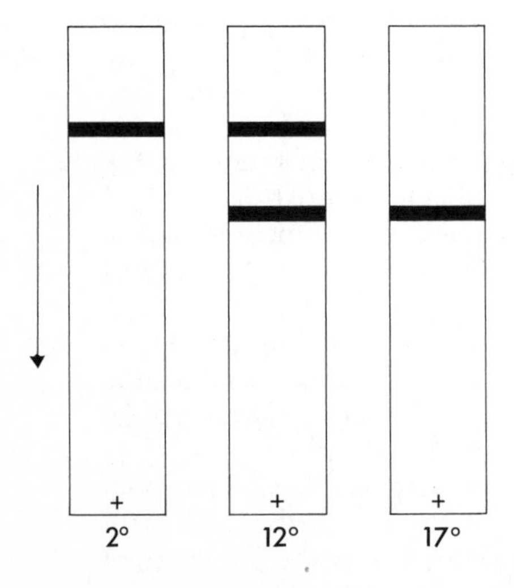

Figure 7–21 Electrophoretic patterns of rainbow trout brain acetylcholinesterase isozymes from 2°, 12°, and 17°C acclimated fish. (Data from Baldwin and Hochachka, 1970.)

12°C both enzymes appear to be synthesized (Figure 7–21). This finding suggests that the regulatory mechanisms which control the synthesis of AChE are capable of providing the trout with precisely the types of isozymes necessary for function over its particular seasonal temperature range. At winter and summer extremes of temperature, only a single AChE is required. At intermediate temperatures, the organism seems to be "hedging its bet" by maintaining both isozymes in its nervous system.

THE MAINTENANCE OF COMPLEX ISOZYME SYSTEMS AS A MEANS FOR INSURING OPTIMAL E-S AFFINITY RELATIONSHIPS

There are other examples of enzyme systems which, like the AChE system of 12°C-acclimated rainbow trout, contain simultaneously several isozymes with different K_m-versus-temperature characteristics. One such system is the isocitrate dehydrogenase (IDH) system of rainbow trout. A given population has been found to contain several phenotypic expressions of IDH, and the overall thermal properties of the IDH reaction are strongly influenced by the exact ratios of the different isozymes present in the cells (Figure 7–22). Individuals having only the A_2 isozyme do not display positive thermal modulation, whereas individuals with the B_2 and C_2 isozymes, in addition to the A_2 variant, do exhibit a rate-compensatory drop in E-S affinity as the temperature rises above 10°C. At least for this locus in metabolism, it appears that individuals possessing all three isozymes are at an advantage, since they will not experience as sharply fluctuating rates of IDH activity during the summer as will those organisms having only the A_2 isozyme. For the latter individuals, an alternate strategy for regulating IDH activity in the face of varying summer temperatures may possibly be employed.

HOW COMMON ARE SEASONAL ISOZYME CHANGES?

The question heading this paragraph has two different meanings. Firstly, within a single species, we can ask how many enzyme systems exhibit the sorts of changes which have been observed for systems like rainbow trout AChE. Secondly, are the isozyme changes found in rainbow trout unique to this species, or do other ectotherms also change their isozymes on a seasonal basis? In considering these two questions we should keep a third question in mind: What properties of an enzyme will raise the need for season-specific isozymes?

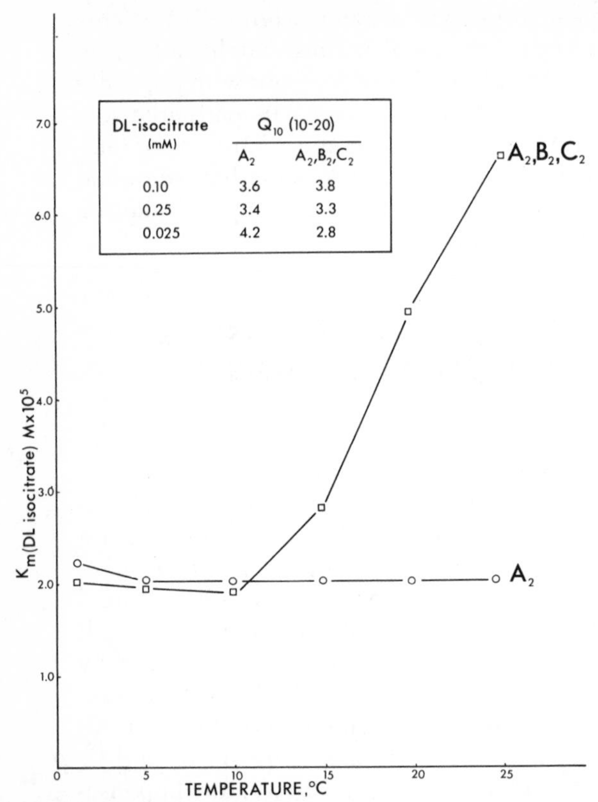

Figure 7–22 The effect of temperature on substrate (DL-isocitrate) affinities of different isozyme forms of isocitrate dehydrogenase (IDH) of the rainbow trout. Within a single hatchery population, individuals were found to contain varying ratios of the A_2, B_2, and C_2 IDH isozymes. Individuals having the greatest amounts of the B_2 and C_2 isozymes (all individuals had the A_2 isozyme) exhibited the most positive thermal modulation. Seasonally, some changes were noted in the isozyme composition of the fishes: 37 per cent of the winter population sampled had only the A_2 isozyme, whereas only 23 per cent of the spring population had only the A_2 isozyme. (Data from Moon and Hochachka, 1971.)

To answer the first question, we must really answer the third question as well. Those enzymes which do occur in season-specific forms in the rainbow trout are those which show relatively sharp changes in E-S affinity with temperature. In particular, isozyme changes seem of great importance when negative thermal modulation occurs well within the normal temperature range of the organism (Figures 7–14 and 7–15). Enzymes which do not exhibit season-specific isozymes in the rainbow trout include LDH (Figure 7–16) and malate dehydrogenase (MDH). Both of these enzymes have relatively flat K_m-versus-temperature curves in the biological thermal range, and neither enzyme exhibits negative thermal modulation.

The question concerning the general occurrence of isozyme changes among different ectothermic species cannot be definitively answered for lack of experimental data. Only a small number of ectotherms have been studied from the standpoint of seasonal isozyme changes, and no seasonal differences have been found except in rainbow trout. However, the enzyme systems examined in other species have usually been enzymes which, in trout, do not show highly temperature-sensitive K_m values or negative thermal modulation. Furthermore, when complex isozyme patterns are present at all seasons, the organism may have in its cells, at all times, differently temperature-sensitive enzymes which perhaps have functional properties comparable to the AChE isozymes found in 12°C-acclimated trout.

There is also the likelihood that not all species variants of a given enzyme need display similar K_m-versus-temperature properties. For example, the PK of *Gillichthys*, which experiences temperatures at least over the range from 4° to 30°C, is markedly temperature-insensitive in its binding of substrate (PEP). Contrast this seemingly "eurythermal" enzyme with a single PK isozyme of the rainbow trout. As we speculated earlier, extremely eurythermal organisms may have enzymes with stable ligand-binding properties. The PKs of intertidal molluscs and of a squid species which migrates diurnally through a 10°C thermocline exhibit temperature-independent K_m values like the *Gillichthys* enzyme. Thus two evolutionary strategies may lead to the same important end result: in some species, such as the rainbow trout, E-S affinity values are held within an optimal range by means of the joint actions of two or more isozymes, while in other cases, including some trout enzymes, selection has promoted the development of highly temperature-insensitive E-S affinities and, as a result, a single set of enzymes that can function well over the entire thermal tolerance range of the organism.

"ALLOZYMES": ENZYME POLYMORPHISM AND TEMPERATURE ADAPTATION

The term "allozymes" is used to refer to the gene products of a heterozygous locus, whereas the term "isozyme" is usually restricted to systems in which more than one gene locus codes for a given enzyme. On the basis of our discussion of isozyme changes during temperature acclimation, we might predict that allozymes could play important roles in temperature and, indeed, other environmental adaptations.

The hypothesis which has been forwarded, concerning the advantages of protein heterozygosity in temperature adaptation, proposes that organisms inhabiting unpredictable thermal regimes will tend to have more genetic variability—more allozymes—than organisms living in stable environments. This hypothesis is predicated on the assumption that different allozymes of a given catalyst will work particularly well over different temperature ranges. Thus, much as we found that trout have two gene loci to produce "warm" and "cold" AChE variants, we might expect to find situations in which one allele of a single gene is a "warm" allozyme and the other allele is a "cold" allozyme.

Testing of this interesting hypothesis has only just begun. And, while there is some evidence supporting the hypothesis that genetic heterozygosity is correlated with temperature variability (e.g., the critical enzyme involved in insect flight metabolism, α-glycerophosphate dehydrogenase, is highly polymorphic in temperate species but not in tropical species of *Drosophila*), many more data must be gathered before this hypothesis can be considered definitively tested. One interesting conclusion which has emerged from studies of enzyme polymorphism, however, is that the level of polymorphism is highest for those enzymes which catalyze essentially irreversible reactions. As we have seen, most regulatory "valves" are located at these kinds of reactions, and, therefore, these reactions may be unusually sensitive sites of natural selection.

CONFORMATIONAL VARIANTS: "INSTANTANEOUS ISOZYMES"

In all of the enzyme systems so far discussed, including isozyme and allozyme systems, changes in the functional characteristics of the enzyme system have been due to changes in the primary structures of the enzymes involved. Effecting these changes is highly time-consuming. Genetic change, of course, involves at least one generation and, more likely, a very large number of generations. Even the induction of new enzyme variants during thermal acclimation appears to take at least one or two weeks. Thus, if the production of new enzyme variants is to play a role in the process of immediate temperature compensation, mechanisms must be found for generating the required variants over much shorter time intervals than are available to organisms compensating to more gradual temperature changes.

The one known mechanism for the rapid, nearly instantaneous formation of new functional variants of an enzyme appears to involve a direct, temperature-mediated change in protein conformation such that above and below a certain temperature a single protein, in terms of primary structure, can act in a manner analogous to the "warm" and "cold" trout isozymes discussed earlier. We term these functionally distinct conformers of a single protein "instantaneous isozymes" to indicate that their formation occurs essentially as rapidly as the habitat (body) temperature is altered. An example of this type of enzyme variant generation is the PK system of the Alaskan king crab (*Paralithodes camtschatica*) (Figure 7–23).

The upper frame of Figure 7–23 illustrates the complex sub-

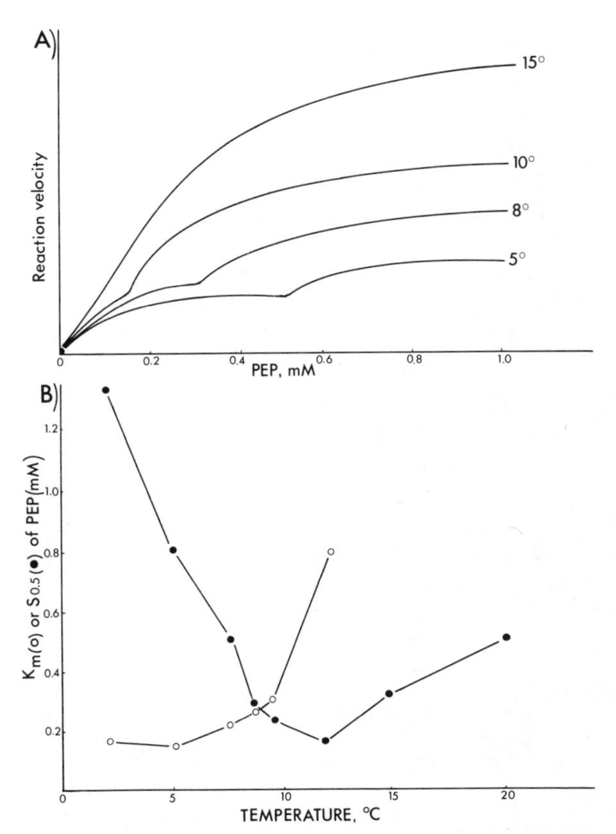

Figure 7–23 The effects of temperature on the kinetics of Alaskan king crab (*Paralithodes camtschatica*) leg muscle pyruvate kinase. (a) Substrate (PEP) saturation curves at different temperatures. Note the leftward migration of the "bump" in the saturation curve as temperature is raised, and the increasing share of activity which is due to the enzyme responsible for the "bumpy" part of the curve. By 15°C, all of the enzyme has been converted to the sigmoidal form, responsible for the "bump" at lower temperatures. (b) The effect of temperature on PK-PEP affinity for the two forms of king crab PK. The hyperbolic form of the enzyme displays maximal PEP affinity at low temperatures (approximately 5°C); the sigmoidal form of the enzyme has highest PEP affinity near 12°C, the approximate upper temperature of the organism's habitat. (Modified after Somero, 1969.)

strate (PEP) saturation curves characteristic of the PK of this large marine crustacean. The "bumpy" shapes of these curves are readily explicable on the assumption that two different PK variants are operative in the system and that the ratios of the activities of the two variants are temperature-dependent. The kinetic differences between the two variants can be seen clearly when we plot the E-S affinity of each enzyme as a function of temperature (Figure 7–23b).

It is obvious that the king crab PK system is the *functional* analogue of the rainbow trout PK system (Figure 7–15). However, the similarities between the two systems do not extend to the level of structure or, indeed, to the fourth dimension, time. The king crab system is the result of an apparently immediate, temperature-mediated change in protein conformation; the trout system likely involves the generation of new enzyme primary structures via the time-consuming processes of gene activation, transcription, translation, assembly, and so forth.

In the king crab system, reductions in temperature affect the enzyme in a second manner comparable to a positive modulator: temperature decreases promote a shift from sigmoidal kinetics to hyperbolic kinetics (as well as an increase in E-S affinity *per se*). Low temperatures thus reduce the cooperativity of substrate binding, perhaps by making the interactions between enzyme subunits more difficult.

These observations should impress the reader that the primary structure of a protein is not the unique determinant of its functional characteristics. The interconversion which occurs in the king crab PK system involves only a single protein, in the sense of primary structure, yet generates at least a pair of enzymes in a functional sense.

ENZYME-SUBSTRATE AFFINITIES AND TEMPERATURE COMPENSATION: SOME FINAL COMMENTS

In theory, changes in E-S affinity could effect metabolic rate compensation over all of the three time-courses of compensation we have been discussing. We have already suggested that positive thermal modulation of catalysis might be an important mechanism for stabilizing the rates of enzymic function in the face of short-term temperature fluctuations; i.e., immediate temperature compensation might depend on positive thermal modulation of key enzyme steps in metabolism.

For the two longer-term temperature compensation processes, we can also visualize means by which E-S affinity changes could promote rate compensation: during evolutionary adaptation and winter acclimation, new enzyme variants could be produced which have lower absolute K_m or $S_{0.5}$ values than the enzymes present in warm-adapted or summer-acclimated organisms. If this temperature compensation mechanism is important, then we would expect to find that warm-adapted (warm-acclimated) organisms will have enzymes with relatively high K_m or $S_{0.5}$ values, relative to the enzymes of cold-adapted (cold-acclimated) forms. In fact, this correlation is generally not observed, and the reasons why this is the case should be obvious. In Figure 7–20 we illustrated the disadvantages of having too high an affinity for substrate: the enzyme could be easily saturated with substrate and, in addition, the effects of regulatory metabolites might be hindered. Consequently, selection does not appear to have favored the development of enzymes with different absolute levels of substrate affinity as a mechanism for temperature compensation over evolutionary or seasonal time spans. Again, we can picture a "conflict of interest" between (i) the catalytic potential of the enzyme, which would be enhanced by an extremely high affinity for substrate, and (ii) the regulatory efficiency of the enzyme, which is greatest when the affinity parameter falls into the midrange of physiological substrate concentrations. And, as we have noted in other systems, the maintenance of regulatory function may gain precedence over the ability merely to catalyze a transformation at an exceedingly high rate.

We are therefore left with the conclusion that evolutionary and seasonal compensation to low temperatures cannot be effected by wholesale reductions in K_m or $S_{0.5}$ values. Instead, cold-compensating organisms may have to rely on (i) the "quantitative strategy," the production of higher enzyme concentrations for low temperature function; (ii) the generation of new enzyme variants with greater abilities to reduce the free energies of activation of their reactions, a possibility which remains to be tested experimentally; and/or (iii) the alteration of the local environment in which the enzymes function, such that compensatory changes in reaction rate occur. We will now examine one particularly important manifestation of this third compensatory mechanism, the "modulation strategy."

LIPOPROTEIN ENZYMES: THE EFFECTS OF CHANGES IN THE LIPID MOIETY

The local environment of an enzyme, its "catalytic environment," plays a major role in shaping the functional properties of the enzyme. Positive and negative organic modulators, hydrogen ions, inorganic ions, and the molecules to which the enzyme protein is bound if it possesses a quintinary level of structure may all affect the catalytic and regulatory properties of the enzyme. While changes in

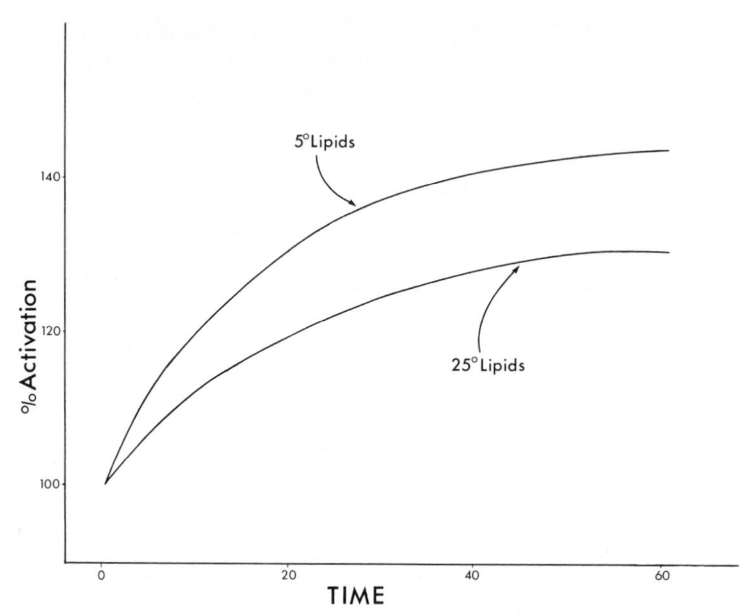

Figure 7–24 The activation of purified succinic dehydrogenase (SDH) protein by purified mitochondrial lipids isolated from mitochondria of 5° and 25° acclimated goldfish. Note that the time-dependent activation process leads to much higher SDH activities when lipids from the cold-acclimated mitochondria are present in the system. (Data from Hazel, 1972.)

any or all of these "catalytic environmental" factors could in theory effect temperature compensatory changes in enzymic activity, the only known example of this type of effect (the "modulation strategy") is in the case of lipoprotein enzymes.

Many of the enzymes involved in mitochondrial oxidative reactions are membrane-associated enzymes which contain a lipid moiety as an obligatory part of their structures. One of these enzymes, succinic dehydrogenase (SDH), can be readily purified and separated into distinct protein and lipid fractions. In an examination of SDH from 5°- and 25°-acclimated goldfish (*Carassius auratus*), Hazel observed a significant degree of temperature compensation in the rate of SDH activity. To determine the possible bases of this change, Hazel purified the enzyme and fractionated it into its protein and lipid components. Electrophoretic analysis of the protein moieties of the 5° and 25° enzymes revealed no differences; i.e., the fish contained the same protein variant at both temperatures. However, when the effects of the lipid component on enzymic activity were compared (Figure 7–24), a striking difference was found in the extent of activation caused by the purified lipids from the 5° and 25° mitochondria. The former activated the protein significantly more than the latter, and thus it appears that at least a major fraction of the rate compensation characteristic of SDH activity is due to differences in the lipid "environments" of the cold- and warm-acclimated enzymes. For lipoprotein enzymes, "isozymic" variation, at least in the all-important functional sense of the term, may derive from lipid changes rather than from protein changes.

Additional support for this hypothesis was obtained by Hazel in experiments where different classes of lipids were added to the purified SDH protein. In these carefully controlled studies, it was found that the extent of activation was, in most cases, directly proportional to the degree of unsaturation of the added lipids.

TEMPERATURE-DEPENDENT CHANGES IN MEMBRANE LIPID SATURATION

Changes in the saturation of the lipid components of membranes are a major homeostatic mechanism in temperature-compensating ectotherms. As indicated by the data of Table 7–11, there is a general trend toward reduced fatty acid saturation in low temperature acclimated organisms, and, in fact, what limited data are available suggest that the same trend may pertain among differently adapted species of ectotherms.

TABLE 7–11 THERMALLY INDUCED CHANGES IN FATTY ACID SATURATION OF MEMBRANE LIPIDS OF THE GOLDFISH (*Carassius auratus*) AND THE YELLOW BULLHEAD (*Ictalurus natalis*)

Gill Mitochondria[1]

	Acclimation Temperature	Fatty Acid Class (mole per cent) Saturated	Monounsaturated	Polyunsaturated
Goldfish	10°	21.57	14.47	48.98
	30°	23.60	15.77	44.62
Bullhead	15°	26.68	19.77	41.52
	28°	30.64	22.54	37.54

Brain[2]

		Fatty Acid Class		
Goldfish	5°	37	36	9
	15°	39	30	9
	25°	42	31	7
	30°	43	32	3

[1]Caldwell, R. S., and Vernberg, F. J. (1970). *Comparative Biochemistry and Physiology 34*:179–191.

[2]Johnston, P. V., and Roots, B. I. (1964). *Comparative Biochemistry and Physiology 11*:303–309.

The functional significance of changes in lipid composition with temperature may involve a number of effects. As noted above, unsaturated fatty acids may serve as more potent activators of the protein component of lipoprotein enzymes and, therefore, a shift to less saturated lipids at low temperatures may effect compensatory changes in enzymic activity without any change in enzyme protein concentration having to occur. A possible mechanism for the differing abilities of various saturation classes of lipids to activate enzymes can best be raised in the context of a second proposed function for the temperature-dependent lipid compositions: the maintenance of a proper physical state of the cell membranes.

MAINTENANCE OF THE PHYSICAL STATE OF CELL MEMBRANES

What is perhaps the most widely accepted model of cell membrane structure is illustrated in the schematic diagram of Figure 7–25. In this model, some proteins are embedded in the lipid phase of the membrane, while other proteins are plated onto the membrane surface. All of the protein-lipid complexes are probably stabilized by weak bonds. Hydrophobic interactions between the long, nonpolar chains of the fatty acids and the hydrophobic residues exposed on the surface of the membrane proteins may be of key importance in stabilizing the "burial" of certain membrane proteins. Ionic interactions between charged groups on the proteins and the charged (polar) ends of the lipids may be important in binding the surface-plated proteins to the membrane.

Because of the intimacy of lipid-protein interactions in the membrane, especially in the case of "buried" proteins, changes in the properties of membrane lipids would be expected to alter the functional properties of the membrane proteins. The likely basis of this effect on protein function involves changes in the physical state

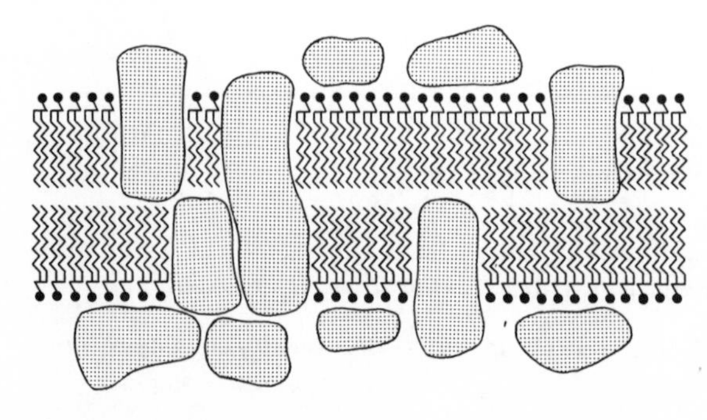

Figure 7–25 A schematic illustration of the structure of a biological membrane. The basic framework of the membrane is a phospholipid bilayer within which and on which different proteins are attached. The structure of phospholipids is shown schematically in Figure 7–26. Proteins are represented by the stippled shapes. (After Fox, 1972.)

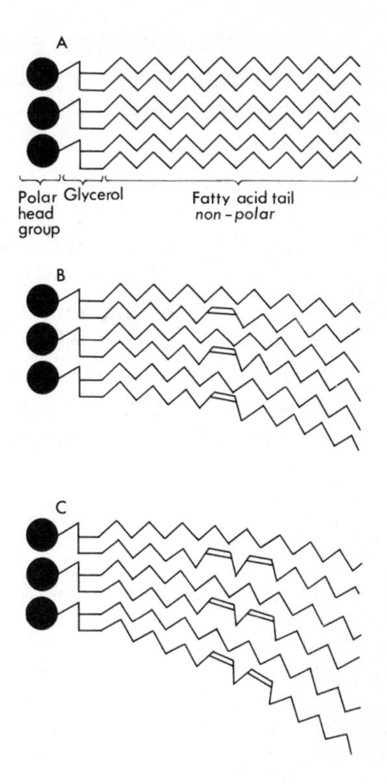

Figure 7–26 A schematic illustration of membrane lipid structure and the effects of varying fatty acid saturation on membrane lipid "fluidity." Each membrane lipid molecule consists of (i) a polar head group, often containing a charged phosphate residue; (ii) a glycerol molecule, which forms a bridge between the polar head and the tail; and (iii) the fatty acid "tail" of the molecule, composed of two hydrocarbon chains indicated by the zig-zag lines. Each angle of the zig-zag represents a carbon atom to which two hydrogens are attached. A double line, as in (b) and (c), represents a double bond (—C=C—).

In (A), all of the fatty acids are saturated; i.e., no double bonds are present between carbon atoms. In (B), one double bond is present in one fatty acid of each pair per molecule, and in (C), one fatty acid of each pair has two double bonds. The addition of increasing numbers of double bonds disrupts the orderly stacking of the hydrocarbon chains illustrated in (a). Consequently, the more double bonding, the more "fluid" is the membrane. (After Fox, 1972.)

of the membrane. The melting points of fatty acids are determined largely by the degree of saturation of the molecules. The higher the degree of unsaturation, the lower the melting point, chain lengths being equal. This difference in melting temperatures stems from the effects of double bonds on the geometry of aliphatic chains. As illustrated in Figure 7–26, fatty acids which lack double bonds can readily align and form solid, crystalline structures. Double bonds create bends in the aliphatic chain and make crystallization more difficult. Only at quite low temperatures are the weak bonds which stabilize lipid-lipid interactions strong enough to promote solidification of highly unsaturated lipids.

As the data of Table 7–10 indicate, one function of lipid changes noted during thermal compensation is the maintenance of a more or less "liquid" state of the cellular membranes. The addition of double-bond-rich fatty acids to the membrane lipids enhances the abilities of the lipids to remain in a liquid or quasi-liquid state at low temperatures. The formation of rigid, crystalline membranes appears to be inimical to successful membrane function, a conclusion which has recently been supported by a number of elegant studies of temperature-dependent membrane phenomena.

MITOCHONDRIAL ENZYME FUNCTION IN "HOMEOTHERMIC" AND "POIKILOTHERMIC" ORGANISMS

In comparative studies of the effects of temperature on certain membrane-bound mitochondrial enzymes, Lyons and Raison observed a most interesting relationship between the source of the enzyme and its thermal responses. As illustrated in Figure 7–27, the activities of mammalian variants of the enzyme decrease sharply below a critical temperature of approximately 23° C, whereas the ectothermic enzymes show no break in their activity-versus-temperature curves. Similar differences were observed when chill-resistant and chill-sensitive plants were compared and, most strikingly

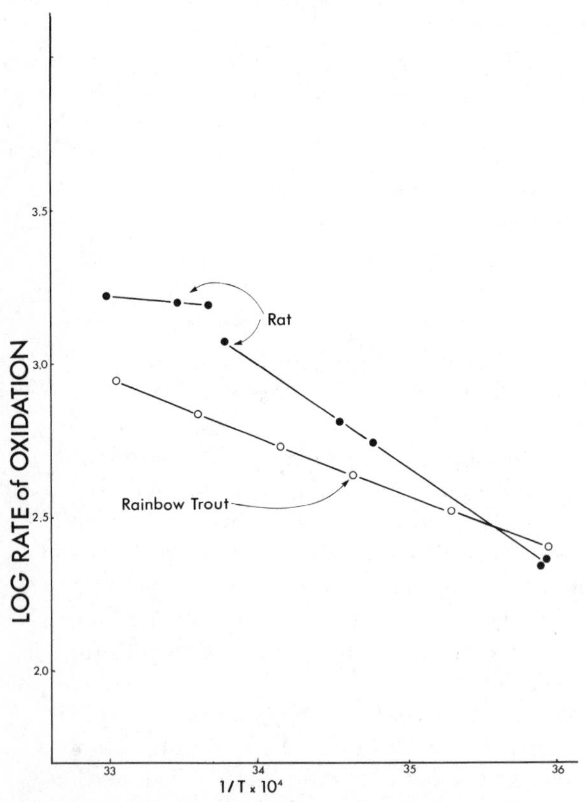

Figure 7–27 Arrhenius plot for oxidation of succinate by mitochondria from rat liver and rainbow trout liver. "Breaks" in the slope of the Arrhenius plot are characteristic of enzymes of mammals (excluding hibernators in the hibernating phase of their annual cycle) and chill-sensitive plants. Ectotherms, such as fishes, have enzymes which display no "breaks" in Arrhenius plots.

of all, when enzymes from the dormant and active states of a hibernator were compared. In the latter case, in which the active (homeothermic) hibernator's enzyme resembled a typical mammalian enzyme, whereas the hibernating (poikilothermic) organism's enzyme responded to temperature like a typical ectothermic enzyme, an acclimation process similar to that observed with goldfish SDH may be responsible for the observed differences in enzymic activity.

Indeed, even though the latter possibility was not examined, comparisons of the membrane physical state of the different systems did reveal that the sharp breaks in the Arrhenius plots were probably due to lipid-based changes involving liquid-to-solid phase transitions. In the membranes of the mammal, the frost-sensitive plant, and the active hibernator, a phase change was observed at the same temperature at which enzymic activity displayed a sharp decrease. This lipid-based phase change may have induced a conformational change in the enzyme, reducing its activity. Or, alternatively, the more rigid membrane structure which exists below the critical temperature may render the passage of substrates, cofactors, and other molecules more difficult, thereby reducing the rate of enzymic function. Whatever the ultimate "damage" in molecular terms, the change which occurs at the critical temperature normally proves lethal for the organism.

DIETARY CONTROL OF MEMBRANE LIPID COMPOSITION: EXPERIMENTAL MANIPULATION OF PHASE TRANSITION TEMPERATURES

Perhaps the most elegant demonstration of the importance of membrane lipid saturation in setting the thermal tolerance limits of organisms has come from studies of the microorganism *Mycoplasma laidlawii*, in which the lipid composition of the cell membrane can readily be varied experimentally. In this particular organism, the lipid composition of the cell membrane is largely determined by the fatty acid composition of the diet, i.e., of the culture medium.

Steim and colleagues cultured *M. laidlawii* in media of different fatty acid composition and examined the resulting cultures to determine whether their thermal tolerances differed. As expected, the organisms grew normally only at temperatures exceeding the membrane phase transition temperature. Organisms grown in culture media with highly unsaturated fatty acids grew at much lower temperatures than specimens grown in media rich in saturated fatty acids.

To demonstrate unequivocally that the observed effects were due to phase transitions in the lipids of the membranes, Steim *et al.*

determined the phase transition temperatures of cell membranes *in situ*, of isolated cell membranes, and of purified membrane lipids. For organisms from any single culture medium, all three phase transition temperatures were the same. These results thus suggest that in microbes, as well as in multicellular organisms, an essential feature of the temperature-compensatory response is the close regulation of the physical state of the cell membranes, an adaptation which to a large degree may be "directed" towards providing the proper local environment for enzymic function.

THE "MODULATION STRATEGY": OTHER CHANGES IN THE "CATALYTIC ENVIRONMENT" OF ENZYMES

The lipid changes just discussed are the only clear example of how changes in the immediate surroundings of enzymes can effect compensatory changes in their catalytic properties. Nonetheless, there may well be a multitude of other "uses" of the "modulation strategy." For example, the marked ion dependencies of many enzymes might serve as an important locus for temperature-compensatory regulation of enzymic activity. Thus, an enzyme which is strongly activated by a divalent cation such as Mg^{++} might be activated during cold exposure by elevations in the intracellular Mg^{++} concentrations. Similarly, most enzymes display characteristic pH optima, and any temperature-dependent pH change which occurs with the correct vector could lead to compensatory changes in enzymic activities.

All of these other possible examples of the "modulation strategy" remain to be rigorously tested. While there have been numerous studies of ionic changes in ectotherms during thermal acclimation, these studies have dealt almost exclusively with blood. Thus, we know little about the changes in ionic content or composition which might occur in the local environments of enzymes. The only good argument that the blood ion changes which have been observed might reflect tissue ion changes has come from experiments in which blood serum from differently acclimated ectotherms has been added to *in vitro* tissue preparations. These studies have shown that serum from cold-acclimated, but not warm-acclimated, specimens can have a marked stimulatory effect upon tissue metabolism. Unfortunately, the blood constituents responsible for this metabolic stimulation have not been determined.

While modulatory changes such as those which could be caused by inorganic ion or hydrogen ion concentrations seem of potential importance in temperature compensation, we must not forget that changes of this nature have one major disadvantage: changes in any ion or in pH will affect different enzymes differently. For example, a change in the concentration of Mg^{++} may sharply activate some enzymes, mildly activate others, not affect some enzymes, and even inhibit certain enzymes. pH changes would probably exert even more dramatically differential effects. Thus, such an apparently simple and economical way of compensating enzymic activities to temperature changes may have only limited potential.

FREEZING RESISTANCE AND FREEZING TOLERANCE: GENERAL CONSIDERATIONS

The physical states of aqueous systems, like the lipid-based systems discussed previously, are determined by the efficacies of large numbers of weak bonds. The hydrogen bonding among water molecules becomes increasingly stabilized as temperature is reduced and, as a result, water tends to become more and more structured as heat is removed from the system. Below a certain critical temperature, the freezing point, water exists as a solid. As in the case of liquid-to-solid phase transitions in lipid-based systems such as membranes, the solidification of water often is inimical to life, especially if ice formation occurs within the cell. Thus, we find that organisms normally exposed to temperatures at or below those at which their body fluids freeze employ a wide variety of adaptations, either to prevent the formation of ice within their bodies ("freezing resistance") or, alternatively, to shield the structures of the cell from the potentially damaging or lethal effects of ice formation in the extracellular fluids ("freezing tolerance").

As in the cases of many other environmental adaptations, the avoidance of freezing may be accomplished by behavioral, anatomical, physiological, and biochemical means. Many organisms, endothermic and ectothermic, migrate away from freezing conditions. These migrations may involve latitudinal displacements of thousands of miles, or they may entail the seeking of a warmer microclimate within the organism's normal geographical locale. Thus, hibernators may burrow into the soil and avoid the extremes of temperature which occur at the land-air interface. Aquatic species may also migrate away from regions where the danger of freezing exists. Some marine fishes, such as the Arctic char (*Salvelinus alpinus*), leave the ocean in winter, where the temperature may fall below the freezing point of the body fluids of most teleosts, and migrate into freshwater rivers and lakes, where the temperature will not fall below 0°C. Certain shallow water marine fishes of high latitude waters may spend the winter season near the bottom to avoid contact with ice crystals which could "seed" ice formation in their body fluids. These fishes thus spend the winter in a supercooled state.

Although many organisms succeed in reducing or eliminating the dangers of freezing by migrating to regions where temperatures are relatively high and/or no ice is present to "seed" the body fluids, for many organisms there is no escape from freezing stress short of biochemical adaptation. The most acute freezing resistance/tolerance problems are, of course, encountered by terrestrial ectotherms and some intertidal invertebrates, the latter being exposed to sub-freezing air temperatures during low tide. Aquatic species never encounter temperatures lower than approximately −2°C (unless, of course, their habitat freezes). However, terrestrial ectotherms may experience environmental temperatures lower than −50°C. These latter organisms furnish us with some of the most dramatic examples of the types of biochemical adaptations which permit life to exist at extraordinarily low temperatures.

In order to understand the types of biochemical changes which can lead to freezing resistance and freezing tolerance, we must first appreciate the actual causes of cellular damage during freezing. Firstly, under almost all circumstances, *intracellular* ice formation leads to cell death due to the irreversible disruption of the cellular ultrastructure. The only known instances in which intracellular ice formation is not lethal have been observed under laboratory conditions in which massive levels of cryoprotectant substances were added to cells, and the preparations were cooled and then thawed under precise conditions. At present there appear to be no data showing that similar tolerance of intracellular ice formation occurs in nature.

Whereas intracellular ice formation is normally, if not always, lethal to the cell, many organisms display the ability to tolerate a significant amount of *extracellular* ice formation. When extracellular ice formation is lethal to the organism, the likely cause of death is cellular dehydration. As the extracellular water is changed to ice, a concentrated solution of salts, macromolecules, and so forth is left behind in the extracellular space. This concentrated fluid establishes a strong osmotic gradient which draws water from the cell. For most organisms which have been studied, death of the cells occurs when approximately two-thirds of the cellular water has been lost to the extracellular space. The causes of death by cellular dehydration need not be dwelt on at this point; recall the discussion of the cell's "solvent capacity" given earlier (pp. 97–103).

To avoid these damaging effects of ice formation, three biochemical approaches seem possible. Firstly, the organism can reduce the freezing point of its body fluids. Secondly, the organism can enhance the abilities of its body fluids to supercool. Thirdly, if it is not possible to prevent ice formation in the extracellular fluids, cellular damage may yet be avoided if means can be found for reducing the osmotically driven outflow of cellular water into the extracellular space. Terrestrial and intertidal ectotherms provide examples of all three approaches to freezing resistance/tolerance.

FREEZING RESISTANCE/TOLERANCE IN INSECTS

The prevention of ice formation could be effected by increasing the concentration of any non-toxic solute, at least in theory. In fact, only a relatively small number of compounds appear to be utilized as biological "antifreezes." Of particular importance are organic polyhydroxyl compounds such as glycerol. These molecules lower the freezing point and extend the supercooling point of biological fluids by interacting with water molecules, via hydrogen bond formation between the hydroxyl groups of the antifreezes and the polar water molecules. The hydroxyl-water interactions thus reduce the number of water-water interactions and, consequently, the formation of ice is retarded. A further biologically important attribute of glycerol and like molecules is their ability to permeate the cell membranes. They can therefore serve as extra- and intracellular antifreezes.

The role of glycerol in freezing resistance/tolerance is nowhere better documented than in the cases of Arctic insects such as the carabid beetle, *Pterostichus brevicornis*. Miller and Baust followed the freezing point, the supercooling point, and the glycerol content of *P. brevicornis* over a full year period. The temperatures of freezing and supercooling dropped over the period from late summer to late winter as the concentration of glycerol in the body fluids and tissues increased. Winter beetles could withstand freezing (probably only of the extracellular fluids) at temperatures as low as −35°C. Summer beetles died at −6.6°C. The former specimens contained approximately 22 per cent glycerol, while the summer organisms had only about one per cent glycerol.

In spite of the close correlation between glycerol content and freezing resistance/tolerance observed in this and other similar studies of cold-hardy insects, there exists a certain degree of ambiguity in the precise importance of glycerol as a cryoprotectant in insects. For example, Miller and Baust found that the supercooling point of the beetle fell significantly in late summer before the glycerol content of the body increased. Furthermore, during midwinter a substantial drop in glycerol content occurred, but freezing resistance/tolerance was not reduced at this time. In other insect species, high concentrations of glycerol are not always accompanied by cold hardiness. In yet other species, a high degree of freezing resistance/tolerance is observed where no large concentrations of glycerol are found. In conclusion, although glycerol no doubt can exert an important cryoprotectant influence in the organism, other biochemical changes may well play important roles in insect freezing resistance/tolerance.

FREEZING RESISTANCE/TOLERANCE IN INTERTIDAL MOLLUSCS

Although most marine organisms do not encounter temperatures below about −1.9°C, some intertidal forms which (i) are normally exposed on the surface of rocks and pilings, and (ii) have limited mobility, may experience temperatures below −10°C in winter, during low tides. These exposed invertebrates display much greater capacities to resist or tolerate freezing than do related species which either remain under water all of the time or live beneath several centimeters of insulative substrate, such as sand or mud. The common mussel, *Mytilus edulis,* was found by Williams to freeze only at −10°C, whereas the clam *Venus mercenaria,* which normally is not exposed to the air, died of freezing at −6°C. In the case of both species, freezing death occurred when 64 per cent of the cellular water was withdrawn into the ice-laden extracellular space.

The ability of *Mytilus* to withstand temperatures down to approximately −10°C lies in part in the osmotic properties of its intracellular fluids. Twenty per cent of the cellular water is osmotically inactive; i.e., this fraction of the cell's water cannot be drawn from the cell under the severe osmotic stresses which occur during extracellular ice formation. In contrast, the cellular water of *Venus* is not "tied up" to such a large extent.

The factors responsible for "tying up" the cellular water of *Mytilus* are not known. Williams postulates that common, low molecular weight solutes such as inorganic ions or small organic molecules (amino acids and sugars) cannot be responsible, since during dehydration of the *Mytilus* cell the osmotic properties of the cellular water change in manners which appear inconsistent with what one would predict on the basis of colligative relationships: the water remaining behind in the cell becomes osmotically inactive at such a rate that more than mere numbers of solute molecules must be involved. Williams suggests that unusual compounds are present in *Mytilus* which exert an osmotic influence out of proportion to their numbers. While the identities of these hypothetical compounds have not been determined, there is one well-studied example of a biological antifreeze which appears to have unique abilities to depress the freezing points of solutions to a much higher extent than would be predicted on the basis of classical colligative relationships. This is the class of freezing-point-depressing glycoproteins found in certain Antarctic and, probably, Arctic fishes.

GLYCOPROTEIN ANTIFREEZE COMPOUNDS IN POLAR FISHES

The biochemical freezing resistance/tolerance adaptations discussed up to this point have been examples of what we term the "quantitative strategy." All of these freezing adaptations have been effected by increasing the concentrations of types of molecules normally present in the cells.

In the case of polar fishes we find examples of most elegant "uses" of the "qualitative strategy" of biochemical adaptation. In both Antarctic and Arctic fishes, classes of glycoprotein molecules are found which appear to have as their sole function the lowering of the freezing point of biological fluids.

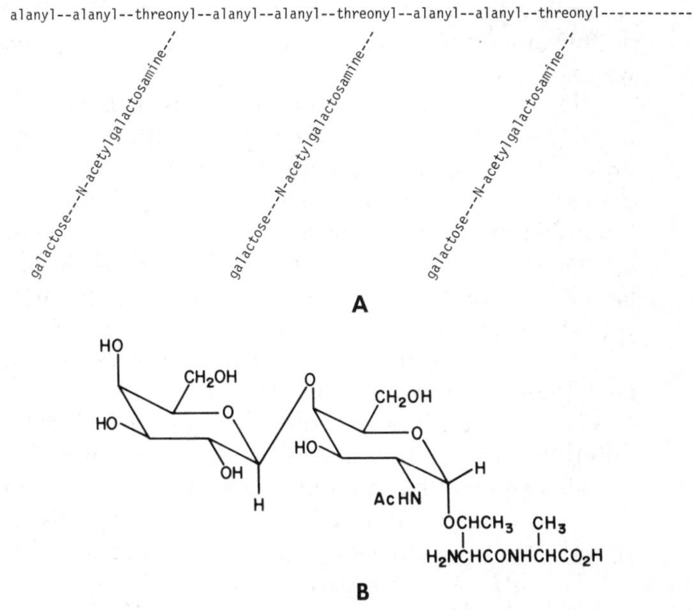

Figure 7–28 The structure of the freezing-point-depressing glycoproteins of Antarctic Nototheniid fishes. (A) The basic repeating unit of the larger (molecular weight greater than 10,500) glycoprotein antifreezes. In the smaller glycoproteins, which have molecular weights of 2600 and 3500, proline residues can replace the alanine residues at certain sites along the chain. (B) A stereochemical representation of the carbohydrate moities of the glycoprotein antifreezes, illustrating the essential hydroxyl groups of the molecule. (After Lin *et al.*, 1972.)

The chemical structure of a typical freezing-point-depressing glycoprotein of an Antarctic Nototheniid fish is illustrated in Figure 7–28. These molecules, which range in molecular weight up to approximately 21,500, can be seen to contain large numbers of hydroxyl groups. As in the case of glycerol and related compounds, the hydroxyl groups of the freezing-point-depressing glycoproteins are essential for antifreeze activity. Chemical blocking of these groups via acetylation and acetonation led to a complete loss of the marked antifreeze potential of these molecules; i.e., following the loss of active hydroxyls, the glycoproteins were no better in depressing the freezing points of solutions, on a per mole basis, than a typical solute such as sucrose, which obeys the colligative rules of freezing point depression.

The mechanism by which the freezing-point-depressing glycoproteins lower the freezing point of solutions more than would be predicted strictly on the grounds of concentration is not fully known. One hypothesis forwarded by DeVries proposes that the glycoproteins coat small ice crystals and somehow weaken the ice structure sufficiently to cause melting. Thus, even though ice crystals may enter the body fluids of the fish (e.g., at the gills), the glycoprotein antifreezes could prevent the survival and growth of this ice within the body. In addition, the hydroxyl-rich glycoproteins no doubt function like other polyhydroxyl molecules in lessening water-water interactions.

The ecological and evolutionary aspects of the freezing-point-depressing glycoproteins are as fascinating as their chemical properties. For example, among different members of the Antarctic genus *Trematomus*, the concentration of antifreeze found in the blood is directly proportional to the freezing dangers encountered by the species. *T. borchgrevinki*, a pelagic fish which encounters ice crystals in its habitat, has the highest concentration of the glycoprotein antifreeze molecules. Shallow water, benthic species, which also experience some ice in their sea-floor habitat, have the next highest concentrations. Deeper water fish have much lower levels of the antifreeze molecules, a trend which is noted among different species and among populations of a single species. DeVries has also shown that Antarctic fishes from very deep water may lack the glycoprotein antifreezes entirely. These fishes are likely to spend their entire lives in a supercooled state.

Evolutionally, the glycoprotein antifreezes of the Antarctic Nototheniid fishes represent an especially good example of an "exploitative" adaptation. No doubt the origin of the freezing-point-depressing glycoproteins enabled these fishes to colonize the food-rich waters of the Antarctic and thereby exploit a vast, untapped food source. This exploitation is especially striking in the case of *T. borchgrevinki*, which lives much of the time among the loose ice platelets that form beneath the solid sea ice, swimming down beneath the ice layers only to capture planktonic prey.

Among various Arctic fish species, similar glycoprotein antifreeze compounds appear to exist, although their chemical structures remain to be detailed. One interesting aspect of the freezing resistance adaptation of Arctic fishes is the seasonal change in glycoprotein content, which has been observed in recent laboratory studies by Duman and DeVries. Like the glycerol content of Arctic insects, the glycoprotein concentration of the Arctic fishes' blood varies seasonally.

OSMOTIC CONFORMITY AS A FREEZING-RESISTANCE STRATEGY

In the case of marine invertebrates, which are osmoconformers, freezing represents little threat, unless the medium itself solidifies. For marine fishes, which are hypotonic to their environment, problems of freezing resistance are obviously more acute. However, one might ask, "Why don't polar fishes merely conform, osmotically, to their environments? Isn't this an easier solution to the freezing resistance problem than the evolution of a new class of glycoprotein antifreeze substances?"

The answer to this question is obviously negative. When we consider the number of physiological and biochemical processes and structures which are dependent on specific ionic conditions, both qualitative and quantitative, we can appreciate the enormity of the task which would result if the fish should "try" to become an osmotic conformer. For example, many (perhaps most) enzymes have specific ion requirements. To alter the intracellular ion concentrations would therefore entail "redesigning" a vast number of proteins. Similarly, nucleic acids and nucleic acid-containing structures, such as ribosomes, have precise ionic requirements. Osmotic conformity would again entail the "redesign" of complex macromolecular ensembles. Perhaps most obvious are the effects on membrane potentials. The enzymes which are instrumental in maintaining ionic disequilibria are "designed" to function optimally only within restricted ionic concentration ranges. Any large scale increase in ionic concentrations would thus prove inimical to the maintenance of proper ionic gradients across membranes.

These considerations suggest that the evolutionary development of unique compounds, such as the glycoproteins of the Antarctic and Arctic fishes, for freezing resistance adaptation does, in fact, represent an "easy" approach to solving a problem in environmental adaptation, as most of the biochemical (i.e., genetic) change involved is located at a few loci.

POLYHYDROXYL COMPOUNDS AND THE CRYOPROTECTION OF ENZYMES

The preceding discussion of freezing resistance/tolerance was framed entirely around questions of ice formation and its consequences to the organism. However, ice crystal formation is but one of the weak bond dependent changes which occur at low temperatures. Another significant "threat" posed by freezing or near-freezing temperatures is the loss of higher orders of protein structure resulting from the destabilization of hydrophobic interactions at low temperatures.

As we indicated in Section I of this chapter, many enzymes lose their 3° or 4° structures (or both) at low temperatures. In many cases, enzymologists have succeeded in preventing these losses of structure, *in vitro,* by incubating the enzymes in high concentrations of glycerol. This molecule acts to stabilize hydrophobic interactions. With this fact in mind, we can readily appreciate that cryoprotectant substances such as glycerol may play a dual role in freezing resistance/tolerance: in addition to preventing ice formation and minimizing cellular damage if and when ice forms in the extracellular spaces, glycerol and similar molecules may exert an important protective influence on cellular proteins. Whether the glycoprotein antifreezes have a similar effect is unknown. Perhaps, if the various cryoprotectants are stabilizers of enzyme structure, they may in fact be acting as temperature compensatory mechanisms by keeping a significant fraction of the cell's enzymes in active conformational and/or aggregational states.

REFERENCES

Books, Review Articles, Symposia

Brett, J. R. (1971). Temperature—animals—fishes. In *Marine Ecology,* edited by O. Kinne, Volume I, pp. 513–560. Wiley-Interscience, London.

Bullock, T. H. (1955). Compensation for temperature in the metabolism and activity of poikilotherms. *Biological Reviews 30*:311–342.

Hochachka, P. W., and Somero, G. N. (1971). Biochemical adaptation to the environment. In *Fish Physiology,* edited by W. S. Hoar and D. J. Randall, Volume VI, pp. 99–156. Academic Press, New York.

Prosser, C. L. (1967). *Molecular Mechanisms of Temperature Adaptation.* American Association for the Advancement of Science, publication number 84, Washington, D.C., 390 pp.

Journal Articles

Baldwin, J. (1971). Adaptation of enzymes to temperature: acetylcholinesterases in the central nervous system of fishes. *Comparative Biochemistry and Physiology 40*:181–187.

Baldwin, J., and Hochachka, P. W. (1970). Functional significance of isoenzymes in thermal acclimation: acetylcholinesterase from trout brain. *Biochemical Journal 116*:883–887.

Baust, J. G., and Miller, L. K. (1970). Variations in glycerol content and its influence on cold hardiness in the Alaskan carabid beetle, *Pterostichus brevicornis. Journal of Insect Physiology 16*:979–990.
Dean, J. M. (1969). The metabolism of tissues of thermally acclimated trout (*Salmo gairdneri*). *Comparative Biochemistry and Physiology 20*:185–196.
DeVries, A. L. (1971). Glycoproteins as biological antifreeze agents in Antarctic fishes. *Science 172*:1152–1155.
DeVries, A. L. (1971). Freezing resistance in fishes. In *Fish Physiology,* edited by W. S. Hoar and D. J. Randall, Volume VI, pp. 157–190. Academic Press, New York.
Fox, C. F. (1972). The structure of cell membranes. *Scientific American 226*:30–38.
Hazel, J. R. (1972). The effect of temperature acclimation upon succinic dehydrogenase activity from the epaxial muscle of the common goldfish (*Carassius auratus* L.): lipid reactivation of the soluble enzyme. *Comparative Biochemistry and Physiology, 43B*:837–861, 863–882.
Hazel, J., and Prosser, C. L. (1970). Interpretation of inverse acclimation to temperature. *Zeitschrift für vergleichende Physiologie* 67:217–228.
Hochachka, P. W. (1967). Organization of metabolism during temperature compensation. In *Molecular Mechanisms of Temperature Adaptation,* edited by C. L. Prosser, American Association for the Advancement of Science, publication number 84, Washington, D.C., pp. 177–203.
Hochachka, P. W. (1968). Action of temperature on branch points in glucose and acetate metabolism. *Comparative Biochemistry and Physiology 25*:107–118.
Hochachka, P. W., and Lewis, J. K. (1970). Enzyme variants in thermal acclimation: trout liver citrate synthases. *Journal of Biological Chemistry 245*:6567–6573.
Hochachka, P. W., and Lewis, J. K. (1971). Interacting effects of pH and temperature on the K_m values for fish tissue lactate dehydrogenases. *Comparative Physiology and Biochemistry 39*:925–933.
Johnson, G. B. (1971). The relationship of enzyme polymorphism to metabolic function. *Nature 232*:347–349.
Lin, Y., Duman, J. G., and DeVries, A. L. (1972). Studies on the structure and activity of low molecular weight glycoproteins from an Antarctic fish. *Biochemical and Biophysical Research Communications 46*:87–92.
Lyons, J. M., and Raison, J. K. (1970). A temperature-induced transition in mitochondrial oxidation: contrasts between cold and warm-blooded animals. *Comparative Biochemistry and Physiology 37*:405–411.
Miller, L. K. (1969). Freezing tolerance in an adult insect. *Science 166*:105–106.
Moon, T. W., and Hochachka, P. W. (1971). Temperature and enzyme activity in poikilotherms: isocitrate dehydrogenases in rainbow-trout liver. *Biochemical Journal 123*:695–705.
Moon, T. W., and Hochachka, P. W. (1971). Effect of thermal acclimation on multiple forms of the liver-soluble $NADP^+$ linked isocitrate dehydrogenase in the family Salmonidae. *Comparative Biochemistry and Physiology 40*:207–213.
Moon, T. W., and Hochachka, P. W. (1972). Temperature and the kinetic analysis of trout isocitrate dehydrogenase. *Comparative Biochemistry and Physiology, 42B*:725–730.
Newell, R. C., and Pye, V. I. (1970). Seasonal changes in the effect of temperature on the oxygen consumption of the winkle *Littorina littorea* (L.) and the mussel *Mytilus edulis* (L). *Comparative Biochemistry and Physiology 34*:367–383.
Newell, R. C., and Pye, V. I. (1971). Quantitative aspects of the relationship between metabolism and temperature in the winkle *Littorina littorea* (L.) *Comparative Biochemistry and Physiology 38B*:635–650.
Precht, H. (1964). Über die Bedeutung des Blutes für die Temperaturadaptation von Fischen. *Zoologische Jahrbücher für Physiologie 71*:313–327.
Roots, B. I. (1968). Phospholipids of goldfish (*Carassius auratus* L.) brain: the influence of environmental temperature. *Comparative Biochemistry and Physiology 25*:457–466.
Somero, G. N. (1969). Pyruvate kinase variants of the Alaskan king crab: evidence for a temperature-dependent interconversion between two forms having distinct and adaptive kinetic properties. *Biochemical Journal 114*:237–241.
Somero, G. N. (1969). Enzymic mechanisms of temperature compensation: immedi-

ate and evolutionary effects of temperature on enzymes of aquatic poikilotherms. *American Naturalist 103*:517–530.

Somero, G. N. (1973). Thermal modulation of pyruvate metabolism in the fish *Gillichthys mirabilis:* the role of lactate dehydrogenases. *Comparative Biochemistry and Physiology, 44B*:205–209.

Somero, G. N., and Hochachka, P. W. (1968). The effect of temperature on catalytic and regulatory functions of pyruvate kinases of the rainbow trout and the Antarctic fish *Trematomus bernacchii. Biochemical Journal 110*:395–400.

Somero, G. N., and Hochachka, P. W. (1971). Biochemical adaptation to the environment. *American Zoologist 11*:159–167.

Somero, G. N., and Johansen, K. (1970). Temperature effects on enzymes from homeothermic and heterothermic tissues of the harbor seal (*Phoca vitulina*). *Comparative Biochemistry and Physiology 34*:131–136.

Steim, J. M., Tourtellotte, M. E., Reinert, J. C., McElhaney, R. N., and Rader, R. L. (1969). Calorimetric evidence for the liquid-crystalline state of lipids in a biomembrane. *Proceedings of the National Academy of Sciences (U.S.) 63*:104–109.

van Handel, E. (1966). The thermal dependence of the rates of glycogen and triglyceride synthesis in the mosquito. *Journal of Experimental Biology 44*:523–528.

Williams, R. J. (1970). Freezing tolerance in *Mytilus edulis. Comparative Biochemistry and Physiology 35*:145–161.

CHAPTER 8 PRESSURE

THE BASIC EFFECTS OF PRESSURE ON BIOLOGICAL SYSTEMS

INTRODUCTION

Hydrostatic pressure impinges directly and unavoidably on the cellular chemistry of all organisms. But even though it is an environmental parameter with which all forms of life must cope, biologists largely have tended to ignore pressure in their studies of environmental physiology and biochemistry. In part this neglect is well justified: the organisms which biologists favor in their studies, for reasons which include ease of capture and maintenance, almost exclusively inhabit the terrestrial environment or the first few meters of the waters of the earth. These are organisms which encounter only modest absolute pressures and essentially no variation in pressure: from sea level to the highest point on earth, atmospheric pressure varies only about 4-fold (Table 8–1). Changes in pressure over this range can have little effect on the chemistry of the cell. Thus, even though adaptation to different altitudes is known to involve major anatomical, physiological, and biochemical changes, these latter must be regarded entirely as adaptations to other altitude-associated factors, especially altered oxygen availability, and not to pressure.

TABLE 8–1 THE RANGE OF PRESSURES IN DIFFERENT POSITIONS IN THE ATMOSPHERE AND IN THE HYDROSPHERE.

Relative Position	Absolute Pressure in Atmospheres
Mt. Everest	0.25
sea level	1.0
5 km below sea level	500.0
10 km below sea level	1000.0

Only in the marine environment and in deep freshwater lakes does pressure assume great importance as an environmental parameter extensively affecting biological systems. Sea water is nearly 1000 times as dense as air, and hydrostatic pressure rises by 1 atm. with each 10 meters of depth. The average abyssal pressures of the ocean floor range between 300 and 500 atmospheres; in the deepest oceanic trenches, pressures exceed 1000 atmospheres. Temperatures are extremely low at these depths, probably averaging less than 2° C.

In addition to encountering severe absolute pressures, marine organisms may also be subjected to wide variations in pressure, either diurnally or at different times throughout the life cycle. For example, fishes such as the Myctophids undergo daily 300 to 500 meter vertical migrations in the water column, depths corresponding to pressure changes of some 30 to 50 atmospheres. Other midwater organisms (squids and other fishes such as *Stomias* and *Ectreposebastes*) undergo vertical migrations which are 2- to 5-fold greater. Many benthic organisms appear to have pelagic larval stages, and through a complete life cycle these species can encounter even greater pressure ranges. Organisms migrating through the water column also face linear changes in temperature, often of 10 to 15°C range. Thus, we find that pressure stress is commonly associated with temperature stress, and as a result, the biochemical machinery of many marine organisms must be adapted to deal simultaneously with two major physical parameters of the environment.

In our examination of temperature effects, we built our discussion on a basic understanding of the fundamental effects on the cellular chemistry of the organism. We shall again use this approach in the case of pressure. As with temperature, the effects of pressure can be grouped into two categories:

(1) *Rate effects,* which are due to the effects of pressure on metabolic reaction rates.

(2) *Weak-bond—structural effects,* which involve the effects of pressure on weak-bond dependent structures and processes, such as those listed in Table 7–2 (p. 180).

We employ these categories for didactic reasons, and do not claim that they represent mutually exclusive sets of phenomena in the real world. Certainly, these basic effects interact with each other. Changes in protein conformation clearly can affect enzyme reaction rates, as can alterations in the physical state of cellular constituents. However, in spite of the overlapping nature of these categories, they may nonetheless help us to appreciate the basic problems posed to marine organisms by extremes of, and changes in, ambient pressure.

RATE EFFECTS: FUNDAMENTAL THEORY OF THE ACTION OF PRESSURE ON ENZYME REACTION RATES

In considering the factors instrumental in establishing the pressure sensitivity of a reaction, it is essential to distinguish carefully between parameters which govern the final equilibrium of the reaction and those which determine how rapidly this equilibrium is attained. When discussing temperature effects, we stressed that the equilibrium which the reaction will attain is set by the free energy change (ΔG^0) which occurs as substrates are converted to products. The *rate* at which this equilibrium is attained is governed by a second free energy function, the free energy of activation, $\Delta G^{\ddagger}$.

We find analogous parameters involved in the kinetics and thermodynamics of pressure effects. As we might expect, the key parameters which establish the direction and magnitude of pressure effects on a reaction involve volume changes. If the volume of the system containing the reactants is greater than the volume of the system containing the products, then pressure will shift the equilibrium toward product formation. If the volume of the reactants is less than the volume of the system containing the products, pressure will favor the accumulation of reactants at equilibrium. Formally, these relationships can be described as follows:

$$\Delta V = 2.3\ RT\ \frac{\log_{10}K_{P_1} - \log_{10}K_{P_2}}{P_1 - P_2}$$

where ΔV is the volume change of the reaction, R is the gas constant with the value 82.07 atm cm^3/mole °K, T is the absolute temperature in degrees Kelvin, and K is the equilibrium constant at pressures P_1 and P_2 atmospheres.

The overall volume change of the reaction, ΔV, has its analogue in the overall free energy change of the reaction, ΔG^0, discussed with reference to temperature (p. 182): *both parameters indicate what the final equilibrium is likely to be, but say nothing about the speed with which this equilibrium is attained.*

The critical factor which determines the effect of pressure on reaction rate, as opposed to reaction equilibrium, is the volume change which accompanies the formation of the transitory activated complex. We can equate this volume change with the volume difference between the system containing nonactivated reactants and that containing activated reactants. As with activation parameters discussed for temperature effects, we symbolize the volume change of activation as $\Delta V^{\ddagger}$.

The equation describing the dependence of velocity on pressure is formally analogous to the Arrhenius equation relating reaction rates to temperature:

$$\Delta V^{\ddagger} = 2.3\ RT\ \frac{\log_{10}k_{P_1} - \log_{10}k_{P_2}}{P_1 - P_2}$$

When the volume of the activated complex exceeds the average volume of its constituents outside the complex, pressure inhibits the reaction. If the volumes become equal, there is no change in the reaction rate under pressure. When the volume of the activated complex is less than that of the reactants, pressure increases the reaction rate. In this latter case, for enzyme-catalyzed reactions, the reaction rate can go through a pressure optimum. Either one of two factors usually determines such an optimum: (1) the reaction can become diffusion limited due to large increases in the viscosity of the medium, or (2) high pressures can lead to enzyme denaturation. As both factors can be influenced by temperature, the actual pressure at which the optimum occurs often varies with temperature.

THE SIGN OF $\Delta V^{\ddagger}$ AND ITS METABOLIC CONSEQUENCES

From the above discussion, it will be evident that with regard to pressure, biochemical reactions can be classified into three categories:

(1) pressure activated reactions, with a negative $\Delta V^{\ddagger}$,

(2) pressure inhibited reactions, with a positive $\Delta V^{\ddagger}$, and

(3) pressure independent reactions proceeding with no net volume change.

The chief determinants of the category of response observed are the structures of the reactants and of the transition "activated" complex. Since in enzyme-catalyzed reactions these can vary with temperature, it follows that pressure can bring about all three effects on a given enzyme, depending upon the temperature.

The combined consequences of (1) differential pressure effects among different enzymes, and (2) differential pressure responses by a single enzyme at different temperatures render metabolism exceedingly "vulnerable" to pressure changes. Where two or more enzymes compete for a single common metabolite, such *differential* effects of pressure can profoundly alter the flow of carbon through critical metabolic branching points, and thus strongly influence the partitioning of carbon and energy between various cellular processes. The problem can be illustrated by examining the potential effects of pressure on a single branch point, such as that involved in the competition for G6P.

In tissues such as vertebrate liver, at least four enzyme pathways compete for the common substrate, G6P (Figure 8–1). In addition, the relative activity of hexokinase strongly determines the net direction of the glucose ⇌ G6P interconversion. Hence, any complete assessment of the effects of pressure upon G6P channelling through this metabolic crossroad must take into account the effects of pressure upon at least five enzyme-catalyzed reactions.

In theory, $\Delta V^{\ddagger}$ for each of these reactions could be estimated from knowledge of the molecular geometry of the initial and transition states of the reacting components. In practice, of course, sufficient information about the configuration of the transition state and its neighboring water and solute molecules is available for only the simplest of reactions: such information is certainly unavailable for the reactions of G6P metabolism. When faced with such problems, the chemist must turn to first approximations. For simple chemical reactions, these approximations usually assume $\Delta V^{\ddagger}$ to be the sum of two terms: a "structural" term, $\Delta V_1^{\ddagger}$, is the change in volume of the reacting molecules during bond-breakage and bond-formation; the "solvation" term, $\Delta V_2^{\ddagger}$, is the accompanying volume change due to interaction with surrounding water molecules. When ionic species are formed or disappear in the transition state of simple chemical reactions, the "solvation" contribution to $\Delta V^{\ddagger}$ usually exceeds the "structural" volume change in the reacting molecules. *For enzyme catalyzed reactions, the opposite is to be expected, and large "structural" contributions to $\Delta V^{\ddagger}$ most probably occur.*

For a branch point such as the G6P crossroads, (1) the reacting compounds involved in each reaction, (2) the reaction mechanism, and (3) the transition "activated" complexes are specific to each of the five competing enzyme pathways. Hence, the "solvation" contribution to $\Delta V^{\ddagger}$ is specific to each reaction. In addition, each reaction is catalyzed by a different kind of enzyme and the "structural" contribution to $\Delta V^{\ddagger}$ will depend on the enzyme involved.

Now, if both "solvation" and "structural" contributions to $\Delta V^{\ddagger}$ are specific to each reaction, we can with complete assurance conclude that *only by coincidence would the quantitative effects of pressure be the same for all five enzymes.* In all probability, the effects of pressure on each of these reactions would be quantitatively very different. Since at least three of these enzymes (G6PDH, G6Pase, and hexokinase) are known to be involved in the regulation of G6P metabolism in the liver, it appears that the flow of carbon through metabolic branch points may be strongly affected by differential pressure effects, and therefore that control requirements at such points may depend critically upon the absolute pressure. It is such differential effects of pressure on biological processes that may indeed represent the most essential biochemical problem in pressure adaptation.

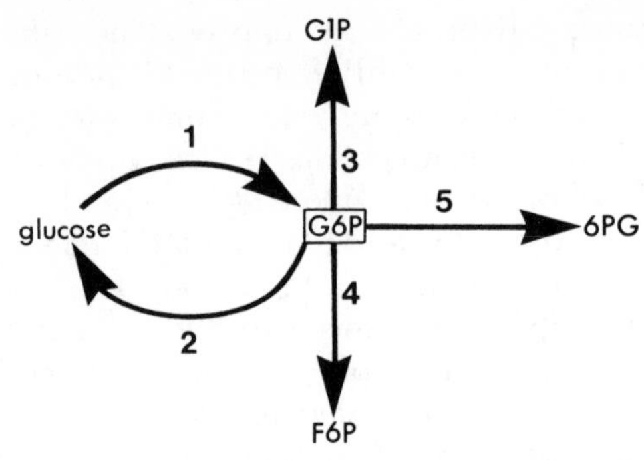

Figure 8–1 The G6P branching point in vertebrate liver metabolism. The enzymes catalyzing each step at the branchpoint are numbered and their properties are summarized below.

No.	Enzyme	Reaction
1	Hexokinase	Glucose + ATP → G6P + ADP
2	G6Pase	G6P → Glucose + P_i
3	PGM	G6P ↔ G1P
4	Isomerase	G6P ↔ F6P
5	G6PDH	G6P + NADP → 6PG + NADPH

No.	Properties	Size (MW)	Number of Subunits
1	equilibrium between bound and unbound states	96,000	4
2	lipoprotein microsomal	?	?
3	soluble, single chain	54,000	1
4	soluble oligomer	140,000?	2?
5	equilibrium mixture of tetramers + dimers	210,000 (tetramer)	4

THE MAGNITUDE OF $\Delta V^{\ddagger}$ FOR ENZYME CATALYZED REACTIONS

Whereas the sign of $\Delta V^{\ddagger}$ determines whether the pressure effect is a stimulation or a retardation in rate, the absolute value of

$\Delta V^{\ddagger}$ will determine how sharply pressure affects reaction velocity. Enzymic reactions characteristically exhibit much larger volume changes during activation than non-enzyme-catalyzed reactions. No doubt this difference is due to volume changes occurring during catalysis and involving the enzyme protein itself.

For non-catalyzed reactions, $\Delta V^{\ddagger}$ values normally are less than 20 cm³/mole. $\Delta V^{\ddagger}$ values for enzymic reactions may surpass 100 cm³/mole. As illustrated in Figure 8–2, $\Delta V^{\ddagger}$ values of this magnitude would lead to approximately ten-fold changes in reaction rate as pressure is increased from 1 atmosphere to 1000 atmospheres, assuming that $\Delta V^{\ddagger}$ does not vary with pressure. In theory, there-

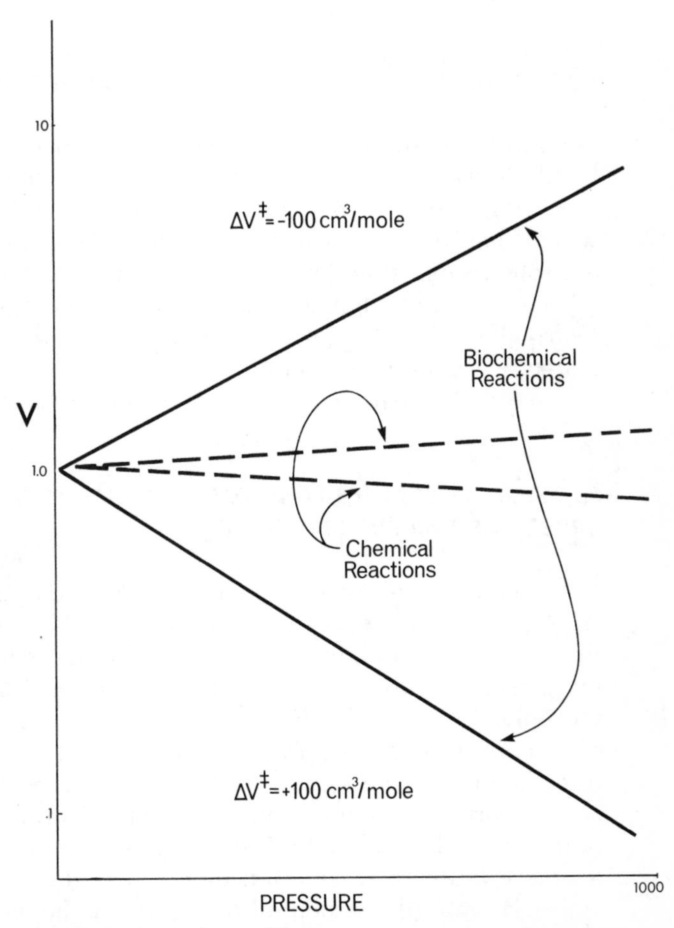

Figure 8–2 Log velocity (V) vs. pressure in atmospheres for pressure-activated and pressure-inhibited biochemical (enzyme-catalyzed) reactions and for non-enzyme catalyzed reactions, assuming that $\Delta V^{\ddagger}$ does not change with pressure to 1000 atmospheres.

fore, reaction rates occurring in deep sea organisms could be altered 3- to 10-fold by pressures of 300 to 1000 atmospheres.

These are large effects of pressure. First of all, they are large enough to suggest that any abyssal organism capable of reducing or minimizing them would gain a strong selective advantage, made all the more significant by the low temperature in abyssal regions of the sea. Secondly, these effects are much larger than are found, or indeed anticipated, for simple chemical reactions. As they are enzyme-catalyzed reactions, the basis for their unusual pressure sensitivity may lie (1) in the large volume changes accompanying the reaction, and (2) in the effects of pressure on enzyme structures and hence their functions. That is, the effects of pressure on the higher structural orders of the main macromolecular component of the system may strongly determine its functional properties.

TWO CLASSES OF WEAK-BOND STRUCTURAL EFFECTS

For heuristic reasons, many of the effects of pressure on weak-bonded systems can be placed into two kinds of transitions:

(1) Transitions between solid, liquid, and gaseous states, which may have a basic but essentially indirect influence upon macromolecular functions such as enzyme catalysis.

(2) Complex changes in macromolecular structure which directly affect the functional characteristics of the particular macromolecule under consideration.

We shall consider the second of these in some detail.

PRESSURE AFFECTS WEAK BONDS AND THEREFORE HIGHER ORDERS OF PROTEIN STRUCTURE

According to current theories, the assumption of 3° and 4° structure of proteins is thermodynamically favored because of the removal of nonpolar, hydrophobic residues from contact with water by "burying" these residues, either within the tertiary structure of the polypeptide or between the faces of adjoining subunits in the oligomeric protein (see Figure 7–4). Hydrogen bonding and ionic interactions between polar or charged residues of the protein are also important in determining the final conformation and aggregation state of the molecule. Knowledge of the influence of pressure on these kinds of weak bond interactions therefore is basic to an appreciation of the structural pressure-dependence of proteins.

In this context, the relevant question concerns the volume changes which accompany the removal of hydrophobic and ionic groups from the aqueous phase of the cell. In other words, when amino acid residues interact among themselves, rather than with cellular water, is the volume of the system (protein plus water) increased or decreased?

The best current estimates suggest that ionic and hydrophobic interactions which lead to the formation of higher orders of protein structure cause a net increase in the volume of the system. In other words, denaturation leads to a decrease in the total volume of protein + water. The most reasonable explanation of this phenomenon involves the effects of hydrophobic and ionic groups on water structure. Both classes of amino acid residue appear to structure water more tightly, in a volumetric sense, than do other water molecules.

The rupture of hydrophobic bonds is accompanied by volume changes of −6 to −20 cm^3/mole. For proteins having higher orders of structure largely stabilized by hydrophobic interactions, the effects of pressure on 3° and 4° structure therefore could be very great. For a polypeptide chain of molecular weight 15,000 (roughly 100 amino acid residues) having 25 hydrophobic residues buried in the interior of the molecule, the ΔV accompanying unfolding of the 3° structure is approximately −250 cm^3/mole, assuming that the 3° structure is stabilized almost entirely by hydrophobic interactions and that the exposure of each hydrophobic residue to the aqueous phase is accompanied by a volume change of −10 cm^3/mole.

We can calculate the equilibrium constant for the denaturation process as follows: in going from the native (N) folded state to the unfolded, denatured (D) state, the equilibrium constant is given by $K = \frac{D}{N}$. If ΔV is independent of pressure,

$$-\ln \frac{K_p}{K_1} = \frac{P\Delta V}{RT},$$

where K_p is the equilibrium constant at P atmospheres pressure and K_1 is the equilibrium constant at 1 atmosphere pressure. At 32°C, RT = 25,000 atm cm^3/mole, and the value of ΔV given above is −250 cm^3/mole. Thus, at 1000 atmospheres,

$$-\ln \frac{K_p}{K_1} = \frac{1000 \times -250}{25{,}000} = -10$$

Therefore, $K_p = K_1 e^{10}$ or $K_p = 2.2 \times 10^4\ K_1$. In other words, at 1000 atmospheres the protein should be essentially all in the unfolded state. Even at less extreme pressures, proteins with 3° and 4° structures stabilized largely by hydrophobic interactions will tend to exist in denatured form. Do available data generally agree with these theoretical predictions? As it turns out, in most cases the results of pressure denaturation experiments are consistent with the hypothesis that extremes of pressure labilize the higher orders of protein structure via disruption of the weak bonds holding 3° and 4° structures intact.

THE EXAMPLE OF RIBONUCLEASE (RNase)

RNase is one of the structurally simplest and, partly for this reason, one of the best understood enzymes. It is a monomeric enzyme but, when treated with the proteolytic enzyme, subtilisin, can be cleaved into two fragments, the "S-protein" containing 104 amino acids and the "S-peptide" containing only 20 amino acids. The two fragments can be separated readily and, in this condition, are inactive. However, when suitable conditions pertain, the two fragments can be non-covalently reunited to yield a functional enzyme, according to the scheme:

$$\underset{\text{(inactive)}}{\text{S-protein}} + \underset{\text{(inactive)}}{\text{S-peptide}} \rightleftharpoons \underset{\text{(active)}}{\text{RNase-S}}$$

In the formation of the active RNase-S, the larger fragment appears to provide a "template" surface for the S-peptide, inducing the formation of (i) a helical region in the S-peptide involving at least 10 of the 20 residues, and (ii) hydrophobic bonds between the S-peptide and the S-protein.

Most likely only three or four hydrophobic bonds (involving phenylalanine, histidine, and methionine residues in positions 8, 12, and 13 of the S-peptide) are responsible for joining the S-peptide and S-protein; consequently, we would expect the RNase-S molecule to be pressure labile.

This expectation is realized: high pressures shift the equilibrium strongly toward the free S-protein and S-peptide. This dissociation is accompanied by a volume change of approximately -40 cm^3/mole, a value consistent with the observation that the two fragments are held together by only a small number of hydrophobic bonds.

THE EXAMPLE OF ACTIN POLYMERIZATION

The formation of enzyme 4° structures is but one example of polymerization processes which depend largely, if not entirely, on the formation of weak chemical bonds between subunits or protomers. In the assembly of muscle contractile elements, monomeric G-actin units assemble to form the fibrous F-actin (Figure 8–3a).

The bonds which stabilize the F-actin structure are hydrophobic interactions, and are therefore favored by neutral salts and high temperature. Formation of F-actin from G-actin involves a volume change of 84 cm^3/mole of G-actin. In light of this large ΔV value, we would expect that high pressures would shift the equilibrium strongly toward G-actin. This, in fact, occurs.

The most striking confirmation of this prediction has come from electron microscopic analyses. Examinations of muscle tissues from deep sea fish, where actin (and myosin) assembly normally occurs under several hundred atmospheres pressure, indicate that the filaments are significantly shorter than those characteristic of muscles in terrestrial vertebrates.

THE EXAMPLE OF MYOSIN POLYMERIZATION

As shown in Figure 8–3b, the assembly of the functional myosin polymer from myosin subunits (which in turn are composed of five polypeptide chains) is also pressure sensitive. Unlike the actin polymer (F-actin), the myosin polymer is largely stabilized by polar interactions. The formation of H-bonds is accompanied by volume increases of about 3 to 7 cm^3/mole. In this case, assembly of the polymer is accompanied by a volume increase of about 300 cm^3/mole, a value of such magnitude as to favor the monomeric state

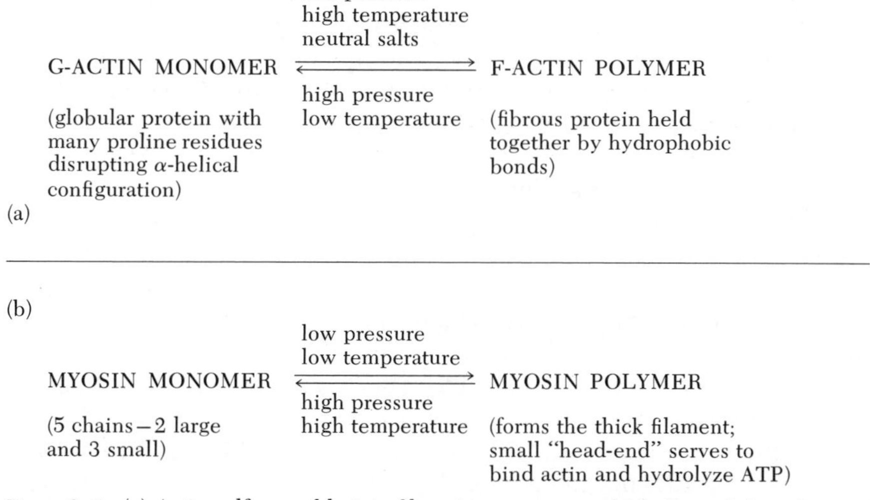

Figure 8–3 (a) Actin self-assembly into filaments, a process which depends largely upon the formation of hydrophobic interactions between actin monomers. (b) Myosin self-assembly into filaments, a process which depends largely upon the formation of hydrogen bonds between the myosin monomers. For discussion of pressure effects on muscle proteins, see studies by Dreizen, Kim, and coworkers in *Amer. Zoologist,* Vol. *11* (1971).

strongly at high pressures. Again, dramatic support for the contention *that myosin filaments assembled under pressure should be shorter than average* comes from direct electron microscopic examinations of muscle of benthic fishes.

ARE ALL MULTI-SUBUNIT PROTEINS UNUSUALLY PRESSURE SENSITIVE?

At this juncture it may be well to ask whether all oligomeric proteins are likely to be inhibited by pressure, due to the tendency for pressure extremes to shift the equilibrium between subunits and aggregated-subunits toward free subunits. At least one author, Penniston, feels that a dichotomy indeed exists among enzymes as regards their pressure sensitivity: multi-subunit enzymes tend to be pressure inactivated, while monomeric enzymes are either pressure insensitive or pressure activated. While all the examples he has presented fit this general scheme (all of his data, incidentally, are from enzyme studies of non-marine organisms), and while he does raise an interesting question concerning deep sea organisms (Can they have multi-subunit enzymes and, if so, are the subunits held together by forces different from those which stabilize the 4° structure of homologous enzymes from terrestrial organisms?), there are exceptions to his general rule. A moment's contemplation will indicate why these exceptions are inevitable.

Enzymes display a tremendous heterogeneity in terms of size, number of chains, number of S-S bridges, phospholipid requirement, and so forth; for this reason, each enzyme system may display specific pressure responses depending upon such factors as (1) the number of weak bonds formed or ruptured by pressure, (2) the kind of weak bonds changed, and (3) their particular contribution to overall protein structure. A well documented exception to the dichotomy suggested by Penniston is the enzyme lactate dehydrogenase.

THE EXAMPLE OF LACTATE DEHYDROGENASE (LDH)

The globular enzyme, LDH, consists of four subunits. In many species and many tissues, at least two kinds of subunits occur, termed here H and M. *In vivo* these self-assemble into five compositionally distinct tetramers, H_4, H_3M_1, H_2M_2, H_1M_3, and M_4. The forces involved in holding together the four subunits of the LDH tetramer are definitely "weak bonds," for the tetramer can be reversibly dissociated by mild conditions which are known to disrupt both hydrogen bonds and hydrophobic interactions. Under appropriate conditions, these dissociated subunits can be shown to reassociate into the native tetrameric enzyme (Figure 8–4). It is possible in this way to mimic the *in vivo* assembly of the LDH enzymes, since the process occurs spontaneously both in the test tube and in the cell. Jaenicke and coworkers took advantage of these properties to study the effects of pressure upon LDH self-assembly. In the mammalian system that they examined, high pressure strongly shifted the equilibrium toward LDH tetramer formation. Since it is probable that the active form of LDH is the tetrameric form, it is evident that such effects of pressure could influence LDH catalysis *in vivo*. And the same conclusion would appear equally valid for all oligomeric enzymes.

This finding suggests that there is more to consider with regard to pressure dependent equilibria than the changes in solvent volume. Perhaps, for at least some enzymes, the assumption of a 4° structure leads to a significant volume reduction due to a tighter compacting of the polypeptide chains. Is it possible that deep sea organisms have multisubunit enzymes which, unlike their homologues in terrestrial or shallow water forms, either show little volume change upon aggregation or, indeed, may be especially tailored by selection to decrease their volumes as 4° structure forms? We do not know. As is so often true in the field of high pressure biology, we can only hope that the present stage of knowledge on this matter will soon be improved.

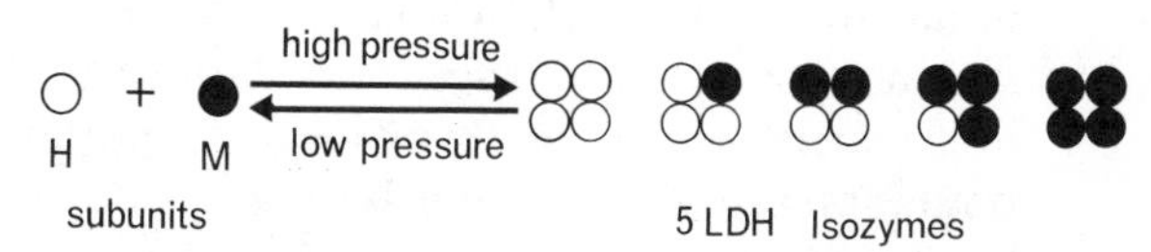

Figure 8–4 Effect of pressure on LDH reassembly from dissociated subunits. The experimental procedure for this type of study is summarized in the form of a flow chart:

$H_4LDH \xrightarrow[\text{conditions}]{\text{dissociating}}$ 4 H monomers

e.g., freeze-thaw cycles at 1M NaCl — reassociating conditions

$M_4LDH \xrightarrow[\text{conditions}]{\text{dissociating}}$ 4 M monomers

Tetramers Formed	Ratio
H_4	1
M_1H_3	4
M_2H_2	6
M_3H_1	4
M_4	1

Starting with pure M and H type subunits, the appearance rate of heterotetramers can be used as a measure of the reassembly rate.

A RECURRING THEME: MANY KEY BIOLOGICAL STRUCTURES AND FUNCTIONS ARE BASED ON WEAK BONDS

In terms of pressure biology, the above studies have yielded some important insights. Firstly, they establish experimentally the range of volume change to be anticipated when weak bonds are formed or are disrupted in various metabolic structures. This information gives the marine biologist interested in pressure problems an idea of the magnitude of pressure effects which may be operative in deep sea species. Furthermore, the kinds of interactions occurring between "S-peptide" and "S-protein" in RNase, between M and H subunits in LDH, and between the actin and myosin subunits are similar to a host of key interactions in the structures and functions of macromolecules in the cell. Included in this list are:

(1) protein-protein interactions, in the formation of complex subcellular structures,
(2) protein-nucleic acid interactions, in the formation of chromosomes, ribosomes, and polyribosomes,
(3) protein-lipid interactions, in the formation of membrane structures, and
(4) enzyme-ligand interactions, in the formation of various enzyme complexes.

Since these vital interactions are dependent on the formation and breakage of weak chemical bonds, the data and arguments raised in our discussion of pressure destabilization of protein structure apply without qualification. Although these sets of interactions have not all been systematically examined in deep sea organisms, we shall briefly discuss examples of the influence of pressure on each of these categories, as the available studies on non-marine organisms provide an important foundation for further developments in pressure biology.

PRESSURE EFFECTS ON SUBCELLULAR STRUCTURES: PHASE CHANGES IN PROTEIN SYSTEMS

Certain proteins exist in solution in either a "sol" or a "gel" phase. In the sol phase the individual protein molecules are in true solution, whereas in the gel phase the proteins aggregate and are no longer able to remain in solution. The sol-gel equilibrium is determined by the chemical and physical properties of the solution, which in turn affect the interactions between the protein monomers.

An elegant example of a pressure-dependent phase transition in a macromolecular aggregation (which may serve as a model for comparable pressure effects on subcellular macromolecular structures) arises from studies of an abnormal human sickle cell Hb

(HbS), which occurs in individuals with sickle cell anemia. In HbS, a valyl residue in position 6 of the β-chain substitutes for the glutamyl residue of normal β-chains. This single amino acid substitution, allowing the formation of a hydrophobic bond between the first and sixth valyl residues in HbS (Figure 8–5), alters the 3-D structure of *deoxygenated* HbS and causes it to aggregate into large molecular "stacks" or filaments. *In vivo,* these filaments are arranged into hollow "cables," each consisting of six HbS monofilaments. These "cables" are insoluble. Sickling is thus a sol-gel transition.

Interestingly, the gel state of HbS is more stable at physiological temperatures than at lower temperatures. When an aggregated (gelled) HbS preparation is placed in an ice bath, the HbS "cables" dissociate to yield free, soluble HbS molecules. Low temperatures thus shift the sol-gel equilibrium toward the sol stage, due to the weakening of the valyl-valyl hydrophobic interactions. Recall that hydrophobic interactions weaken at temperatures near 0°C.

These same valyl-valyl interactions appear to be highly dependent upon pressure. At pressures up to approximately 50 atmospheres, and at temperatures near 37°C, the equilibrium favors the sol state; i.e., pressure reverses the tendency for sickling, presumably by disrupting the valyl-valyl interactions between positions 1 and 6 of the β-subunit. In terms of pressure biology, two points here should be stressed: In the first place, this complex effect

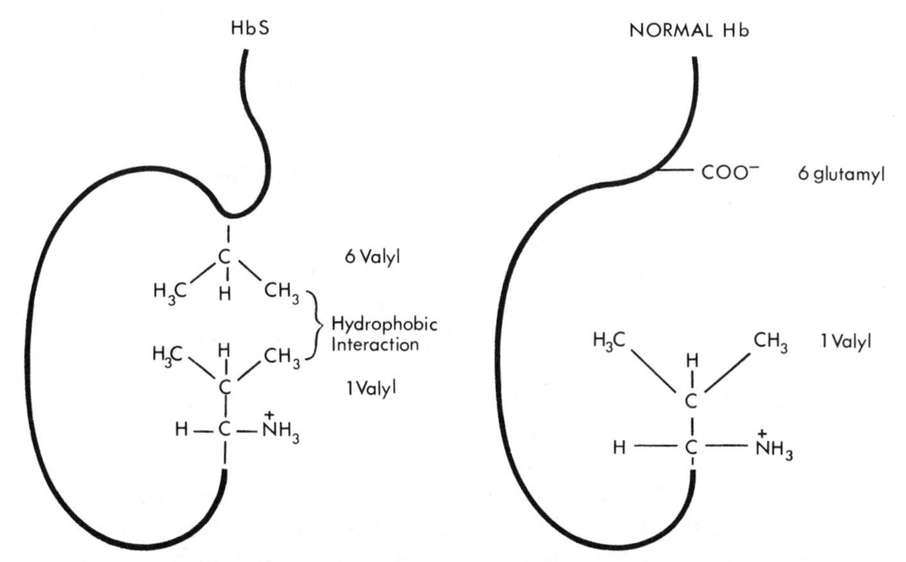

Figure 8–5 Schematic representation of the hydrophobic interaction between valyl 1 and valyl 6 on the beta chain of HbS. In normal Hb, position 6 of the beta chain is a polar glutamyl residue. Modified after Murayama (1966).

on a sol-gel equilibrium is occurring at very modest pressures, well within the biological range for many midwater organisms. Secondly, the effects on macromolecular structure are unusually large. In the case of HbS, the volume change is in the order of 400 cm^3/mole. Should comparable effects occur in other macromolecular structures within the cell, they would represent important adaptational problems to midwater and abyssal organisms. Although critical data on appropriate biological systems are lacking, some highly suggestive information is available from studies of the ultrastructure of protozoans subjected to elevated pressures. In general, there are marked structural changes at high pressures which are related to the duration of the treatment. Alterations in ultrastructure can be seen with 500 atmosphere pressure "pulses" as short as 2 minutes.

The most notable ultrastructural changes are apparent in what appear to be either protein-based structures or membranous strucstures. Thus, control *Amoeba* cells clearly show Golgi complex, mitochondria, and pinocytotic channels; pressure-treated amoebae reveal a loss of Golgi complex and pinocytotic channels. In the heliozoan, *Actinosphaerium,* and in *Tetrahymena,* exposure to pressure leads to a dissolution of the microtubule system. In *Tetrahymena,* the cilium is also highly pressure sensitive. When exposed to high pressures, the central unpaired ciliary tubules as well as the basal body material undergo dissolution or extreme disorganization. Although the lysosomes, the plasma membrane, and the mitochondria appear to be unaffected by pressures of about 500 atmospheres, the endoplasmic reticulum becomes swollen, distended, and clumped. Such effects of pressure on cell ultrastructure may be analogous to the effects of pressure on the HbS sol-gel equilibrium.

DIFFERENTIAL EFFECTS OF PRESSURE ON TRANSCRIPTION AND TRANSLATION

Unlike protein-protein interactions, nucleic acid-nucleic acid interactions may be particularly pressure insensitive. Double-helical DNA, for example, which is stabilized in large measure by extensive hydrogen bonding, is denatured only by exceedingly high (2000 to 3000 atmospheres) pressure. Thus, although it is impossible to predict the quantitative effects of pressure on transcription and translation, we may expect that pressure would alter these processes differentially. Studies with microorganisms such as *E. coli* show this to be the case. When it was established that protein and RNA syntheses are both pressure-inhibited, Landau and his coworkers set out to quantitate these effects with an induced enzyme system in *E. coli.* Induced β-galactosidase in *E. coli* is a particu-

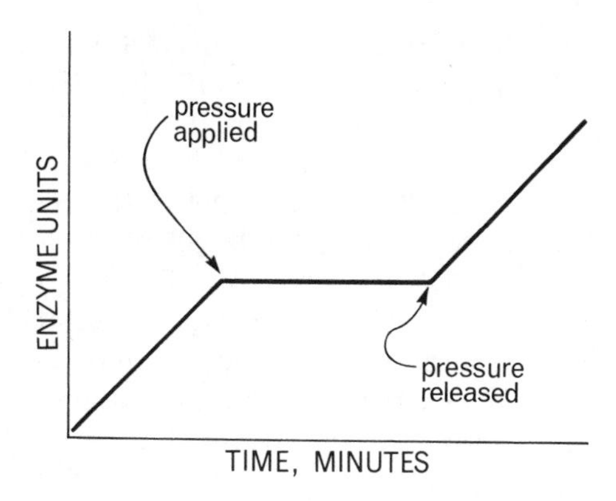

Figure 8–6 The synthesis of β-galactosidase in *E. coli* prior to, during, and after the application of 680 atmospheres of pressure for a period of 10 minutes. The abruptness with which enzyme synthesis returns to control rates after pressure release indicates that mRNA is continuously available for translation. The pressure-block must therefore occur after the synthesis of mRNA; that is, after transcription. See "High Pressure Effects on Cellular Processes," edited by Zimmerman, for further details.

larly good system for sorting out the effects of pressure on transcription and translation, since these operationally distinct phases of protein synthesis can be separated by a variety of experimental techniques. A summary of one key set of experiments is shown in Figure 8–6.

Upon application of high pressures (about 700 atmospheres) to intact *E. coli* cells, induced β-galactosidase synthesis is immediately blocked; with release of pressure, there is an instantaneous return to normal rates of synthesis, suggesting that mRNA is available for immediate translation. The absence of any lag phase on decompression is therefore taken to indicate that the pressure inhibition of enzyme synthesis is at the translational level. This interpretation is supported by the further experimental demonstration that the synthesis of *mRNA for β-galactosidase (i.e., transcription of the β-galactosidase gene)* occurs normally even at very high pressures.

PRESSURE EFFECTS ON RIBOSOMES AND POLYRIBOSOMES

In previous chapters, we have indicated that ribosome structure, and hence function, are highly sensitive to various chemical (p. 106) and physical (p. 195) parameters. In view of those discussions, it is not surprising to discover that ribosomal and polyribosomal structure and function may also be highly pressure sensitive. Zimmerman and his coworkers have carefully investigated the effects of short pulses of pressure upon the formation of ribosomes and polyribosomes in the ciliate protozoan, *Tetrahymena pyriformis.* Their studies clearly indicate that newly forming polysomes are readily disrupted by short pulses of pressure, but that the sedimentation characteristics of the already-functional polysomes are unaffected by pressure. Thus, pressure sensitivity of translation may involve the initiation step rather than the subsequent reactions involved in "reading" the mRNA message.

PRESSURE EFFECTS ON GENE REGULATION: INHIBITION OF INDUCER-REPRESSOR BINDING

Under certain experimental conditions—at pressures too low to inhibit either transcription or translation—it is possible to study the direct influence of pressure upon the formation of the inducer-repressor complex. Under these conditions, a decrease in the rate of β-galactosidase synthesis is observed. These experiments indicate that the binding of the inducer (a low molecular weight sugar) to the repressor (a protein of MW 150,000) is strongly inhibited by pressure:

$$\text{inducer} + \text{repressor} \underset{\text{low pressure}}{\overset{\text{high pressure}}{\rightleftarrows}} \text{inducer-repressor complex}$$

The $\Delta V^{\ddagger}$ for complex formation is about 55 cm^3/mole. As this complex formation undoubtedly involves weak bonds and a conformational change in the repressor upon binding inducer, the observed result perhaps should be expected.

POTENTIAL PRESSURE EFFECTS ON MEMBRANE STRUCTURE

We wish to consider one final example of a subcellular structure that is stabilized by weak bond interactions: the plasma membrane. The basic structure of the plasma membrane (p. 256) consists of a lipid leaflet bylayer, with a strongly hydrophobic interior and a strongly polar exterior. Membrane proteins associate with both the polar and the hydrophobic portions of the phospholipid leaflet. Membranes in many organisms undergo liquid-solid transitions at low temperature (usually between 0° and 20°C), due to the organization of the aliphatic ends of the membrane phospholipids into a crystalline state (p. 257). The functional membrane, in contrast, occurs in a quasi-liquid ("liquid-crystalline") state. If the aliphatic chains of membrane phospholipids exhibit phase transitions comparable to those observed in *in vitro* studies of aliphatic hydrocarbons, then the precise temperature for the liquid-solid phase transition would be dramatically influenced by pressure.

In general, for saturated hydrocarbons, the transition (or melting) temperature increases by about 20°C per 1000 atmospheres. At 1000 atmospheres, for example, a C-12 hydrocarbon (dodecane) undergoes a liquid-solid phase transition at about 15°C, compared

TABLE 8–2 RELATIONSHIP BETWEEN PRESSURE AND THE TEMPERATURE FOR LIQUID-SOLID TRANSITIONS FOR THREE MODEL HYDROCARBON COMPOUNDS. DATA FROM WEALE (1967)

Compound	Approximate Transition Temperature, °C	Pressure, atm
n-C_{12} (dodecane)	−10°	1
	+15°	1000
n-C_{18}	+27°	1
	+50°	1000
n-C_{24}	+50°	1
	+70°	1000

to −10°C at 1 atmosphere; C-18 octadecane melts at over 50°C; C-24 *n*-tetrasocane melts at over 70°C (Table 8–2). These temperatures, of course, are much higher than those of the midwater and benthic environments. In a marine organism even at modest pressures, these kinds of compounds would exist in the solid phase, a situation which is incompatible with continued cell function. It is therefore clear that the solid-liquid phase transitions of membranes must be carefully adjusted in accordance with the depth at which the organism exists.

These phase transitions are strongly determined by such factors as length of the polymer, the number (odd or even) of carbon atoms, and the degree of unsaturation. Aliphatics with an odd number of carbons, for example, have lower "melting points" than would be anticipated by interpolation between the values for their "even" neighbors. This difference, due to the lesser ability of the "odd numbered" molecules to pack in a crystal, *is accentuated at high pressures,* and could supply midwater and abyssal organisms with a means for maintaining their membrane lipids in a liquid state despite the high pressures and low temperatures of their environment. Unfortunately, we know of no studies dealing with the problem of membrane lipid composition as a function of depth in the marine water column; it is an obvious area for examination by students interested in marine biology.

PRESSURE EFFECTS ON MEMBRANE FUNCTIONS

In spite of the key contribution of lipids to membrane structure, most active membrane functions are, of course, carried out by proteins. Many membrane proteins (e.g., G6Pases) display an absolute requirement for a lipid component; these are true lipoproteins. Others (e.g., electron transfer enzymes) may depend upon a hydrophobic environment for maintenance of proper conformation. Still others (e.g., hexokinase) are membrane bound, interacting with the phospholipid components of the membrane largely by hydrophobic forces. Theoretically, the functions of these various membrane-associated proteins could be unusually pressure sensitive for two basic reasons:

(1) The most obvious basis for pressure sensitivity of membrane functions is their dependence upon hydrophobic interactions, which can be readily disrupted by pressure.

(2) Equally important, these functions may become *diffusion-limited because of dramatic effects of pressure on the viscosity of the lipid components of membranes.*

For many organic liquids, the viscosity can change by factors of 10 to 100 as pressure is increased from 1 to 1000 atmospheres. Although the exact viscosity change is highly dependent upon the *mass* and the chemical *structure* of the organic molecule (the presence of ring structure in hydrocarbons, for example, can lead to much larger changes in viscosity with pressure than occur with linear hydrocarbons), it is in all cases large.

Since most of the cell's chemistry occurs in the aqueous phase of the cell—albeit the supply of metabolic "reagents" is highly dependent on the lipid phase, the membranes—it is also essential to understand the effects of pressure on water viscosity. Unlike organic liquids, water exhibits a decrease in viscosity with increasing pressure between 1 and 1000 atmospheres, at least over the temperature range from 0 to 30°C (Figure 8–7). At higher pressures, or in the case where water contains certain electrolytes, viscosity may increase with rising pressure. However, even when increases occur, they are much smaller than the changes found for organic liquids.

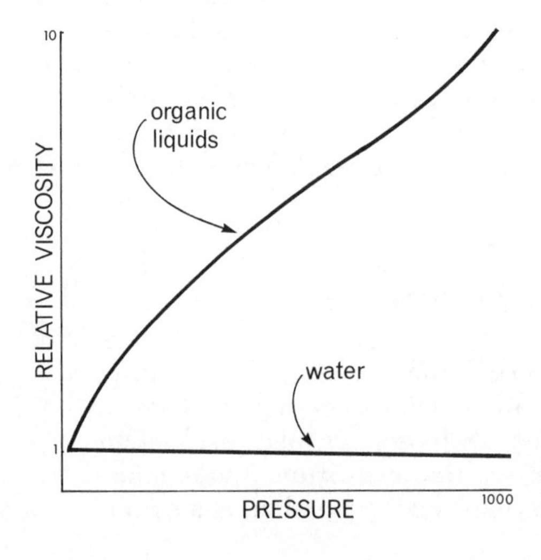

Figure 8–7 The relative changes in the viscosities of organic liquids, such as lipids, and of water. See Weale (1967) for further discussion of this area.

The potential difficulties posed by viscosity changes in the aqueous and lipid phases of the cell stem largely from the fact that increases in viscosity lead to reduced rates of diffusion. Thus, if a reaction rate is limited by the rate at which reactants diffuse onto the enzyme's surface, we might expect an increase in pressure to retard the reaction. For example, bimolecular termination reactions between growing polymer radicals are frequently diffusion-controlled, and we might expect that increases in pressure could significantly alter the molecular weight and structure of a polymer. Diffusion problems might be especially acute for reactions which depend more or less directly on membrane-transported substrates and cofactors, for the viscosity changes which occur in the lipid bilayers of the membranes probably are much greater than those occurring in the aqueous phase. Such differential effects of pressure on these two phases could dramatically influence the coordination of metabolic events occurring in them.

In a recent study of the inward transport and subsequent metabolism of various amino acids in a marine bacterium, for example, it was found that (intracellular) glutamate oxidation was pressure-activated up to about 500 atmospheres, *whereas glutamate transport into the cell was drastically inhibited* (Figure 8–8). These results suggest that the amino acid transport proteins (which are membrane bound) are highly pressure sensitive. Unfortunately, we do not know if the inhibition of glutamate transport is due to viscosity changes in the membrane or to conformational changes in the transport proteins. No comparative data for organisms in different pressure regimes are available for similar cellular functions and no studies have dealt with the isolated, partially-purified, membrane-bound enzyme. However, a small amount of information is available from studies of the effects of pressure on the active transport of ions across membranes and across epithelia.

In these studies, experimental advantage is taken of the observation that isolated gills of fishes can be incubated successfully in saline for fairly long intervals. Under such conditions, the gill cells maintain constant concentrations of intracellular Na^+ by actively pumping it out of the cell. In isolated gills of the European eel, this transport system is completely inhibited by 500 atmospheres of pressure. Studies with isolated Na^+K^+ ATPase indicate an identical pressure sensitivity, and *thereby identify the transport protein rather than the viscosity of the membrane lipids* as the pressure-sensitive "lesion." It is a fair guess that pressure disrupts hydrophobic interactions between the Na^+K^+ ATPase and its lipid milieu; in consequence, the gill cells cannot sustain electrical or chemical concentration gradients across their membranes. As the capacity to sustain such gradients is a property apparently common to all cells, any perturbating effects of pressure upon it would be greatly ampli-

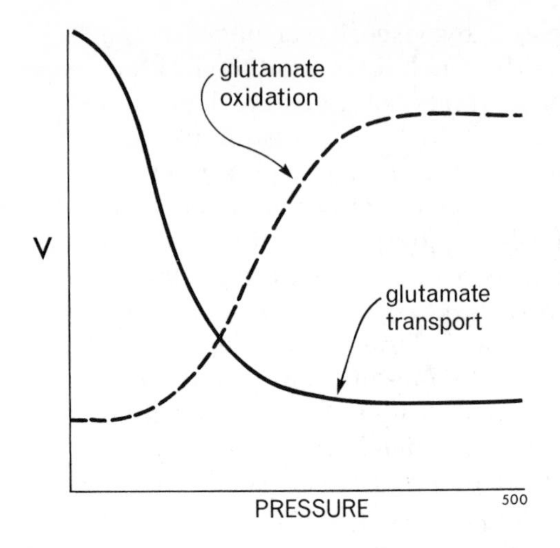

Figure 8–8 The differential effects of pressure upon glutamate uptake into the cell and glutamate oxidation within the cell of a marine microorganism. Modified after Paul & Morita (1971).

fied in the physiology of the organism. In midwater and abyssal organisms, evolution would strongly select against such pressure-sensitive enzymes, and we shall next turn to a consideration of the results of such evolutionary selection, insofar as enzyme adaptations to pressure are currently understood.

STRATEGIES OF ENZYME ADAPTATION TO PRESSURE IN MARINE ORGANISMS

FACTORS DETERMINING CHOICE OF STRATEGY

To appreciate what types of strategies might be employed by marine organisms in adapting to extremes of pressure or widely varying pressures or both, it may help to recall certain of the basic strategic considerations involved in temperature adaptation. We stressed that different ectothermic species range in their thermal tolerance from narrow stenothermality to extremely broad eurythermality. Associated with at least certain representatives of the latter group is a substantial ability to maintain key enzymic parameters relatively independent of temperature. Ectothermic organisms which must cope with large changes in body temperature were found to have enzyme-ligand interactions which were not adversely perturbed by changes in temperatures. In contrast, at least some enzymes of highly stenothermal species, such as acetylcholinesterase of an Antarctic fish, appear adapted to function only over an extremely narrow range of temperatures. However, at the low (−2°C) temperatures at which this fish exists, its acetylcholinesterase appears to attain its optimal efficiency, at least with regard to enzyme-substrate binding.

These findings lead us to predict that the strategies employed in pressure adaptation will likewise be dependent on the absolute value and the variability encountered with regard to this environmental parameter. Abyssal organisms which have no life stages in shallow waters may be "stenobathic," adapted to function well at, but only under conditions of, extremes of pressure. In contrast, organisms which encounter large magnitude pressure changes, either diurnally or during the life cycle, might be expected to exhibit "eurybathic" characteristics. The diurnal migrants may possess enzymes which are insensitive to sudden changes in pressure, much as many eurythermal forms appear to have temperature-insensitive enzymic systems. Species which have abyssal adult stages but mesopelagic larval stages might be expected to show some form of pressure acclimation analogous to the seasonal temperature acclimations found in species such as the rainbow trout.

Thus, the strategies of pressure adaptation may bear striking resemblance to the patterns of temperature adaptation, and this may hold true for all time courses of the adaptation process. We now turn to an examination of what limited data are available, to determine if these strategic similarities do indeed exist and, furthermore, if the enzymic parameters sensitive to selection are the same or different in the cases of temperature and pressure. Because the differential effects of high pressure on enzymes seem to be the most serious problem to controlled cellular chemistry under abyssal and midwater conditions, we shall examine classes of enzymes which show different responses to increases in pressure.

VARIANTS OF A PRESSURE-ACTIVATED ENZYME (FDPase) IN DEEP WATER AND SURFACE FISH

Fructose-diphosphatase (FDPase) is an enzyme which, under alkaline pH conditions, exhibits a large increase in V_{max} at high pressure; that is, it is a reaction which proceeds with a negative $\Delta V^{\ddagger}$. We have previously considered the "anatomy" of the FDPase reaction (p. 42). Briefly, to review the function and metabolic position of this key regulatory enzyme, FDPase is a major valve enzyme in the gluconeogenic sequence. The overall reaction, $FDP \rightarrow F6P + P_i$, requires a divalent cation (Mg^{++} or Mn^{++}), and enzymic activity is largely under the control of intracellular AMP concentrations. When AMP concentrations are low, metabolically equivalent to high ATP levels, the cell can reduce the flow of glucose down the glycolytic channel and "concentrate" on reforming glucose from lower molecular weight precursors, such as pyruvate and glucogenic amino acids. AMP is a potent inhibitor of FDPase, and until AMP levels decrease to low values, FDPase remains in a shut-off position. The opposing effects of the charged and uncharged adenylates on FDPase and phosphofructokinase can thus be appreciated as making eminent sense, physiologically.

In comparative studies of FDPase enzymes from a surface fish, the rainbow trout, and a deep benthic fish, the rattail (genus *Coryphaenoides*), certain similarities in pressure responses were observed. Specifically:

1. pH Profile Effects. The pH dependencies of both enzymes are affected similarly by increasing hydrostatic pressure (Figure 8–9). At alkaline pH values (pH 7.5 and greater), the V_{max} characteristics of both the rattail and trout enzymes are strongly accelerated by increased pressure.

2. Activation Energy Effects. For both enzymes, increases in pressure at neutral pH increase the activation energy (enthalpy of activation) of the reaction. At alkaline pH values this effect is reversed.

3. Reduced Pressure Sensitivity at High Temperatures. For the trout and rattail enzymes, increases in temperature reduce the effects of pressure on enzymic activity. This observation is commonly made in pressure studies. The basis of this effect may lie in the roles of hydrophobic interactions in low temperature and high pressure denaturation phenomena.

4. Pressure Sensitivity of the Native Enzyme. A further similarity between the surface and abyssal enzymes is their extreme pressure lability under conditions where substrate and cofactor are absent. Again, the instability of "naked" enzymes is a frequent observation in studies of pressure and temperature effects.

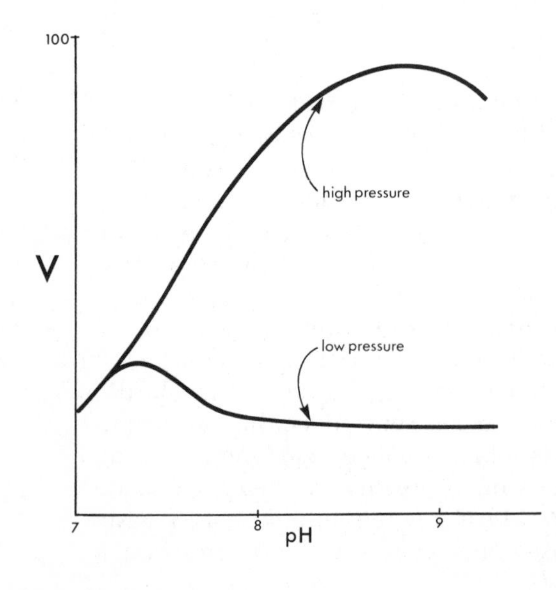

Figure 8–9 Effect of pH on the pressure response of FDPases. The pattern observed is similar for the FDPases extracted from abyssal and surface organisms. See *Amer. Zoologist*, Vol. *11*, for further details.

We list these similarities between two FDPase variants adapted to function at widely different pressures not to suggest that the enzymes are functionally identical—we shall soon demonstrate that some functional characteristics are extremely sensitive to selection on the basis of pressure differences—but rather to indicate that certain characteristics of an enzymic reaction may be inseparable features of a basic chemical transformation. For example, the free energy change which occurs as substrate is converted to product is of similar magnitude and same sign for all variants of an enzymic reaction. Likewise, the steric changes which must occur in the substrate molecule(s) during the activation process may be identical for all variants of a reaction. In short, there are undoubtedly certain basic thermodynamic and steric parameters which are necessary concomitants of a chemical transformation. These parameters are fixed for a given enzymic reaction, regardless of the species or tissue in which the reaction occurs. Selection cannot tailor these parameters in any major way to adapt the enzyme better for function in its particular milieu.

If we can make any generalization about the nature of these relatively "inviolate" parameters, it is that they are involved in the establishment of the basic catalytic capacity of the enzyme. In no case do these parameters appear to establish the key regulatory functions of the enzyme. These latter functions, which determine the extent to which the enzyme will actually utilize its catalytic potential, are not fixed characters, identical in all variants of an enzyme. In fact, these regulatory parameters, which include all enzyme-ligand (substrate, modulator, cofactor) interactions, in essence *define* the variant nature of enzymes, at least in a functional sense. What, then, is the influence of pressure on these important regulatory reactions between enzymes and lower molecular weight ligands? What is their role in enzyme adaptation to pressure?

FDPase-FDP INTERACTIONS

Although the V_{max} for the FDPase reactions of trout and rattail is accelerated by pressure, FDPase-FDP interactions are affected by pressure quite differently in the two systems (Figure 8–10). For trout FDPase, the activity of the enzyme at physiological substrate concentrations is reduced at high pressure due to a sharp rise in the apparent K_m of FDP as pressure increases. For the abyssal enzyme, a small increase in the apparent K_m of FDP occurs as pressure increases. However, because this effect is slight, when the joint influences of the K_m change and the turnover number change are considered, the rates of FDPase activity at physiological substrate concentration are seen to be pressure-independent (Figure 8–10).

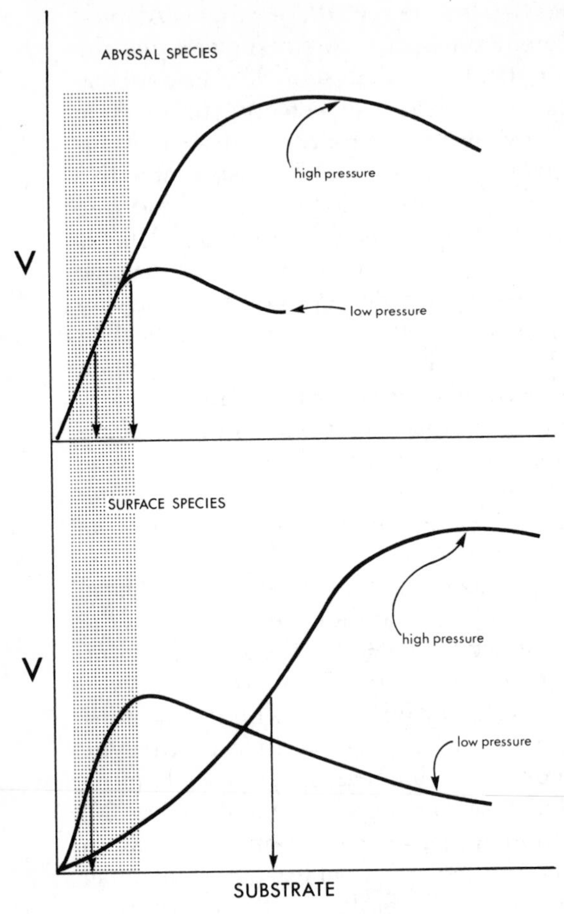

Figure 8–10 Substrate saturation curves for liver FDPases from surface and abyssal fishes. Approximate physiological range of FDP levels is shown by shading. In the surface fishes, pressure has such a large effect on the K_m for substrate that reaction rates are reduced at high pressures because of a reduced affinity for FDP. Of course, when the enzyme is saturated, high pressure activates the reaction velocity (V). In the abyssal species, the K_m still increases with pressure but only to the extent that it counteracts the pressure-activation of catalysis; in consequence, at K_m values of substrate, the reaction velocity is pressure-independent. The effects of pressure on cofactor (Mg^{++}) saturation of these two FDPases are essentially identical to those for FDP saturation shown above. See *Amer. Zoologist*, Vol. *11*, for further details.

Thus, the acceleration of catalysis *per se*, as evidenced by the rise in V_{max} with increasing pressure, is just sufficient to balance the slight decrease in activity at physiological substrate concentrations which stems from a decrease in FDPase-FDP affinity. In consequence, FDPase activity is independent of pressure in the abyssal fish.

This example illustrates a key point of enzymic adaptation: the physiologically important attributes of an enzyme, in this case the rate of the reaction at physiological substrate concentrations, is often the net result of opposing changes in the total catalytic capacity on the one hand, and the regulatory use of this catalytic capacity on the other. For temperature effects, we have already seen examples in which the influence of a temperature change on the turnover number of a reaction is largely or fully counteracted by an opposing change in E-S affinity. We now see that pressure effects, operating at the loci of turnover number and E-S affinity, lead to a similar homeostasis of function. In temperature and pressure adaptations, therefore, it appears that a relatively fixed property of an enzyme—specifically the energy changes involved in activation in the case of temperature, and the volume changes of activation in the case of pressure—of necessity perturbs enzymic function as the temperature or pressure of the environment changes. However, the potentially adverse effects of this change are reduced or eliminated by varying another enzymic parameter which appears more malleable under the influences of natural selection.

FDPase-COFACTOR INTERACTIONS

The effects of pressure on FDP-Mg^{++} interactions for the trout and rattail enzymes mirror the FDPase-FDP results just discussed. The affinity of the trout enzyme for Mg^{++} is highly pressure-sensitive; for the rattail enzyme, the K_a for Mg^{++} is pressure-insensitive. It is important to realize that because of these $K_{a(Mg^{++})}$ changes, the pressure effects on the trout enzyme are differential: at low Mg^{++} levels, increases in pressure inhibit the enzyme, whereas at high Mg^{++} concentrations the enzyme is pressure-activated. This property would be highly maladaptive in an environment of varying pressure.

FDPase-AMP INTERACTIONS

Since the major regulator of FDPase activity is the concentration of AMP in the cell, we might expect that FDPase-AMP interactions would be notably subject to selection. This expectation is realized. Although the shape of the AMP inhibition curve of the rattail enzyme is somewhat altered by pressure (Figure 8–11), the K_i value of AMP remains constant over a wide range of pressures. Thus, AMP control of FDPase activity is pressure-insensitive for the abyssal enzyme.

In contrast, AMP regulation of trout FDPase is highly pressure-sensitive. As pressure increases from 1 atmosphere to approximately 800 atmospheres, the apparent K_i of AMP decreases two-fold. In trout liver, FDPase under these high pressures would be 50% inhibited at AMP concentrations approximating the lower *in vivo* limit. The trout enzyme thus could never be more than 50% active and would be largely nonfunctional in the high pressure environment, even if AMP levels were to rise only slightly. This FDPase variant is thus not well suited for function at the depths at which abyssal organisms abound.

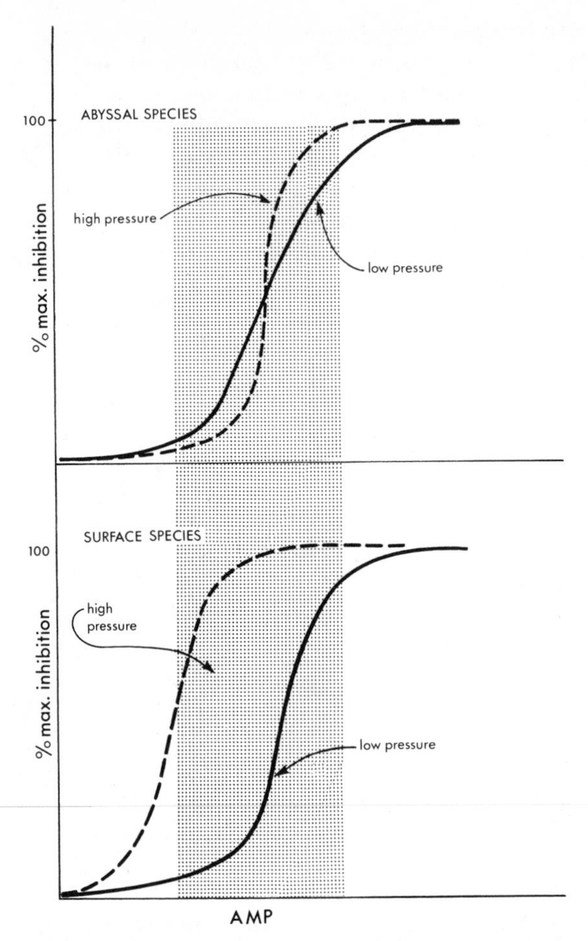

Figure 8–11 Effect of pressure on the percentage maximum inhibition of FDPases by AMP. Approximate physiological range of AMP is shown by shading. In the abyssal species, the enzyme-modulator affinity is pressure-independent and AMP is an equally good inhibitor at all pressures encountered in nature. In the surface species, AMP inhibition is accentuated at high pressures owing to an increase in enzyme-modulator affinity. In consequence, at high pressures (even at the lower limit of AMP concentrations) this FDPase would be partially inhibited. Only slight increases in AMP could lead to complete inhibition of the enzyme. See *Amer. Zoologist*, Vol. *11*, for further details.

In summary, the salient features of pressure adaptation of rattail FDPase involve a closely tailored opposition of turnover number effects and regulatory effects. We can rationalize the advantages of pressure-dependent increases in turnover number (as illustrated by the pressure activated V_{max} of the FDPase reaction): these appear to be useful as a temperature compensatory mechanism (recall that the rattail is living near 2°C). However, if we regard the trout FDPase as closer to the primitive form of the enzyme from which the rattail FDPase evolved, we find that a mere increase in total catalytic potential at high substrate concentrations is not enough to offset the disadvantages of a sharp loss in E-S affinity as pressure rises (Figure 8–10). Instead, it appears that during the long process of evolutionary adaptation to high pressure, selection has tailored the affinity of rattail FDPase for FDP in such a manner that the enzyme can function in a pressure-independent manner at physiological substrate concentrations. This pressure-insensitivity is probably characteristic of pressure-activated enzymes from most deep sea organisms, particularly those species (such as *Coryphaenoides*) which have mesopelagic juvenile stages and those species which are diurnal vertical migrants. But for enzymes which are pressure inhibited, a similar strategy must lead to an opposite set of adjustments. Do these in fact occur?

A PRESSURE-INHIBITED ENZYME: PK FROM TROUT, A MIDWATER SEA BASS, AND THE RATTAIL

Pyruvate kinase is another enzyme which plays a key regulatory role in glucose metabolism. We have earlier discussed the nature of this reaction and the types of regulatory interactions which govern its activity. Briefly, the reaction, which is strongly inhibited by pressure, involves a transphosphorylation,

$$\text{PEP} + \text{ADP} \rightarrow \text{pyruvate} + \text{ATP}$$

and requires both K^+ and Mg^{++} for activity. Regulation is complex. FDP formed in the PFK reaction activates PK in a "feed-forward" manner. ATP is a potent feed-back inhibitor (p. 38), but this inhibition can be reversed by FDP.

PK, like FDPase, appears to have certain properties which are common to all variants, regardless of the pressure to which the enzyme is adapted. These relatively stable properties include:

1. Pressure Retardation of V_{max} (Turnover Number). All variants of PK examined exhibit a decrease in V_{max} as pressure increases. The magnitude of this effect varies somewhat. Abyssal and surface enzymes are more pressure sensitive in this regard than the enzymes from a species of midwater fish (*Ectreposebastes imus*). The pressure sensitivity of reaction rate, as we have stated, is dependent on the volume change which occurs during activation. The $\Delta V^{\ddagger}$ value for the midwater fish is lower than the values characteristic of

TABLE 8–3 The volume change of activation ($\Delta V^{\ddagger}$) for PK from muscle of three fish species living in different positions in the water column. The value of $\Delta V^{\ddagger}$ was calculated from PK maximum catalytic rates at 3° C. For further details, see *Amer. Zoologist*, Vol. *11* (1971).

Organism	$\Delta V^{\ddagger}$, cm^3/mole
surface species (rainbow trout)	47
midwater vertical migrant (*Ectreposebastes*)	29
benthopelagic species (rattail)	44

the *Coryphaenoides* and trout enzymes (Table 8–3). This observation indicates that the $\Delta V^{\ddagger}$ value of an enzyme-catalyzed reaction is not fixed and, in addition, raises the interesting possibility that selection may modify $\Delta V^{\ddagger}$ values in such a way as to minimize the effects of pressure change on reaction rate (recall our earlier discussion of possible adaptive changes in $\Delta G^{\ddagger}$ in temperature adaptation).

2. Pressure Effects on pH Profiles. As in the case of the FDPase variants, the interactions of pH and pressure on enzyme activity were the same in different organisms.

PRESSURE EFFECTS ON PK-LIGAND AFFINITIES

The similarities between PK variants and FDPase variants extend to the effects of pressure on enzyme-ligand affinities. Again, enzymic activity in abyssal forms, as well as in the midwater fish, exhibits key pressure-insensitivities. In particular, the retardation of catalytic rate by high pressures is effectively overcome by other, opposing (i.e., activating) effects. These include:

(1) an increase in enzyme-substrate affinity with pressure, but only in the presence of FDP;

(2) reversal of pressure inhibition by the feedforward activator, FDP;

(3) pressure-independence of the K_a of FDP, insuring retention of regulatory control at different pressures; and

(4) FDP reversal of ATP inhibition, *but only under conditions of high pressure.*

As the data summary of Table 8–4 indicates, the PK variants of the deep sea and midwater fish are adapted for pressure-insensitive function at physiological substrate concentrations. In contrast, the catalytic and regulatory properties of the trout enzyme are extremely pressure-sensitive; this enzyme could not function in a

TABLE 8–4 Summary and comparison of various catalytic and regulatory properties of muscle pyruvate kinases of benthic *Coryphaenoides*, midwater sea bass, and the rainbow trout. From studies summarized in *Amer. Zoologist*, Vol. *11* (1971).

Selected Enzyme Characteristics	Effect of High Hydrostatic Pressure		
	RATTAIL	SEA BASS	TROUT
$K_{m(PEP)}$	small increase	no effect	U-shaped K_m-pressure curve
$K_{m(PEP)}$ in presence of FDP	small decrease	no effect	U-shaped K_m-pressure curve
$K_{m(ADP)}$	small increase	no effect	large increase
$K_{a(Mg^{2+})}$	no effect	(no effect?)	large increase
$K_{i(ATP)}$	no effect	no effect	substantial increase
FDP reversal of ATP inhibition	large increase	no effect	no effect

physiologically satisfactory manner at either high pressures or varying pressures.

The PK-FDP interactions of the rattail enzyme deserve special attention. Firstly, the magnitude of FDP activation of the rattail enzyme is much greater than that found for the trout enzyme. The rattail PK is activated up to 5-fold by physiological FDP concentrations, whereas the trout enzyme is activated less than half this amount. Such strong activation could be rate-compensatory for the deep water reaction, where high pressure and low temperature both promote decreases in substrate turnover.

A second important aspect of PK-FDP interactions in the rattail system is the obligatory role played by high pressure in effecting FDP reversal of ATP inhibition. The requirement for high pressure to bring about this regulatory effect implies that the rattail enzyme is adapted for controlled function *only* at high pressures. This observation raises an important question concerning PK function in mesopelagic juvenile stages: does the juvenile survive with this barophilic enzyme, possibly via other regulatory interactions which reverse the ATP inhibition, or might this life stage use a different, "low pressure" isozyme?

Finally, in consideration of these PK variants, it is important to emphasize the extremely high degree of pressure-independence characteristic of PK from the vertical migrant, *Ectreposebastes imus*. As we would expect on the basis of the analogy between temperature adaptation and pressure adaptation, this fish possesses enzymes which enable it to be tolerant of a wide range of pressures—here, PK is a "eurybathic" enzyme. The pressure-independence of enzymic characteristics pertains both to properties associated with turnover number and to characteristics involving enzyme-ligand affinities (Table 8–3), but there appears to be a definite limit to the extent to which characteristics like $\Delta V^{\ddagger}$ and activation energy can be reduced. Thus, the most vital features of enzyme adaptation to pressure, as to temperature, tend predominantly to involve enzyme-ligand interactions.

SOME UNANSWERED QUESTIONS

Most available data dealing with high pressure effects accomplish at least one thing: they dramatize the types of problems which high and varying pressures impose on many marine organisms. What these data do not do, by and large, is reveal how marine forms have "solved" these problems. The failure of most existing data to reveal true adaptive mechanisms in marine organisms stems from two facts. Firstly, most pressure studies have been performed on species which have never encountered pressures of more than one atmosphere; i.e., many, perhaps most, pressure studies have been done with systems (tissues, enzymes, etc.) from terrestrial or shallow water species. A second reason why most past studies of pressure effects have faltered when it comes to explaining the mechanisms whereby deep sea organisms have adapted to pressure is the tendency to place attention on enzymic parameters which may have only a minor role in governing the function of the enzyme under *in vivo* conditions. Enzymes do not operate at maximal velocity. Enzymes do not operate in a purified, "naked" form within the cell. The enzymic properties which appear most malleable under the influence of selection have often been the parameters most neglected in pressure studies. If the biologist has no other contribution to make to the work of the pressure-oriented chemist or biochemist, he at least can indicate the true nature of the enzyme's intracellular and extracellular environments.

The field of pressure biology is thus full of lacunae. In one sense this is good; there are a multitude of interesting problems awaiting exploration. Among these unanswered questions we would list the following:

(1) Do marine organisms encountering large pressure changes at different stages in their life cycles pressure acclimate, much as ectothermic species temperature acclimate?

(2) Are the 4° structures of enzymes from deep sea organisms as pressure-labile as those of terrestrial organisms? If not, do the structures of enzymes from deep sea organisms depend more heavily on, for example, hydrogen bonding for stabilization of 4° structure?

(3) Are the functions of transcription and translation pressure

sensitive? Are specific pressure adaptations required to stabilize the active conformations of tRNA species, of polyribosomes, and of chromatin?

(4) Are the structures and functions of membrane-bound proteins unusually pressure sensitive in pressure-adapted organisms? Have membrane proteins in deep sea organisms become barophilic? Are the phospholipid components of membranes varied as a function of the depth at which the organism lives?

(5) Lastly, what biochemical mechanisms of pressure adaptation occur in deep-diving homeotherms such as whales, seals, and sea birds? Are there unique problems associated with the tolerance of large scale pressure changes in a system held at a high and constant temperature?

Answers to these questions will undoubtedly prove interesting *per se*, and, equally importantly, they may reveal whether the basic strategies of biochemical adaptation, which characterize other organism-environment interactions, are also utilized in adaptation to pressure.

SELECTED READINGS

Books and Proceedings

Chemical Reactions at High Pressures (1967) by K. E. Weale. Spon Ltd., London. 349 pp.

Effects of High Pressure on Organisms (1972). *Soc. Exp. Biol. Symposium 26*, Cambridge University Press, 516 pp.

High Pressure Effects on Cellular Processes (1970) (Ed. A. M. Zimmerman). Academic Press, N.Y. 324 pp.

Pressure Effects on Biochemical Systems of Abyssal Fishes (1971) (Ed. P. W. Hochachka). *Amer. Zoologist 11*, 401–576.

Reviews and Articles

Hochachka, P. W., D. E. Schneider, and A. Kuznetsov (1970). Interacting pressure and temperature effects on enzymes of marine poikilotherms: catalytic and regulatory properties of FDPase from deep and shallow-water fishes. *Marine Biology 7*, 285–293.

Jaenicke, R., R. Koberstein, and B. Teuscher (1971). The enzymatically active unit of lactate dehydrogenase. *Eur. J. Biochem. 23*, 150–159.

Murayama, M. (1966). Molecular mechanism of red cell "sickling." *Science 153*, 145–149.

Paul, K. L., and R. Y. Morita (1971). Effects of hydrostatic pressure and temperature on the uptake and respiration of amino acids by a facultatively psychrophilic marine bacterium. *J. Bacteriol. 108*, 835–843.

Penniston, J. T. (1971). High hydrostatic pressure and enzymic activity: inhibition of multimeric enzymes by dissociation. *Arch. Biochem. Biophysics 142*, 322–332.

BUOYANCY

CHAPTER 9

THE ADVANTAGES OF NEUTRAL BUOYANCY

Most marine organisms are adapted to live in a finite region of the water column. The particular depth may, as we have seen, vary greatly during the life history of the organism; e.g., benthic adult organisms may have larval stages living in the first few meters of the sea. Diurnal migrants, on the other hand, may course through hundreds of meters of the water column each day. For any given life stage and, in the case of diurnal migrants, for any given time, it appears to the organism's advantage to maintain a particular body density: the density required for neutral buoyancy. This, in turn, allows the organism to remain at a given depth with only a minimal requirement for active locomotion as a means of position maintenance. For essentially non-motile forms, such as many plankters and marine eggs, the correct buoyancy can be achieved only by regulation of cellular/body density.

For motile forms, a measure of the relative significance of this problem can be obtained by considering the amount of energy that is saved by maintaining neutral buoyancy. In the case of bony fishes, where sufficient information is available for appropriate calculations, the possession of neutral buoyancy enables the organism to save energy in two ways: (1) it can remain motionless in midwater without expenditure of energy, and (2) the energy expended in swimming horizontally is reduced. Alexander has calculated that when a fish moves at a velocity of one body length per second, the power needed to overcome the sinking tendency is 60% of the total power of movement. When it moves at 10 lengths per second, only 5% of the total power is used to keep it at the same vertical position. At more normal cruising speeds, 3 or 4 body lengths/sec, about 20% of the total power is expended to overcome sinking. The maintenance of neutral buoyancy therefore gives the motile organism a significant energetic advantage, since this same power expenditure is avoided and can be put to other uses.

DENSITIES OF BIOLOGICAL MATERIALS AND POTENTIAL STRATEGIES OF DENSITY-REGULATION

Table 9–1 lists the densities of the environment and the body constituents of marine organisms. These physical parameters in effect circumscribe the problem of density-regulation. From such a list, it takes little engineering insight to realize that the potential strategies of density adaptation are few in number. One strategy involves the *reduction in quantity of heavy constituents.* Thus the organism can:

(1) Reduce the total ionic concentration of its body fluids.
(2) Reduce the amounts of body protein.
(3) Reduce the amount of bone or its extent of ossification.

Alternatively, the organism may change the types of materials which compose its tissues, *replacing a dense material with one having a lower density.* Examples of this second strategy include:

(1) Replacement of heavy salts with lighter ones.
(2) Deposition of large amounts of fat in the body.
(3) Substitution of a less dense type of fat.
(4) Elaboration of air-filled buoyancy tanks.

Thus density adaptation, like adaptation to other environmental parameters, involves changes both in the total *amounts* of certain molecular species and in the *types* of chemicals present in the system. Let us now examine the manners in which the "raw material" available for density adaptation has been "utilized."

TABLE 9–1 DENSITIES OF SOME COMMON ENVIRONMENTS AND BIOLOGICAL CONSTITUENTS AT 1 ATMOSPHERE PRESSURE AND 20°C

Substance	Density
Air	0.00125
O_2 at 700 atm	0.7
Pure Water	1.00000
Sea Water	1.026
"Protoplasm"	1.02–1.10
Fish Oil	0.93
Triacylglycerol	0.922
Diacylglycerol ether	0.908
Wax esters	0.90 (approximate value*)
Squalene	0.86
Protein	1.33
"Skeleton"	2–3

*J. Nevenzel and A. Yayanos, personal communication.

MARINE BONY FISHES: BUOYANCY ADVANTAGES OF THE HYPOTONIC STATE

Often, when attempting to rationalize "why" an organism has a particular structure or "why" the organism has a particular physiological or biochemical trait, it is difficult to state which of several advantages conferred by the feature in question was the primary selective reason for the appearance of the feature. This is certainly true in the case of body fluid hypotonicity in marine bony fishes. From the standpoint of physiological function, e.g., the generation of membrane potentials, the particular ionic concentration of the marine teleost's tissues and fluids may represent optimal conditions for physiological activity. In addition, however, we can argue that, from the standpoint of buoyancy, maintenance of the hypotonic state is another key selective advantage. Thus, even though it is energetically expensive to maintain a hypotonic state, the marine teleost may recover a considerable portion of this energy expenditure via a reduction in the amount of energy it need expend in maintaining its position in the water column. Hypotonicity might be even more critically important in the cases of fish eggs and young larval stages, which often must develop near the ocean surface. Thus, the capacity to maintain a hypotonic state confers more than one advantage on the organism. The particular advantage which served as the primary impetus for the evolution of hypotonicity must remain a matter of debate.

CONTROL OF IONIC COMPOSITION: SQUIDS

In contrast to fishes, marine invertebrates exhibit only limited abilities for osmoregulation (Chapter 4). Thus, the maintenance of all body fluids in a hypotonic state is not a possible strategy in buoyancy adaptation. Hypotonicity, when it occurs, involves a fluid tank which is largely cut off from the rest of the body fluids. In spite of being osmotic conformers, all marine invertebrates can regulate the *types* of ions in their fluids and tissues. At least some of this qualitative ion regulation appears to be directed toward control of body density. Diverse groups of invertebrates are known to lower the concentrations of heavy ions such as Mg^{++}, Ca^{++}, and $SO_4^{=}$, and to replace these with lighter ions, e.g., Na^+ and NH_4^+. In many cases a special region of the body is used as a buoyancy tank, wherein these ionic substitutions occur. The ionic composition of other body regions may not change; i.e., aside from the buoyancy tank, physiological processes may require the presence of the heavier ionic species.

Especially good examples of the use of relatively non-dense, fluid-filled buoyancy tanks are found in the cases of deep-sea squids. Firstly, it is essential to realize that a fluid-filled buoyancy tank is probably essential for these organisms. A fluid-filled tank is virtually incompressible, a necessary feature for deep-sea organisms. The body location which serves in buoyancy regulation varies among different families of squid. In the Cranchidae a single buoyancy chamber exists: an enlarged coelom. In other families, the buoyant liquid is contained in vacuoles situated mainly in the mantle and arms.

The chemical composition of the buoyant fluid is adjusted to give the animal as a whole neutral buoyancy. For example, in the Cranchidae, in which the coelomic fluid forms 65 per cent of the body mass, Mg^{++} and Ca^{++} are replaced by ammonium ion. Even the concentration of Na^+ is reduced. The ammonium chloride-rich coelomic fluid (480 mM NH_4Cl) has a density of 1.010. This density reduction is just sufficient to give the squid neutral buoyancy.

The retention of these vast amounts of NH_4^+ within the coelomic fluid is made possible only by the acidic nature of the fluid. By maintaining the acidity of the coelomic fluid near pH 5, the organism succeeds in maintaining sharp concentration gradients of NH_4^+ and Na^+ between the blood and coelomic fluid. The blood contains only 3 mM NH_4^+ but 460 mM Na^+, and has a pH between 7 and 8. By thus charging ammonia and rendering it nondiffusible, the squid can trap large quantities of this protein breakdown product within its coelomic fluid. Again we find an example of the use of ammonia in an adaptation process. As in the case of osmoregulation by euryhaline crustaceans, the ammonia formed during protein catabolism is not treated as a waste product, but instead is utilized in an important adaptational process.

Because squid can regulate the density of their coelomic fluid during vertical migration, the rates and directions of ion flow must be subject to close regulation. While these mechanisms are totally unknown, it would seem likely that ion pumps similar to those discussed for fish gills (p. 134) might provide an efficient control scheme. A sodium-ATPase which couples Na^+ to H^+ as a counter-ion seems a particularly attractive candidate for the squid system.

REDUCTIONS IN PROTEIN CONTENT AND BONE OSSIFICATION IN DEEP-SEA FISHES

One of the difficulties in maintaining neutral buoyancy stems from the fact that protein, which is a relatively dense material, is a major constituent of the body. The same is true of the skeleton (Table 9–1). One way around this dilemma involves extensive reduction of these heavy body constituents. In certain deep-sea fishes which are not active swimmers, body ossification may be significantly reduced. Even more dramatically, the amount of body protein may be reduced to approximately 5% – a 4-fold reduction from the protein content characteristic of many shallow-water fishes. The loss of protein occurs largely in the trunk and caudal musculature of bathypelagic fishes, and renders many of these species "floating jaws."

LIPIDS AS DENSITY REGULATORS

Of all the major organic constituents of the body, lipids have the lowest density. In addition, different types of lipids have different densities (Table 9–1). Organisms appear to have "taken these facts into account," and we find numerous examples of lipid-based buoyancy control. Lipid-dependent density regulation is very important in the eggs of marine organisms which float in the surface waters. The same is true for many planktonic organisms with only limited abilities to maintain buoyancy by other means. However, even in the largest of marine organisms, the whales, we find that lipids make a major contribution to buoyancy control. Thus, the strategy of utilizing large quantities of lipids for reduction of body density seems to be phylogenetically widespread.

One of the more refined utilizations of lipids is found in elasmobranch fishes. Sharks have extremely fatty livers (which may account for up to 25% of body volume). The precise lipids found in the liver vary among different species. In the *Centrophorus* elasmobranchs, squalene (Figure 9–1) may account for as much as 90% of the deposited lipids. As indicated in Table 9–1, the use of this relatively non-dense lipid yields a large decrease in density. Squalene thus seems to furnish us with another example of a "better" molecule being used to fulfill a given function.

In dogfish sharks the primary low-density material found in the livers is a class of diacylglycerol ethers (Figure 9–1). The relative proportions of these compounds and of triacylglycerols vary. Malins and coworkers artificially increased the body weight of dogfish by suspending lead weights from their pectoral fins. Compared with control fish, the artificially-weighted sharks exhibited two-fold increases in the ratio of diacylglycerol ethers to triacylglycerol. Although the density differences between these two kinds of lipids is not great (Table 9–1), 1 gm of diacylglycerol ether gives 14% more lift in sea water than 1 gm of triacylglycerol. This advantage presumably accounts for the elaboration of a regulatory mechanism favoring the selective synthesis of diacylglycerol ethers for body density regulation in these species.

Glycerol

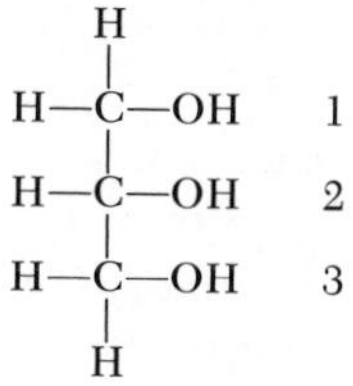

$$\begin{array}{l} \quad\; H \\ H-C-OH \quad 1 \\ H-C-OH \quad 2 \\ H-C-OH \quad 3 \\ \quad\; H \end{array}$$

Nonesterified glycerol ethers

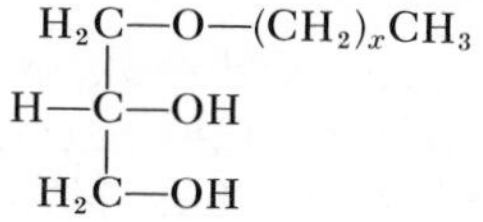

$$\begin{array}{l} H_2C-O-(CH_2)_xCH_3 \\ H-C-OH \\ H_2C-OH \end{array}$$

1-Monoacylglycerol

$$\begin{array}{l} \quad\; H \\ H-C-O-\underset{\underset{O}{\|}}{C}-R_1 \\ H-C-OH \\ H-C-OH \\ \quad\; H \end{array}$$

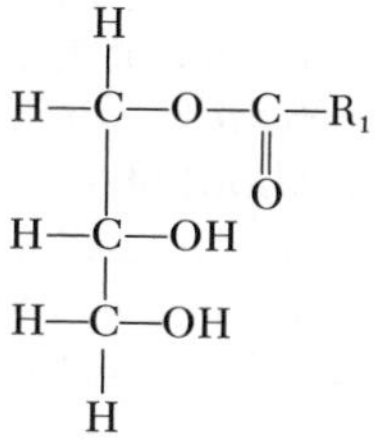

Monoacyl glycerol ethers

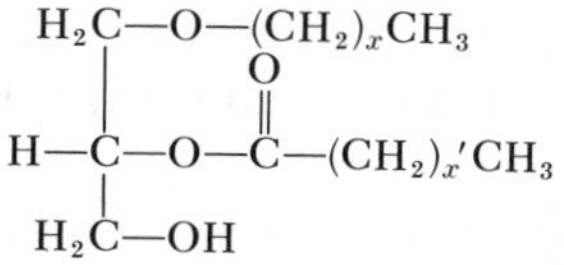

$$\begin{array}{l} H_2C-O-(CH_2)_xCH_3 \\ H-C-O-\overset{\overset{O}{\|}}{C}-(CH_2)_x{'}CH_3 \\ H_2C-OH \end{array}$$

1,2-Diacylglycerol

$$\begin{array}{l} \quad\; H \\ H-C-O-\underset{\underset{O}{\|}}{C}-R_1 \\ H-C-O-\underset{\underset{O}{\|}}{C}-R_2 \\ H-C-OH \\ \quad\; H \end{array}$$

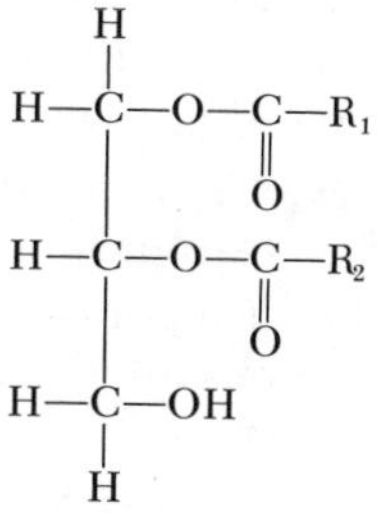

Diacyl glycerol ethers

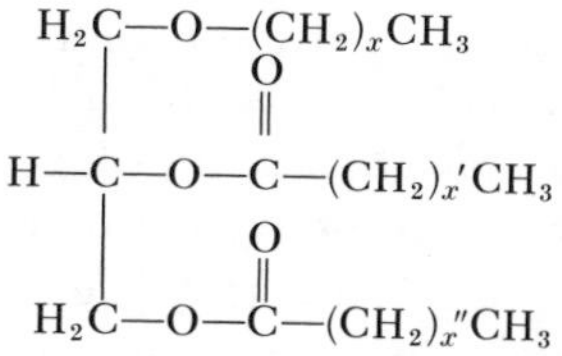

$$\begin{array}{l} H_2C-O-(CH_2)_xCH_3 \\ H-C-O-\overset{\overset{O}{\|}}{C}-(CH_2)_x{'}CH_3 \\ H_2C-O-\overset{\overset{O}{\|}}{C}-(CH_2)_x{''}CH_3 \end{array}$$

Triacylglycerol

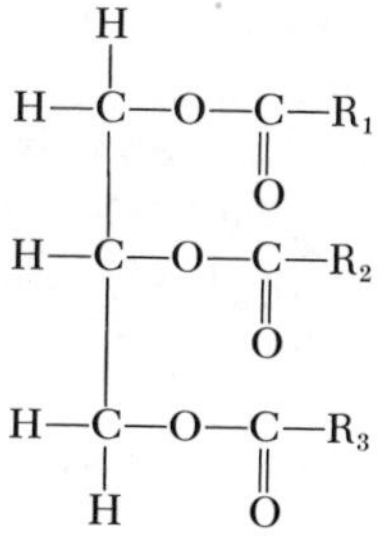

$$\begin{array}{l} \quad\; H \\ H-C-O-\underset{\underset{O}{\|}}{C}-R_1 \\ H-C-O-\underset{\underset{O}{\|}}{C}-R_2 \\ H-C-O-\underset{\underset{O}{\|}}{C}-R_3 \\ \quad\; H \end{array}$$

Wax Esters

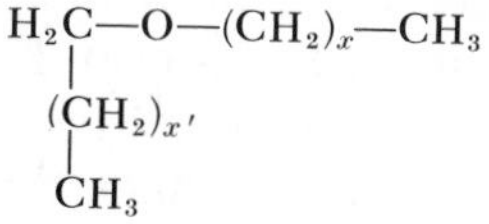

$$\begin{array}{l} H_2C-O-(CH_2)_x-CH_3 \\ (CH_2)_{x'} \\ CH_3 \end{array}$$

Squalene

$$H_3C-\overset{CH_3}{C}=\underset{H}{C}-\overset{H_2}{C}-\underset{H_2}{C}-\overset{CH_3}{C}=\underset{H}{C}-\overset{H_2}{C}-\underset{H_2}{C}-\overset{CH_3}{C}=\underset{H}{C}-\overset{H_2}{C}-\underset{H_2}{C}-\overset{H}{C}=\underset{CH_3}{C}-\overset{H_2}{C}-\underset{H_2}{C}-\overset{H}{C}=\underset{CH_3}{C}-\overset{H_2}{C}-\underset{H_2}{C}-\overset{H}{C}=\underset{CH_3}{C}-CH_3$$

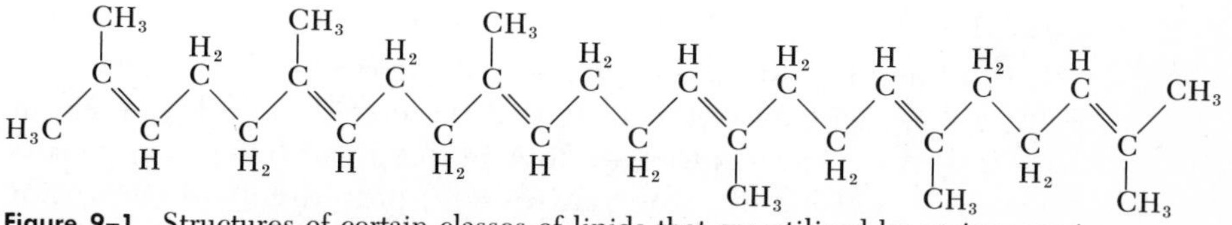

Figure 9–1 Structures of certain classes of lipids that are utilized by various marine organisms in body density regulation.

WAX ESTERS MAY PLAY A ROLE IN DENSITY-REGULATION

In contrast to the elasmobranchs discussed above, deep-sea sharks as well as other marine organisms appear to regulate body density with wax esters. "Wax" is the generic name given to naturally occurring fatty acid esters of any alcohol other than glycerol. In the case of many marine organisms, the alcohols found in wax esters themselves derive from fatty acids; hence, a wax ester is in effect an aliphatic complex held together by an ester linkage (Figure 9–1). In some species, very high proportions of these compounds are present. In midwater lantern fishes, for example, Nevenzel and his coworkers have shown that about 90% of the lipids are wax esters, ranging from C_{30} to C_{38} in chain length. Similarly high concentrations are found in midwater crustaceans. In shallow water crustaceans, the wax esters are found as droplets around the intestine, but in deep water forms they are found as intracellular inclusions in most tissues of the body. In midwater fishes, the wax esters are located predominantly between the muscle fibers as well as in unusually high concentrations in the lining of the swimbladder. In some of these fat-invested swimbladders, the gas space is completely filled by wax esters. These appear to be synthesized *de novo* in all cases; the rate-limiting step in biosynthesis appears to be the reduction of the fatty acid to the alcohol. Once the latter is formed, biosynthesis appears to be a thermodynamically "downhill" process.

At room temperature, wax esters of marine organisms are all in the liquid state, mainly because they are highly unsaturated. However, at the reduced temperatures and elevated pressures of the marine environment, they almost certainly are at least in a semisolid state. Their density appears to be similar to that of most lipids—about 0.90 (Table 9–1). For wax esters, this value is remarkably temperature-sensitive, but the density of wax esters does not appear to be much affected by pressure. Perhaps it is for this reason that many marine organisms "chose" wax esters as a mechanism of density regulation.

THE POTENTIAL ROLE OF OLEIC ACID IN DENSITY REGULATION

Brief mention should be made in this context of an observation made some years ago indicating that oleic acid varied from about 15% in surface species to over 70% in deep water fishes and crustaceans. Unfortunately, the classes of lipids involved were not examined. It is now known that wax esters are the source of the oleic acid in midwater fishes and crustaceans; in deep sea fishes, phospholipids and triacylglycerols are the sources of oleic acid. Thus, although oleic acid (a C-18 unsaturated fatty acid) may be involved in density regulation in some marine organisms, its precise role cannot as yet be specified.

GAS FILLED BUOYANCY TANKS

If the technical problems can be solved, a tank of compressed gas is by far the most efficient kind of mechanism for maintenance of neutral buoyancy. The density of air at sea level, for example, is only 0.00125, a value about three orders of magnitude less than the density of aqueous or lipid liquids. In consequence, a small gas chamber would be as effective in buoyancy as would a much larger chamber that was filled with liquid. These differing efficiencies are readily observed in nature. The gas filled swimbladders of teleosts, for example, constitute less than 5% of their body volume, while the coelomic fluid of cranchid squids occupies approximately 65% of the body. It should not be too surprising, then, to discover that gas-filled chambers are in fact fairly commonly utilized as buoyancy machines by animals at many levels in phylogeny. The tiny rhizopod protozoan, *Arcella* (about 1 mm in diameter), adjusts its buoyancy with bubbles of oxygen. The giant Portuguese man-of-war operates a buoyancy chamber (or a float) which may have a liter capacity. It possesses a gas-generating layer of cells and a pore through which gas can be emitted to adjust the volume. The gas mixture is about 20% O_2, with the remainder consisting of variable amounts of CO_2 (presumably of metabolic origin), N_2 (presumably from solution), and CO (presumably from a specialized serine metabolism).

There is perhaps no more elegant example of this sort of adaptation than the cuttlebone of the cuttlefish, *Sepia*. The cuttlebone is a two-phase buoyancy tank. The gas phase, which is largely nitrogen, is of course primarily responsible for density reduction. The liquid phase regulates the volume of the gas phase and, thereby, the density of the organism.

The mechanism of buoyancy regulation in this vertical migrant again involves active ion transport. The osmotic pressure of the cuttlebone fluid is controlled by the active pumping of ions from the fluid into the outside body fluids. Removal of ions from the tank fluid reduces the tonicity of the fluid and, consequently, water leaves the tank, increasing the volume of the gas phase. When sufficient fluid is removed from the cuttlebone, the cuttlefish rises to the surface where it feeds.

Regulation of the volume of the gas phase occurs in response to change in illumination. Thus, as dawn approaches the cuttlefish adjusts its tank density and sinks; in the evening the density of the tank is reduced and the animal rises to feed.

While an air-filled tank of this sort is extremely efficient as a buoyancy device, it has a major drawback: at more than a few hundred feet of depth, pressure is sufficient to implode the cuttlebone. Hence, it is not surprising that more deep-dwelling cephalopods, such as the Cranchidae, employ less compressible fluid-filled tanks to regulate density.

THE TELEOST SWIMBLADDER

By far the most impressive example of all gas-filled buoyancy tanks is the teleost swimbladder, which is able to develop and sustain gas pressures within the chamber of up to about 500 atmospheres! This spectacular ability to concentrate gas (largely in the form of O_2) in the swimbladder, to tensions that are hundreds of times higher than those present in the arterial blood entering the swimbladder, is achieved by combining the properties of Hb as an O_2 carrier with (1) a specific kind of acid-generating glycolysis in the swimbladder epithelial cells, and (2) a unique "counter-current multiplier" arrangement of arterioles and venules supplying the organ (Figure 9–2).

The architecture of the blood circulatory system to the swimbladder supplies the key to its high O_2 concentrating abilities. The artery which brings oxygenated blood to the swimbladder forms a capillary bed, with the venous return a short distance from the swimbladder epithelium. This capillary system of venous and arterial origin intertwines in such a way that arterial capillaries are surrounded primarily by venous ones and vice versa. Such a capillary bed is called a *rete mirabile* (literally meaning wonderful net!). Blood on its way to the swimbladder must therefore pass the venous return in such a way as to optimize diffusional exchange of gases. If the system were entirely passive, it would be referred to as a *counter-current exchanger,* but because the blood entering the rete is under pressure and its O_2 is in part unloaded into the bladder by acid dissociation of HbO_2, the system can be utilized as a remarkably efficient *counter-current multiplier*: a good portion of the O_2 remaining in the venous blood can be transferred to the entering arterial blood, and, repeated, the process leads to a continuous gradual increase in O_2 tension in the blood on the bladder side of the rete. The rete, therefore, serves two obvious functions: (1) it leads to a piling up of O_2 in the blood on the swimbladder side of the circulation, and (2) it serves to prevent any major loss of O_2 via the venous return. But this in itself would not account for the

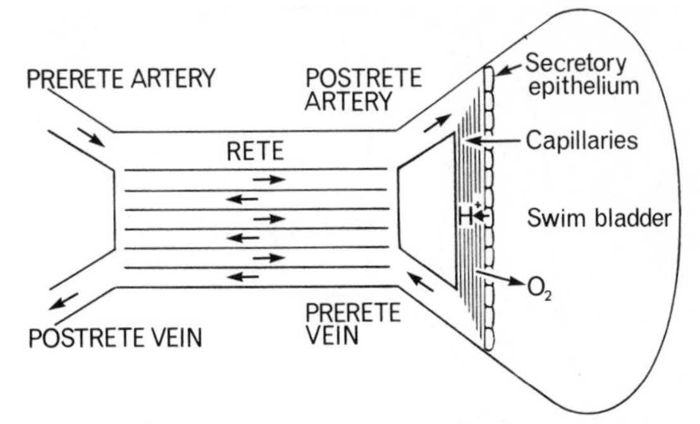

Figure 9–2 A simplified drawing of the gas gland with rete and swimbladder, modified after Steen (1970). The arrangement of vessels is called a hairpin counter-current system. This term is not strictly an anatomical one, but it expresses the loop nature of the circulation. In 1929, August Krogh counted 88,000 venous and 116,000 arterial capillaries in the two retia of an eel. These had an aggregated length of 353 and 464 meters, respectively. The total capillary surface was 106 meters^2 for the venous and 105 meters^2 for the arterial vessels. All this surface was contained in a volume of 0.064 cm^3, of which two-thirds was occupied by blood. The ratio of the total diffusion area to the rete volume is therefore some 1700 cm^2/cm^3. By comparison, the ratio of diffusion area/volume in the alveoli of the human lung is about 100 cm^2/cm^3.

accumulation of such remarkably high O_2 tensions in the swimbladder.

The mechanism of depositing the O_2 in the swimbladder occurs in a second capillary bed which is formed in the epithelium itself (Figure 9–2). Here, blood enters an environment which sustains a high acidity arising from an active aerobic glycolytic system in the epithelial cells. The glycolytic enzymes of this tissue function efficiently in the presence of high O_2 tensions. The Pasteur effect (inhibition of glycolysis by high O_2 tensions) does not occur, either because of a special form of PFK insensitive to feedback inhibition by metabolites of aerobic metabolism, or because of a very sluggish aerobic metabolism. In any event, the lactate produced can fully account for the acidification that occurs in the blood entering the epithelial capillary bed. In addition, carbonic anhydrase occurs in high activity in this epithelium and presumably contributes to H^+ production.

In many vertebrates, H^+ is known to decrease the Hb affinity for O_2, and this is true in teleost fish as well. In the swimbladder this would possibly assist the unloading of O_2 at low pressures, but it cannot account for unloading at high pressures (see Figure 9–3).

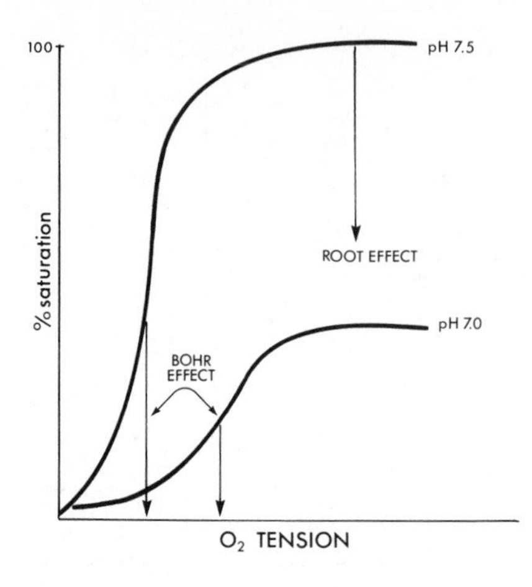

Figure 9–3 The Root Effect and the Bohr Effect on O_2 loading and unloading of fish blood.

It is not surprising, therefore, that in fishes not only is the affinity parameter pH-sensitive, but so also is the *maximum O_2 capacity:* under acidic conditions, the maximum amount of O_2 which Hb can bind is dramatically decreased (Figure 9–3). By this process, the well known Root Effect, Hb unloads a significant fraction of its O_2 into the blood during passage through the epithelial capillaries.

The final transfer of O_2 from the water phase of the blood to the gas phase of the bladder is facilitated by a decrease in solubility of O_2 in the blood as it passes through this capillary bed. The decreased solubility of O_2 stems from the "salting out effects" of increased numbers of particles in solution; the osmotically active particles which bring this about are lactate molecules. If the source of lactate is glycogen (an osmotically inactive substrate), it is evident that increased glycolysis would contribute not only H^+ for inducing the Root Effect, but also osmotically active molecules which decrease the O_2 solubility and hence facilitate O_2 transfer into the swimbladder. According to Kuhn and his coworkers, this kind of system could be effective in developing O_2 pressures of up to 1000 atmospheres—tensions which are higher than would ever be required by the deepest abyssal benthic fishes.

DEEP-SEA FISHES HAVE LONGER RETIA THAN SHALLOW WATER FORMS

The central position of the rete in swimbladder function is clearly demonstrated by a comparison between the extent of its development and the depth at which the fish lives. The efficiency of the swimbladder, whether in maintaining or generating gas tensions, depends directly upon the length and interchange surface

TABLE 9–2 RETE LENGTH AND THE DEPTH RANGES OF DEEP-SEA TELEOSTS DATA FROM MARSHALL (1972).

Zone	Depth in meters	Rete Length in mm
upper mesopelagic	200–600	1–2
lower mesopelagic	600–1200	3–7
bathypelagic	1000–4000	15–25
benthic (true bottom fish)		no swimbladder

area of the retial capillaries and indirectly upon the rate of blood flow. Both functions become harder to perform with increasing depth. Hence, it is not surprising to find a graded increase in retial length in species ranging from shallow to deep reaches of the oceans (Table 9–2).

SWIMBLADDER FUNCTION AT GREAT DEPTHS

Although many fishes which live near the bottom in very deep waters possess swimbladders, in most cases these organs are filled largely with lipids rather than gas. Interestingly, the juvenile stages of these species, which are mesopelagic, have gas-filled swimbladders. During metamorphosis and migration to the sea floor, the gas volume is continuously filled by lipids, mainly cholesterol and phospholipid. The adult swimbladders are not regressed in any sense; they remain highly functional in buoyancy control.

The utilization of lipids in swimbladders "makes sense" on a number of grounds. Firstly, as already mentioned, gas-filled buoyancy tanks are so highly compressible at great depths that they lose much of their efficiency. Secondly, the secretion of oxygen into a gas-filled swimbladder may be difficult, if not impossible, at great depths. Thus, the movement of oxygen into the swimbladder at great depths is impeded by the high oxygen back pressure from the bladder. However, this back pressure likely will not occur when the bladder is filled with lipids. Under high pressures, gases are more soluble in lipids than in aqueous solution. Consequently, in deep-water fishes the swimbladder oxygen, which is dissolved in a mixture of cholesterol and phospholipid, will tend to remain in the bladder due to its higher solubility there than in the blood.

If this hypothesis of Phleger is correct, the oxygen back pressure at the swimbladder surface at, for example, 7000 meters will not be the "expected" 630 atmospheres, but instead may be less than 50 atmospheres. Under these conditions the Root Effect could readily enable oxygen secretion into the bladder to occur. This combined system then could constitute the "magic trick" used by deep-sea fish to secrete O_2 into their swimbladders effectively despite the presence of great pressure gradients. If so, we can, with the wisdom of hindsight, appreciate the wisdom of nature in filling these swimbladders with cholesterol instead of with gas.

SELECTED READINGS

Reviews and Articles

Alexander, R. N. (1972). The energetics of vertical migration by fishes. In *Effects of High Pressure on Organisms. Soc. Exp. Biol. Symp. 26*, 273–294.

Denton, E. J. (1961). The buoyance of fish and cephalopods. *Prog. Biophys. 11*, 177–234.

Gilpin-Brown, J. B. (1972). Buoyancy mechanisms of cephalopods in relation to pressure. In *Effects of High Pressure on Organisms. Soc. Exp. Biol. Symp. 26*, 251–259.

Malins, D. C., and J. C. Wekell (1970). The lipid biochemistry of marine organisms. In *The Chemistry of Fats and Other Lipids.* (Ed. R. T. Holman). Pergamon Press, New York, Vol. 10, 339–363.

Marshall, N. B. (1972). Swimbladder organization and depth ranges of deep-sea teleosts. In *Effects of High Pressure on Organisms. Soc. Exp. Biol. Symp. 26*, 261–272.

Phleger, C. F. (1971). Ph.D. Thesis, University of California, San Diego.

Steen, J. B. (1970). The swimbladder as a hydrostatic organ. In *Fish Physiology.* (Ed. W. S. Hoar and D. J. Randall). Academic Press, New York, Vol. 4, 413–443.

CHAPTER 10

HEMOGLOBIN

INTRODUCTION

In this chapter we reverse our presentational approach. Instead of examining the effects of a single environmental parameter on diverse biochemical systems, we shall examine a single class of molecules, the vertebrate hemoglobins, as they are affected by and adapted to a variety of environmental factors. By treating in outline fashion many of the known manners in which hemoglobins appear to be adapted for function in particular environments, we hope to provide the reader with an overview of much of what has been said in earlier chapters. Specifically, we hope to accomplish two objectives. First, we wish to leave the reader with a sharply focused image of the nature of the basic strategies of biochemical adaptation to the environment. Second, we wish to provide at least some valid rules of thumb for predicting under which circumstances a given strategy of biochemical adaptation is likely to be used.

To achieve these goals, no better study system is available than the hemoglobin (Hb) molecules of vertebrates, for no macromolecules are so thoroughly understood, at all levels of physiological and biochemical resolution. The adaptation of Hb systems to environmental changes which affect oxygen transport occurs via a great variety of mechanisms, and these serve to illustrate vividly the basic "adaptational strategies" which form the theme of this volume.

THE BASIC FUNCTIONS OF RESPIRATORY PIGMENTS

The respiratory pigments (Hb, hemocyanin, chlorocruorin, and hemerythrin) are a chemically and functionally diverse array of macromolecules which have in common one chemical trait: all are complexes of protein with metal atoms which exhibit a characteristic color (hence the name "pigment"). Most of the respiratory pigments located in circulating fluids of the body display physiologically important abilities to bind or "load" oxygen at the respiratory surfaces and lose or "unload" oxygen at the actively respiring tissues.

Large concentrations of respiratory pigments may be found in tissues other than the blood. For example, vertebrate muscle may contain enormous quantities of myoglobin (muscle hemoglobin). These non-circulating respiratory pigments have higher oxygen affinities than the circulating pigments (Figure 10–1); however, the highest oxygen affinities are those of the oxidative enzymes, such as the cytochromes, which are the ultimate recipients and users of molecular oxygen. The non-circulating pigments such as myoglobin thus facilitate the transfer of oxygen from the blood to the oxidative enzymes. In addition, when large amounts of myoglobin or analogous pigments are present in a tissue, the pigment can act as an important storage depot for oxygen, thus insuring an adequate supply of oxygen under conditions where uptake of oxygen from the external environment is restricted, such as during prolonged dives by marine mammals.

Detailed discussions of the properties and the phylogenetic distributions of the different respiratory pigments are found in any comprehensive physiology textbook and need not be elaborated here.

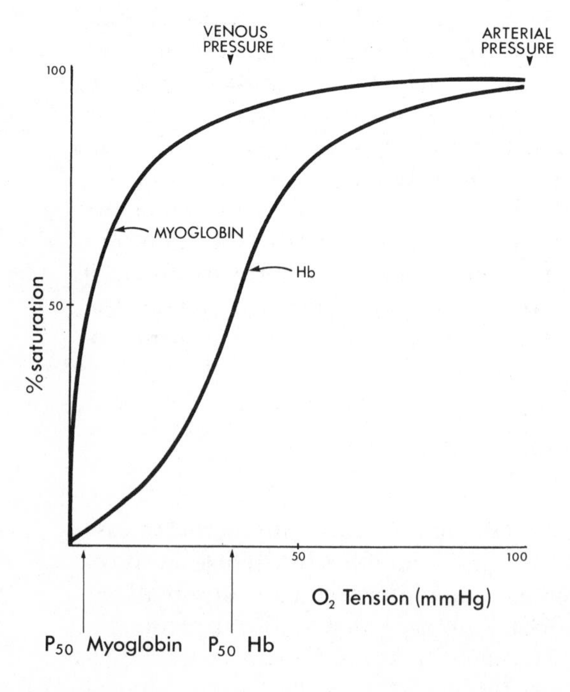

Figure 10–1 Oxygen saturation curves for adult mammalian hemoglobin (Hb) and muscle myoglobin. Note the differences in oxygen affinity, as measured by the P_{50} values, and the shapes of the saturation curves.

HEMOGLOBIN—THE "HONORARY ENZYME"

Hemoglobin is the most ubiquitous respiratory pigment, occurring in vertebrates and some invertebrates, and is the one which appears to be adapted to the widest range of physiological and environmental conditions. Hemoglobin is frequently called an "honorary enzyme," a distinction it merits for both structural and functional reasons (Figures 10–1 and 10–2). Indeed, the similarities between Hb and enzymes—especially complex regulatory enzymes—are so extensive that Hb biochemistry and enzymology have enjoyed a most successful cross-fertilization.

Like regulatory enzymes, most vertebrate hemoglobins have a quaternary structure (Figure 10–2). Furthermore, much as regulatory enzymes frequently are composed of heterogeneous (regulatory and catalytic) polypeptide subunits, so do most Hbs have two subunit types. The most common Hb in adult vertebrates is a tetramer composed of two α and two β subunits, with a heme moiety attached to each of the four subunits (Figure 10–2). The presence of heterogeneous subunit types in the Hb molecule appears to be essential for the cooperativity in oxygen binding illustrated in Figure 10–1. Homotetramers display hyperbolic oxygen saturation curves similar in shape to the myoglobin saturation curve. The assumption of complex, higher orders of structure has allowed the development, in Hbs and regulatory enzymes, of high degrees of regulatory efficiency. In particular, both classes of proteins are able to vary their rate of function sharply over small ranges of variation in substrate or gas concentration. For Hbs this translates into an ability to unload large quantities of its bound oxygen against a sharp oxygen concentration gradient in the tissues. In addition, concentrations of free oxygen in the blood are kept low.

The similarities between Hb and regulatory enzymes include the basic mechanisms by which the functions of the macromolecules are regulated. For both classes of protein, the fine control of function is vested in affinity parameters. The Hb affinity parameter, which is analogous to the K_m or $S_{0.5}$ parameter of enzymes, is termed the P_{50} value. As illustrated in Figure 10–1, the P_{50} characteristic of a Hb is the concentration of oxygen (usually expressed as partial pressure of oxygen in mm Hg) required to half-saturate the Hb under the given conditions of temperature, pH, and ionic strength. P_{50} values, like K_m and $S_{0.5}$ values, generally are of the same order of magnitude as the concentration of substrate (or gas) available to the protein. And, as in the case of enzyme-substrate affinity parameters, P_{50} values are extraordinarily sensitive to changes in the physical and chemical state of the cellular milieu.

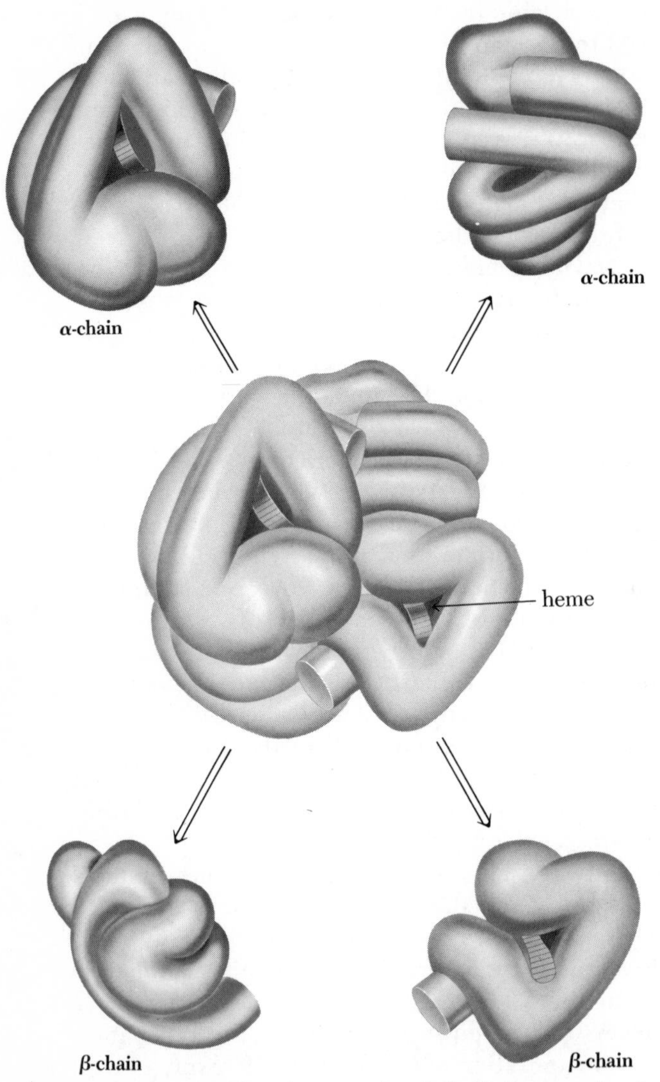

Figure 10–2 (a) A highly schematic, "exploded" diagram showing the higher orders of structure of the most common adult vertebrate Hb. Each Hb molecule contains four polypeptide chains and four heme units. The structure of the heme moiety (b) is invariant among species. The primary sequences of the polypeptide chains exhibit considerable variation. (Diagram from McGilvery, R. W., *Biochemistry: A Functional Approach.* W. B. Saunders Co., 1970.)

(Fig. 10–2 continues on the opposite page.)

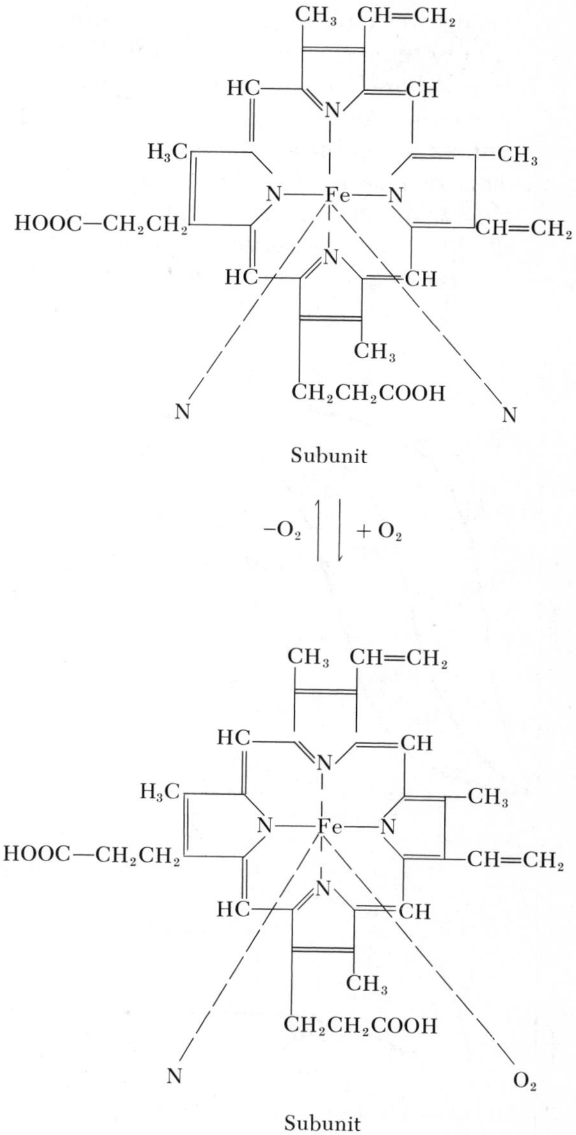

CH_3 $CH{=}CH_2$
HC CH
N
H_3C $-CH_3$
N—Fe—N
$HOOC{-}CH_2CH_2$ $CH{=}CH_2$
N
HC CH
CH_3
CH_2CH_2COOH
N N

Subunit

$-O_2$ $+ O_2$

CH_3 $CH{=}CH_2$
HC CH
N
H_3C $-CH_3$
N—Fe—N
$HOOC{-}CH_2CH_2$ $-CH{=}CH_2$
N
HC CH
CH_3
CH_2CH_2COOH
N O_2

Subunit

(b) The chemical structure of the heme moiety associated with each Hb (and myoglobin) polypeptide chain (subunit). The upper structure illustrates the state of the heme unit in deoxyhemoglobin or deoxymyoglobin; the lower illustration shows the oxygenated state.

REGULATION OF Hb-OXYGEN AFFINITIES

One of the most important fine level modulators of Hb function is the hydrogen ion concentration (pH) of the cell. For the large majority of Hbs, decreases in pH promote reductions in Hb-oxygen affinity (increases in the P_{50} value). Thus, at actively respiring tissues where lactate, CO_2, and other acidic or potentially acidic metabolic end products decrease the pH of the blood, Hb tends to unload a significant fraction of its bound oxygen (Figure 10–3a). This inverse relationship between pH and P_{50} is termed the *Bohr Effect.* By analogy with regulatory enzymes, we can term the hydrogen ion a "negative modulator of Hb function."

Hydrogen ions can negatively modulate Hb function in a second manner which, again, is similar to the way in which some nega-

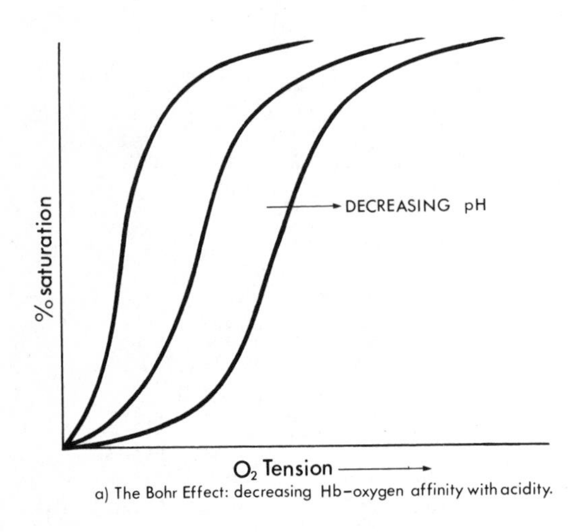

a) The Bohr Effect: decreasing Hb–oxygen affinity with acidity.

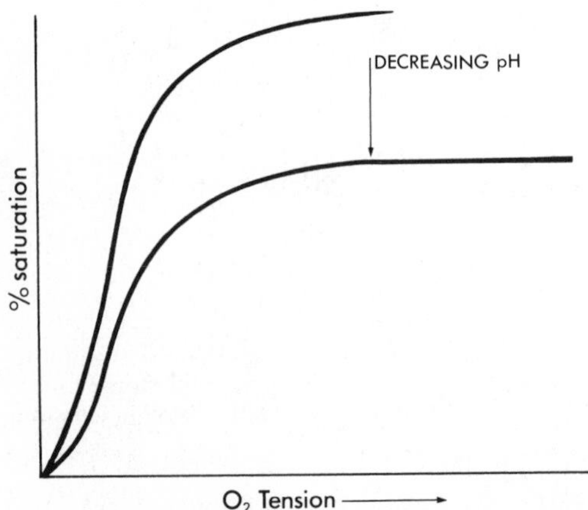

b) The Root Effect: decreasing oxygen carrying capacity with rising acidity.

Figure 10–3 pH modulation of Hb function.

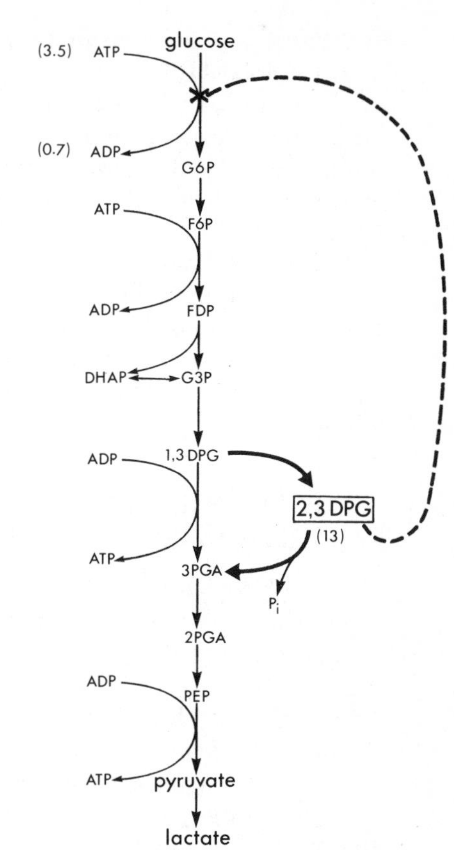

Figure 10–4 The modified glycolytic sequence utilized for 2,3-diphosphoglycerate (2,3-DPG) synthesis in mammalian erythrocytes. Major modifications of the standard glycolytic pathway include: (i) The presence of the enzyme diphosphoglycerate mutase (DPGM), which catalyzes the transformation of 1,3-diphosphoglycerate (1,3-DPG) to 2,3-DPG. Note that this reaction leads to the *removal* of a potential ATP molecule, since the 1,3-DPG no longer is channeled through the phosphoglycerate kinease (PGK) reaction. (ii) The rate of glycolysis is subject to the control of 2,3-DPG, which acts as a feedback inhibitor of hexokinase. Of historical interest is the fact that this particular example of enzyme modulation was the first case of feedback inhibition to be discovered.

The concentrations of ATP, ADP, and 2,3-DPG are given in μmoles of phosphorylated intermediate per gram of Hb. (Figure modified after Brewer and Eaton (1971).)

tive modulators affect rates of enzymic catalysis. The total binding capacity of Hb can be reduced by a fall in pH; i.e., hydrogen ion can lower the "V_{max}" characteristic of Hb function as well as the affinity for oxygen (Figure 10–3b). The variation in oxygen carrying capacity with pH is termed the *Root Effect.* A large Root Effect is often associated with a large Bohr Effect.

Modulation of Hb-oxygen affinity may also involve low molecular weight metabolites *formed within the erythrocytes.* In mammalian erythrocytes, for example, Hb-oxygen affinity is reduced as concentrations of 2,3-diphosphoglycerate (2,3-DPG) rise (Figure 10–4). 2,3-DPG, another "negative modulator" of Hb, is formed in a truncated version of the glycolytic sequence which is unique to mammalian erythrocytes (Figure 10–4). 2,3-DPG is a highly negatively charged molecule, and it fits into a positively charged "pocket" of the Hb tetramer, formed in part by the NH_3^+ terminal of the β subunits and by serine, lysine, and histidine residues. One mole of 2,3-DPG is bound per mole of Hb tetramer.

There is evidence suggesting that the intra-erythrocyte concentrations of 2,3-DPG vary between venous and arterial blood. Measurements of 2,3-DPG titres in freshly oxygenated blood leaving the lungs and in venous blood which has passed through the deep body tissues reveal substantially higher 2,3-DPG levels in the venous erythrocytes. Thus, the composition of the intracellular medium in which Hb functions may vary, in a physiologically adaptive manner, as the blood travels from the lungs to the deeper tissues. In other vertebrate groups, the role of 2,3-DPG appears to be assumed by other organophosphate compounds. In bird and reptile erythrocytes, inositol hexaphosphate is thought to modulate Hb-oxygen affinity. In fish, ATP is probably the key organophosphate modulator.

ADAPTATION OF Hb SYSTEMS: BASIC DEMANDS AND RESPONSES

The basic demands placed on the oxygen transport system are seen to be two-fold: the respiratory pigments must capture sufficient oxygen to meet the tissues' needs and, in addition, the pigments must be able to unload this oxygen where it is needed. Again, to reiterate our analogy between Hb and enzymes, the oxygen supply system must possess a sufficient *capacity* and a tight *control* over the utilization of this capacity.

Adaptations of Hb systems generally relate to two basic variables: altered demands for oxygen by the organism and changes in the oxygen available in the external environment. In some cases, the requisite changes in Hb function can be brought about by what we have termed the "quantitative strategy," the compensatory variation of macromolecular concentrations. In other cases, the changes in oxygen demand or oxygen supply may be such as to render the Hb variants already present in the organism poorly suited for continued function. Under these circumstances the "qualitative strategy" may be employed: new Hb variants may be produced which function better than the old variants. Lastly, the organism may instead adapt by altering the "functional environment" in which the Hb molecules bind and release oxygen. This, we will see, is an especially important strategy in Hb adaptation.

HIGH ALTITUDE ADAPTATIONS OF Hb FUNCTION

As an organism moves from a relatively oxygen-rich lower elevation to an oxygen-poor higher elevation, two important modifications of its oxygen transport-delivery system probably will be necessary. On the one hand, since the external oxygen tensions at high altitude may be too low to saturate the blood, some addition to the overall oxygen transport *capacity* must be made. In other words, since the oxygen transport capacity per Hb molecule is reduced, an increase in the total amount of circulating respiratory pigment may be required.

On the other hand, it would appear to be advantageous for the Hb to increase the facility with which it can release oxygen to the oxygen-requiring tissues. Thus, an increase in the type of regulatory response noted with 2,3-DPG would seem to be to the organism's advantage.

The means by which high altitude adaptations are effected are multifarious, and the "choice" of adaptive strategy is strongly influenced by time considerations. As in many other adaptations to environmental stresses, behavioral and physiological adaptations may form the first line of defense. When a human climbs to a high elevation, the unpleasant sensations arising from reduced oxygen availability—headaches, dizziness and, perhaps, nausea—tend to reduce the subject's level of activity and, thereby, his need for oxygen. At a physiological level, the rate of breathing and the heartbeat rate may rise, thus tending to increase the amount of oxygen which is transported to the tissues by a given amount of Hb. However, as basic molecular adaptations ensue, dependence on these behavioral and physiological responses largely ceases.

The initial change which leads to rises in the oxygen carrying capacity of the blood is an increase in the number of circulating erythrocytes, which results from a contraction of the spleen, where significant quantities of erythrocytes may be stored. To complement this increase in oxygen carrying capacity, the ability of Hb to unload its oxygen at the tissues is elevated via an increase in 2,3-DPG concentrations in the erythrocyte, an adaptation which occurs during the first two days of high altitude adaptation. During prolonged stays at high altitudes the proliferation of erythrocytes may also be increased as a means for maintaining high levels of circulating Hb. In conclusion, the adaptive changes which occur in the oxygen delivery system during high altitude adaptation are analogous to many of the enzymic adaptations we have previously discussed, e.g., during temperature and salinity adaptations. In all of these instances, adaptation involves both the adjustment of the *overall capacity* of the system to perform its function and, equally importantly, the *fine control* of this function, which remains under stringent physiological regulation.

Hb ADAPTATIONS DURING ONTOGENY IN MAMMALS

For many organisms, the availability of oxygen varies markedly throughout the life cycle. The *in utero* development of mammalian embryos occurs under low "environmental" oxygen tensions compared to those experienced by adults. Similarly, among non-mammalian vertebrates, the pre-adult stages often may live under conditions of low oxygen content. In most of these instances, the transition from oxygen-poor to oxygen-rich environments involves a shift from aquatic to terrestrial existence, e.g., during amphibian metamorphosis, hatching in birds, and parturition in mammals.

In each of these cases, the Hbs characteristic of the adult and pre-adult life stages differ in their functional properties. These differences are clearly of adaptive significance, in that they insure that the oxygen transport system will function satisfactorily under the particular oxygen concentrations available at the respiratory surfaces of the organism. In these Hb adaptations a different strategy is employed from the ones just discussed under the heading of altitude adaptations. Ontogenetic changes in Hb function involve the appearance of new Hb variants with different primary structures and, consequently, new functional characteristics. These changes are thus analogous to the adaptive changes which occur in the enzyme systems involved in nitrogen excretion during water-to-land transitions.

The basic differences between the pre-adult (or aquatic-functioning) and adult (or terrestrial-functioning) Hbs are illustrated by the saturation curves of fetal and adult mammalian Hbs (Figure 10–5). In placental vertebrates, the fetus must compete with the maternal tissues for oxygen. The efficiency with which the fetal blood can extract oxygen from the maternal blood, across the placental barrier, is greatly facilitated by the presence in the fetal

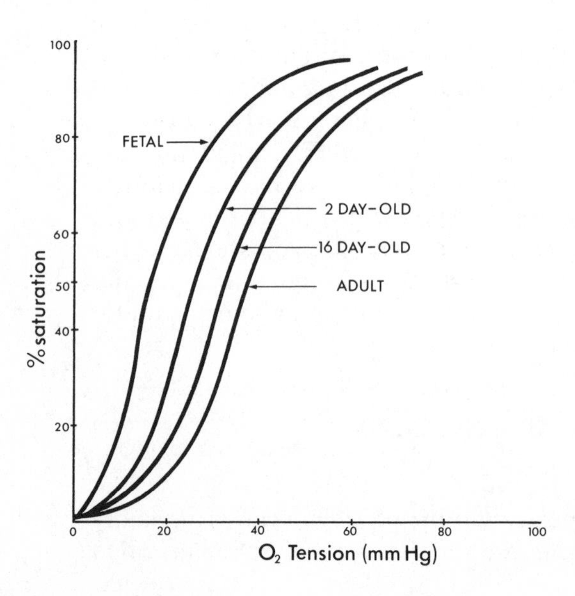

Figure 10–5 Oxygen saturation curves for blood of fetal and newborn sheep. Note the much higher affinity (lower P_{50}) value for the fetal blood, containing the fetal Hb variant, which must extract oxygen from an aqueous medium, i.e., the maternal circulation. The low affinity Hb in the adult blood extracts oxygen from an oxygen-rich medium, i.e., air.

erythrocytes of an Hb variant ("fetal Hb") which has a greater oxygen affinity than adult Hb. Fetal Hb, with an oxygen affinity intermediate between the affinities of the adult Hb and the oxidative enzymes of the fetal tissues, thus functions analogously to myoglobin in terms of transporting oxygen from the blood to the respiring tissues.

ONTOGENETIC CHANGES IN AMPHIBIAN Hbs: ADAPTATIONS TO AEROBIC AND ACTIVE MODES OF LIFE

In many amphibian species the transition from aquatic (juvenile) to terrestrial (adult) existence involves major changes in the oxygen demands of the organism, as well as in the amount of oxygen available in the environment. Adult amphibians normally are much more active than juveniles, such as tadpoles, and thus require large amounts of oxygen to be unloaded in their tissues.

In some amphibians, the aquatic-to-terrestrial transition is accompanied by the appearance of new Hb species which, unlike the juvenile Hbs they replace, have strongly sigmoidal (cooperative) oxygen saturation curves. In other amphibians, both juvenile and adult Hbs display cooperative oxygen binding kinetics. However, in this latter group, only the adult Hb displays a Bohr Effect. The presence of high cooperativity in oxygen binding and a pronounced pH dependency of oxygen binding permit the adult amphibian Hbs to unload large fractions of their bound oxygen against a high oxygen concentration gradient at the actively respiring tissues.

INTERSPECIFIC DIFFERENCES IN OXYGEN BINDING COOPERATIVITY AND BOHR EFFECT

A positive correlation between the degree of cooperativity in oxygen binding and the general activity level of the species is also found among other classes of vertebrates. More active species also tend to have the largest Bohr Effects. As we have just argued in the case of amphibian Hbs, these molecular properties suit the Hbs for function under conditions where demands for oxygen by rapidly respiring tissues are very high.

The importance of adaptation in Hb regulatory function—as opposed to adaptation strictly in the total oxygen carrying capacity of blood—is attested by the fact that among various fish species there may be little difference in the total oxygen carrying capacity of the blood, whereas there are marked differences in the manner in which use of this capacity is regulated. The more active species have a much greater ability to unload bound oxygen against a steep oxygen gradient. Manwell and Baker have proposed that the "hybrid vigor" noted for a certain sunfish hybrid results from the increased sigmoidicity (i.e., oxygen unloading potential) of the organism's Hb. Here, then, is another instance in which environmental adaptation involves the presence of an "improved" molecular variant, rather than a mere adjustment in the total concentration of the molecule in question.

Hbs WITH NO BOHR EFFECT: "BACK-UP" SYSTEMS IN HIGHLY ACTIVE FISHES

While it is advantageous for rapidly metabolizing organisms to have Hbs with high Bohr Effects, situations can arise where too high a level of negative modulation by hydrogen ion can work to the organism's disadvantage. Some highly active fishes accumulate very high levels of blood lactate during strenuous exercise. Under these conditions the pH of the blood may fall to such an extent that oxygen loading at the gills is severely impaired owing to the Bohr Effect. It is known, in fact, that fishes may die of apparent suffocation during or following vigorous exercise.

In some highly active fishes, Hb variants which lack a Bohr Effect have been observed. These hydrogen-ion-insensitive Hbs are normally a minor component of the total Hb population. However, they most likely represent an important "back-up" system of oxygen delivery which can be utilized whenever the major Hb components are under strong negative modulation from metabolically generated acidity.

Powers (personal communication) has found that the molecular differences between "normal" Hbs and the "non-Bohr Effect" Hbs lie in the terminal amino acids of the α and β Hb chains. In mammals it is known that the Bohr Effect is due largely to the activities of the C-terminal histidine residues of the β chains and the free amino groups at the N-terminal end of the α chains. In one fish Hb variant which lacks a Bohr Effect, the C-terminal amino acid of the β chains was not histidine and, in addition, the N-terminal amino groups of the α chains were blocked.

TEMPERATURE ADAPTATION OF ECTOTHERMIC Hbs—ACCLIMATION EFFECTS

Because Hb is an "honorary enzyme," we might expect that most of the temperature effects on enzymic structure and function already discussed would apply closely to Hb. This in fact is true. In addition, temperature affects Hb function in a second vital way: in the case of aquatic species, it largely determines the concentration of Hb's "substrate." Since the amount of gas dissolved in a solution is inversely proportional to temperature, warm water species have less oxygen available at their respiratory surfaces than colder water species. Temperature influences on Hb function are likely to be particularly acute for aquatic organisms which encounter large fluctuations in temperature over time periods of a few minutes or hours. It is known that, for active Salmonid species (trout and Pacific salmon), the reduced oxygen availability which occurs at temperatures above approximately 18°C places an upper limit on the rate of active metabolism.

A second major effect of temperature on the blood transport of gases involves Hb-oxygen affinity. In light of our earlier discussion of temperature-dependent changes in E-S affinity and the analogy made above between Hbs and enzymes, it should not come as a surprise to learn that the P_{50} values of most Hbs which have been studied decrease as temperature falls. The direction of this temperature-dependent change in the P_{50} value is the *opposite* of what we would rationalize as being adaptive to the organism, since increases in temperature at once reduce the amount of oxygen available at the respiratory surface(s) and the efficiency with which a given Hb molecule can load oxygen.

The physiological and biochemical changes which enable ectothermic species to overcome the adverse effects of temperature on oxygen transport are multifarious and, again, the adaptation strategy employed often reflects the time course of the adaptation process. On a short-term basis, the organism may compensate for an increase in water temperature by means of increasing its ventilation and circulatory rates. These two adaptations provide at least some relief from the acute effects of reduced oxygen tension. However, as in the case of high altitude adaptation by mammals, these physiological changes are superseded by basic cellular and molecular adaptations.

The oxygen carrying capacity of the blood may be increased by elevating the number of circulating erythrocytes during acclimation to warm temperatures. In addition to this increase in oxygen carrying capacity, the regulatory function of the circulating Hb may also be altered. Thus, warm acclimation of the bullhead is marked by an increase in Hb-oxygen affinity. The molecular basis of this change appears to reside in an alteration of the composition of the erythrocyte cytosol. The P_{50} differences between the blood of warm- and cold-acclimated fish are observed only when unruptured erythrocytes are studied. When the Hbs of warm- and cold-acclimated fish were isolated from their normal intracellular environment and oxygen saturation curves were estimated using solutions of free

Hbs, no difference in P_{50} values were noted. Further evidence that the functional differences between the Hbs of warm- and cold-acclimated fish resulted from changes in the milieu in which the Hbs function came from studies of the electrophoretic patterns of the Hbs: in both acclimation groups, seven Hb zones were detected on electrophoretic media, and the relative quantities of these different Hbs were approximately the same in both warm- and cold-acclimated specimens. Thus, it is very likely that seasonal adjustment in oxygen affinities of certain ectothermic Hbs may occur via a mechanism comparable to that employed in high altitude adaptation by humans: the overall capacity of the oxygen transport system is adjusted through changes in the amounts of circulating Hbs, while the loading and unloading properties of the Hbs are adaptively adjusted via changes in the erythrocyte cytosol.

In the case of ectothermic Hbs, we do not know what types of modulators are involved in effecting seasonal changes in Hb-oxygen affinity. However, the recent demonstration that ATP negatively modulates fish Hbs in a manner similar to 2,3-DPG modulation of mammalian Hbs makes ATP a promising candidate for this role.

Another important consideration concerning the modulation of ectothermic Hbs is that regulatory substances may bind to Hb more tightly as temperature is reduced. This is known to occur in the case of 2,3-DPG binding to mammalian Hbs. In an ectothermic system such an effect would be compensatory: the influence of the tightly-bound negative modulator would counteract at least some of the "positive thermal modulation" of Hb-oxygen interaction which, we have said, most likely works to the organism's disadvantage.

EVOLUTIONARY ADAPTATION OF ECTOTHERMIC Hbs TO TEMPERATURE

The seasonal changes in Hb-oxygen affinity just discussed have their counterparts in ectothermic species evolutionally adapted to different thermal regimes. The Hb(s) of an ectothermic species is (are) capable of loading and unloading oxygen under the particular gaseous and thermal conditions of the species' habitat due to strong selective pressure for P_{50} values which are low enough to permit efficient loading at the respiratory surfaces, yet not too low for oxygen to be efficiently released at the deeper tissues. These evolutionary patterns are illustrated in Figure 10–6. As in the case of enzymic adaptation to temperature over evolutionary time spans, in which case selection favors enzymes with functional K_m or $S_{0.5}$ characteristics, the elaboration of well-adapted Hb species seems to require changes in the primary structure of the polypeptide chains

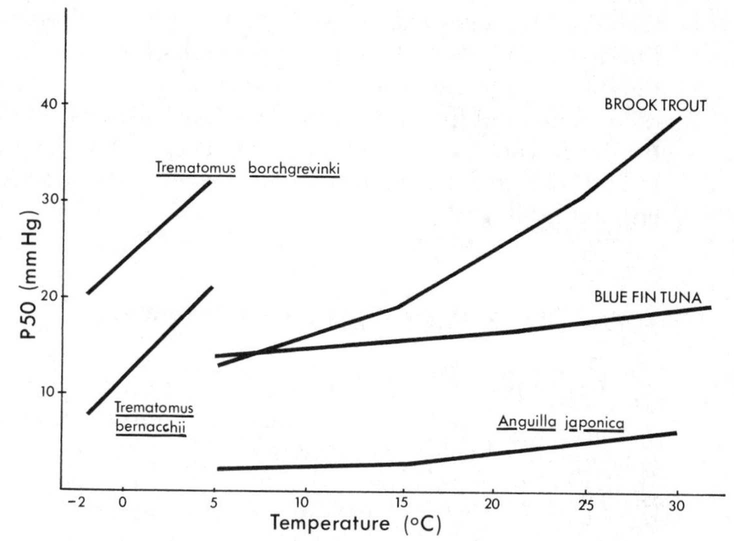

Figure 10–6 Change in P_{50} value with temperature for the Hbs of several fishes. Several important features of Hb adaptation are illustrated by these data: (i) As in the case of enzyme-substrate affinity, the P_{50} value of the Hb is under strong selection for having the correct affinity to extract and deliver oxygen. (ii) The P_{50} values of highly active species tend to be higher than in the case of less active forms. For example, of the two Antarctic species illustrated, *T. borchgrevinki* is an active, pelagic form, while *T. bernacchii* is a fairly sluggish benthic species. (iii) The absolute value of Hb-oxygen affinity may also reflect the oxygen content of the organism's habitat. Thus, we find that the Antarctic fishes, which live in oxygen-saturated, −1.9°C water, have higher P_{50} values than, for example, the sluggish, bottom-dwelling Japanese eel (*Anguilla japonica*) which lives in a relatively low-oxygen environment. (iv) Lastly, note the relatively temperature-independent P_{50} value of tuna Hb. We discuss this adaptation later in the chapter. (The data are from Grigg and from Rossi-Fanelli and Antonini.)

of the Hb molecule. Parenthetically, it should be noted that the heme moiety is invariant among all Hbs which have been studied. The heme unit of Hb, which we might term the "honorary coenzyme" of Hb, is thus analogous to the "real" coenzymes, such as NAD and NADP, which also display the same chemical and functional characteristics regardless of the changes which occur in the protein chain(s) to which they are bound.

As in the case of seasonal changes in the oxygen transport system, evolutionary adaptation of Hb systems to temperature is marked by major changes in Hb concentration. For example, many Antarctic fishes have either no detectable erythrocytes or greatly reduced levels of them. Yet these fish display metabolic rates not markedly lower than those of warm water fishes (p. 214). The ability to remain metabolically active under these anaemic circumstances stems from the high oxygen carrying capacity of −1.9°C solutions. The sea water itself likely is saturated with oxygen, and the quantity of oxygen which can be added to, and carried by, the blood under these conditions is extremely high. In addition, there is little variation in oxygen content of the sea throughout the year and, therefore, the role of Hb in buffering free oxygen content is minimized. Lastly, the exponential increase in fluid viscosity which occurs as temperature is lowered may make it advantageous for Antarctic fishes to reduce the cellular content—and, thereby, the viscosity—of their circulating fluids.

TUNA Hb: A WAY AROUND THE BENDS

In Chapter 7 we discussed the counter-current heat exchanger systems which, in fishes like large tunas, enable a significant quantity of metabolically produced heat to be retained in certain tissues. In light of the temperature dependence of P_{50} values for most Hbs, the heat exchanger system in tunas would seem to pose extreme problems for the gas transport system. Thus, as the relatively cool and highly oxygenated blood courses into the heat exchanger, a temperature dependent change in P_{50} would create two serious problems. Firstly, large amounts of oxygen would be released from the Hb before the most rapidly respiring tissues receive the arterial blood. Secondly, and of even greater seriousness, the rate of oxygen unloading as the blood is warmed in the heat exchanger might be great enough to supersaturate the warmed plasma. Oxygen could then bubble out of the blood, and as these bubbles were carried into the finer vessels, the "bends" would result, likely with a fatal outcome.

Clearly, tuna do not suffer this fate. The molecular basis of their ability to avoid this danger lies in the fact that, at least for bluefin Hbs, temperature has only a minimal effect on P_{50} values (Figure 10–6). The Hb of this, and no doubt other, tuna species therefore will not release enormous amounts of oxygen as it enters the warmed tissues. Physiologically, oxygen release is most likely caused by a strong Bohr Effect, as in the case of most other Hbs of highly active vertebrates.

THE BASIC STRATEGIES OF BIOCHEMICAL ADAPTATION

In this chapter we have briefly sketched the process of environmental adaptation as observable at the level of Hb function. The wealth of data available on Hb structure and function has allowed us to present clear examples of what we feel are the key attributes of biochemical adaptation.

In the case of Hb function, as with other macromolecular functions, biochemical adaptations frequently can be grouped into three general categories:

1. Quantitative adaptations, in which the concentrations of macromolecular components are adaptively altered.

2. Qualitative adaptations, in which new types of macromolecules are incorporated into the system in response to an altered environment.

3. Modulation-type adaptations, in which the activities of existing macromolecules are adaptively adjusted.

These three types of biochemical adaptation, acting together or separately, are either *compensatory* or *exploitative* in nature. Thus, the various biochemical changes involved in altitude adaptation all serve an essentially homeostatic, compensatory function—they compensate for the reduced availability of atmospheric oxygen by increasing the organism's abilities to extract oxygen from the environment. The Hb adaptations which occur during developmental processes are likewise compensatory for alterations in oxygen availability. In contrast, the "back-up" Hb systems present in highly active, fast fishes can be viewed as exploitative adaptations: these Hbs likely enable the fishes to enter and exploit new habitats previously unavailable to them. To quite a large extent, therefore, compensatory adaptations represent an organism's successful retention of the status quo in the face of an environmental change which tends to perturb this equilibrium, whereas exploitative adaptations bear the mark of true evolutionary "innovation."

THE ROLE OF TIME IN DETERMINING ADAPTIVE STRATEGY

Another important generalization to emerge from studies of biochemical adaptation concerns the relationship between the element of time and the adaptive strategy which is—or can be—utilized by the organism. Each time-course of environmental adaptation draws from a different pool of adaptational "raw material." Evolutionary adaptation may proceed by mechanisms of change in the DNA base sequences. Shorter-term adaptations which occur during the life-span of the individual have a more limited pool of raw material at their disposal. The information coding for new types of macromolecules or for the alteration of transcriptional control circuitry cannot be added to the genome. However, by means of differential gene activation, new types of macromolecules can be brought into the system, at least in cases where the organism has hours to make its adaptational response.

It is worth noting that many of the phenotypic adaptations we have studied appear to "recapitulate" evolutionary adaptations; i.e., these two temporally diverse adaptation processes may be effected by almost identical biochemical mechanisms. For example, the water-to-land transition which occurs during amphibian development and, in mammals, during parturition, involves Hb changes which probably mirror those which first occurred during the colonization of land by the vertebrates. Similarly, the changes which take place in the nitrogenous waste disposal pathways during the water-to-land transitions of amphibians must reflect the biochemical adaptations which occurred during the dawn of terrestrial vertebrate evolution.

When environmental adaptations must be nearly instantaneous, we frequently observe that the pool of adaptational raw material shrinks to eliminate all but the modulation strategy. When a rapid biochemical adjustment must be made to an environmental change, the adaptive regulation of macromolecular function is often vital. We noted many examples of this particular strategy in the case of Hb systems, e.g., the modulation of Hb-oxygen binding by 2,3-DPG. We also noted, especially in the case of enzymes, that changes in physical parameters could lead to immediate and apparently adaptive changes in the functional properties of the macromolecules. Thus, even though the pool of raw material available for these nearly instantaneous adaptations is relatively small, the organism is not without a significant armamentarium for defending itself against certain effects of sudden changes in its environment.

THE ESSENTIAL ROLE OF MACROMOLECULAR-LIGAND AFFINITIES

One common characteristic of biochemical adaptations of all time-courses is the fundamental importance of macromolecular-ligand affinities. The efficiencies and the specificities of biochemical processes are due, in large measure, to the existence of precise intermolecular affinities which permit macromolecules to bind lower molecular weight substances. And, because these affinities are dependent on weak chemical bonds such as hydrogen bonds, these affinity characteristics are highly susceptible to disruption by changes in temperature, pressure, ionic concentrations, etc. Thus, a significant number of biochemical adaptations involve the intricate "tailoring" of weak-bonded systems to yield macromolecules with the necessary affinity characteristics to permit function in particular microenvironments.

SOME UNANSWERED QUESTIONS

The biochemical adaptations we have discussed are expressions of the genotypic diversity existing among various organisms. Many of the questions raised, and left unanswered, by the data presented in this volume concern the genetic foundation on which the phenotypic changes—the environmental adaptations—are erected. For example, what is the relationship between genomic complexity and the adaptability of an organism? Will those organisms found to be most eurytolerant turn out to be organisms having highly complex genetic and, thereby, protein systems? Will eurytolerant organisms have the most complicated allozyme and isozyme systems?

The element of time also appears when we consider the questions involved in the genetics of adaptation. How rapidly can organisms acquire new molecular variants? Does an increase in ploidy further an organism's chances for evolving new and adaptive proteins? And what is the cost of maintaining complex genetic (protein) systems?

Lastly, questions concerning the regulation of gene expression must be raised. What are the primary environmental "cues" which trigger a biochemical adaptation? How are genes "turned on" and "turned off" in response to an environmental change? Do eurytolerant organisms possess the most complicated transcriptional control circuits? How rapidly do control circuits evolve?

The single, unifying thread which ties these and related questions together is a potential for examining the mechanisms of biological evolution at the molecular level. As we gain the ability better to understand the types of molecular changes which allow organisms to survive in and exploit their environments, we are further charting the course of evolution at its most fundamental level.

SELECTED READINGS

Books and Review Articles

Hoar, W. S. (1966). *General and Comparative Physiology.* Prentice-Hall, Inc., Englewood Cliffs, N. J.

McGilvery, R. W. (1970). *Biochemistry: A Functional Approach.* W. B. Saunders Co., Philadelphia.

Manwell, C., and C. M. A. Baker (1970). *Molecular Biology and the Origin of Species.* University of Washington Press, Seattle.

Prosser, C. L. (1973). *Comparative Animal Physiology* (3rd edition). W. B. Saunders Co., Philadelphia.

Riggs, A. (1971). Properties of Fish Hemoglobins. In *Fish Physiology*, Vol. IV, pp. 209–252 (W. S. Hoar and D. J. Randall, eds.). Academic Press, New York.

Journal Articles

Brewer, G. J., and J. W. Eaton (1971). Erythrocyte metabolism: interaction with oxygen transport. *Science 171*, 1205–1211.

Grigg, G. C. (1967). Some respiratory properties of the blood of four species of Antarctic fishes. *Comparative Biochemistry and Physiology 23,* 139–148.

Grigg, G. C. (1969). Temperature-induced changes in the oxygen equilibrium curve of the blood of the brown bullhead (*Ictalurus nebulosus*). *Comparative Biochemistry and Physiology 28,* 1203–1223.

Meschia, G., A. Hellegers, J. N. Blechner, A. S. Wolkoff, and D. H. Barron (1961). A comparison of the oxygen dissociation curves of the bloods of maternal, fetal, and newborn sheep at various points. *Quarterly Journal of Experimental Physiology 46,* 95–100.

Powers, D. (1972). Hemoglobin adaptation for fast and slow water habitats in sympatric catostomid fishes. *Science 177*:360–362.

Rossi-Fanelli, A., and E. Antonini (1960). Oxygen equilibrium of hemoglobin from *Thunnus thynnus. Nature 186,* 895–896.

INDEX